Contents

Participants

M. D. Beecher   Departments of Psychology and Zoology, Animal Behavior Program, University of Washington, Seattle, WA 98195, USA

T. Bouchard   Minnesota Center for Twin and Adoption Research, Department of Psychology, University of Minnesota, Elliott Hall, 75 East River Road, MN 55455–0344, USA

D. Buss   Department of Psychology, University of Texas, Austin, TX 78712, USA

L. Cosmides   Center for Evolutionary Psychology, Department of Psychology, University of California, Santa Barbara, CA 93106, USA

H. Cronin   Centre for Philosophy of Natural and Social Sciences, The London School of Economics and Political Science, Houghton Street, London WC2A 2AE, UK

M. Daly *(Chairman)*   Department of Psychology, McMaster University, Hamilton, Ontario, Canada L8S 4K1

R. Dawkins   Oxford University Museum, Parks Road, Oxford OX1 3PW, UK

S. W. Gangestad   Department of Psychology, University of New Mexico, Albuquerque, NM 87131, USA

S. J. C. Gaulin   Department of Anthropology, University of Pittsburgh, Pittsburgh, PA 15260, USA

G. Gigerenzer   Center for Adaptive Behavior and Cognition, Max-Planck-Institute for Psychological Research, Leopoldstrasse 24, D-80802 Munich, Germany

M. D. Hauser   Departments of Anthropology and Psychology, Program in Neuroscience, Harvard University, Cambridge, MA 02138, USA

A. Kacelnik   Department of Zoology, South Parks Road, Oxford University, Oxford OX1 3PS, UK

J. Maynard Smith   School of Biological Sciences, Biology Building, University of Sussex, Falmer, Brighton BN1 9QG, UK

L. Mealey   Department of Psychology, University of Queensland, Brisbane 4072, Australia

G. F. Miller   Economic and Social Research Council Research Centre for Economic Learning and Social Evolution, University College London, Gower Street, London WC1E 6BT, UK

A. P. Møller   Laboratoire d'Ecologie, CNRS URA 258, Université Pierre et Marie Curie, Bât. A, 7ème étage, 7 quai St. Bernard, Case 237, F-75252 Paris Cedex 5, France

R. Nesse   The University of Michigan, 5057 ISR, PO Box 1248, Ann Arbor, MI 48106–1248, USA

R. Nisbett   Institute for Social Research, University of Michigan, Ann Arbor, MI 48109, USA

S. Pinker   Department of Brain and Cognitive Sciences, Massachusetts Institute of Technology, Cambridge, MA 02139, USA

A. R. Rogers   Department of Anthropology, University of Utah, Salt Lake City, UT 84112, USA

J. Scheib *(Bursar)*   Department of Psychology, University of California, Davis, CA 95616, USA

R. N. Shepard   Department of Psychology, Building 420, Stanford University, Stanford, CA 94305–2130, USA

D. F. Sherry   Department of Psychology, University of Western Ontario, London, Ontario, Canada N6A 5C2

D. Sperber   CREA, Ecole Polytechnique, 1 Rue Descartes, 75005 Paris, France

R. Thornhill   Department of Biology, University of New Mexico, Albuquerque, NM 87131, USA

J. Tooby   Center for Evolutionary Psychology, Department of Psychology, University of California, Santa Barbara, CA 93106, USA

M. Wilson   Department of Psychology, McMaster University, Hamilton, Ontario, Canada L8S 4K1

CHARACTERIZING
HUMAN
PSYCHOLOGICAL
ADAPTATIONS

Ciba Foundation Symposium 208

CHARACTERIZING HUMAN PSYCHOLOGICAL ADAPTATIONS

1997

JOHN WILEY & SONS

Chichester · New York · Weinheim · Brisbane · Toronto · Singapore

Published in 1997 by John Wiley & Sons Ltd,
Baffins Lane, Chichester,
West Sussex PO19 1UD, England

National 01243 779777
International (+44) 1243 779777
e-mail (for orders and customer service enquiries): cs-books@wiley.co.uk
Visit our Home Page on http://www.wiley.co.uk
or http://www.wiley.com

Other Wiley Editorial Offices

John Wiley & Sons, Inc., 605 Third Avenue,
New York, NY 10158-0012, USA

WILEY-VCH Verlag GmbH
Pappelallee 3, D-69469 Weinheim, Germany

Jacaranda Wiley Ltd, 33 Park Road, Milton,
Queensland 4064, Australia

John Wiley & Sons (Canada) Ltd, 22 Worcester Road,
Rexdale, Ontario M9W 1L1, Canada

John Wiley & Sons (Asia) Pte Ltd, 2 Clementi Loop #02-01,
Jin Xing Distripark, Singapore 129809

Ciba Foundation Symposium 208
viii+295 pages, 35 figures, 7 tables

Library of Congress Cataloging-in-Publication Data

Characterizing human psychological adaptations.
p. cm.–(Ciba Foundation symposium ; 208)
Includes bibliographical references and index.
ISBN 0-471-97767-5 (alk. paper)
1. Genetic psychology. 2. Adaptability (Psychology) I. Series.
BF701.C49 1997
155.7–dc21
97–24261
CIP

British Library Cataloguing in Publication Data

A catalogue record for this book is available from the British Library

ISBN 0 471 97767 5

Typeset in 10/12pt Garamond by Dobbie Typesetting Limited, Tavistock, Devon.
Printed and bound in Great Britain by Biddles Ltd, Guildford and King's Lynn.
This book is printed on acid-free paper responsibly manufactured from sustainable forestation, for which at
least two trees are planted for each one used for paper production.

Introduction

Martin Daly

Department of Psychology, McMaster University, Hamilton, Ontario, Canada L8S 4K1

'Is it not reasonable to anticipate that our understanding of the human mind would be aided greatly by knowing the purpose for which it was designed?'

(Williams 1966, p 16)

'Evolutionary psychology' has enjoyed a recent vogue, at least in the popular press. For most psychologists, however, the case that they should invest the effort to become sophisticated evolutionists still has to be made.

The workings of the psyche are obviously organized to achieve various ends, and effective psychological scientists have always been adaptationists, partitioning the psyche into component processes with putative functions. This is perhaps clearest in the study of perception, where proposed mechanisms and processes are labelled in terms of the information-processing tasks that they address: movement detection, sound localization, face recognition and so forth. There is greater uncertainty and controversy about the best way to subdivide 'central' (or 'higher') processing, but cognitive psychologists are also adaptationists, concerning themselves with how the mind achieves such tasks as memory encoding and retrieval, categorization and selective attention. Even social psychologists often characterize hypothesized mental processes in terms of what they achieve, although proposed functions in this field (the maintenance of Heiderian balance, the minimization of cognitive dissonance, etc.) often seem arbitrary. Moreover, recent advances in neuroscience and artificial intelligence have reinforced the view of psyches as bundles of modules dedicated to distinct tasks.

But if psychologists are already adaptationists, they are seldom selectionists: psychological science has made scant use of the Darwinian insight that the ultimate criterion of adaptive functional organization is not health or happiness or homeostasis, but contribution to fitness. Over 30 years have passed since George Williams made the rather wistful suggestion that I quoted above, but many psychologists have been slow to grasp the point.

Evolutionary approaches in psychology and other social sciences encounter not just apathy, but also antipathy, founded in ignorance and false dichotomies. An unfortunate side-effect is that Darwinists are often tempted to wave off more serious sceptical challenges, dismissing them as more of the same old hostility, ignorance and foolishness. The challenges to which I refer are to some extent challenges not simply to

1

evolutionary psychologists, but also to the scientific pretensions and aspirations of psychology as a whole. Is the concept of 'mental organs' a stretched metaphor, for example, and the concept of discrete mental adaptations an excessive reification of processes? Must we be neurological reductionists to justify such materialistic language as psychological 'mechanisms'? Can algorithmic/procedural characterizations of mental adaptations be made sufficiently rigorous and testable that they will command consensus and become the foundations of further discovery?

Evolutionary psychologists have generally adopted the stance that psychology is an ordinary branch of biology and that our task in identifying and studying psychological adaptations is essentially like that facing anatomists or physiologists. But psychology is tricky because its objects of study are ephemeral processes, less tangible than bones and muscles whose functional designs an adaptationist anatomist strives to understand. It may also be especially tricky because the workings of the mind are in some sense more holistic and less amenable to componential analysis than the 'mental organ' metaphor would imply. In any event, for whatever reason, the history of psychology seems to indicate that the recognition of adaptations in this branch of biology is, at the least, peculiarly difficult. Consider Freudian theory, for example. Many intelligent people were once persuaded that its constructs (id, ego and superego, parricidal motives, thanatos and all the rest) constituted a valid dissection of the human psyche, while others were convinced that they were fantastic. The latter view has prevailed, although the former lingers in pop psychology and literary criticism. It is difficult even to imagine an analogous case arising in functional morphology, that is, a theory that attracts many adherents by partitioning the body into components which are eventually seen to be non-existent when the theory is discredited! Must we conclude that psychologists are simply parroting the language of materialistic science while unable to deliver its substance?

There is another, more hopeful view, and it is the one to which I subscribe. The reason why psychologists have wandered down so many garden paths is not that their subject is resistant to the scientific method, but that it has been inadequately informed by selectionist thought. Had Freud better understood Darwin, for example, the world would have been spared such fantastic dead-end notions as Oedipal desires and death instincts. And why has social psychology been in large measure a succession of fads without cumulative progress? Could it be because social psychologists have repeatedly postulated shallow intrapsychic functions, such as the defence of one or another sort of mental 'balance', which are unrelated to the basic social information-processing tasks around which any Darwinian would organize the subject? While mainstream social psychology has gone around in circles, theory and research on non-human animal social psychology and behaviour have achieved real progress, and I believe the main reason for this difference is not that human beings are particularly mysterious. Behavioural ecologists and sociobiologists have built cumulative understandings because they have partitioned the subject along the lines of discrete, real-world problems of social information processing, such as kin recognition, maintenance of reciprocity balance sheets, allocation of parental affection and investment, and mate value assessment. This approach will also work for the human animal.

There is simply no question that the mind/brain owes its complex functionality to a history of selection, and it follows as a corollary that an understanding of how the natural selective process works can be a valuable aid to the practising psychological scientist. Williams was right, as the chapters in this symposium volume demonstrate.

Reference

Williams GC 1966 Adaptation and natural section. Princeton University Press, Princeton, NJ

The concept of an evolved adaptation

Randy Thornhill

Department of Biology, University of New Mexico, Albuquerque, NM 87131, USA

Abstract. A Darwinian adaptation is an organism's feature that was functionally designed by the process of evolution by selection acting in nature in the past. Functional design rules out explanations of drift, incidental effect, phylogenetic legacy and mutation. Elucidation of the functional design of an adaptation entails an implicit reconstruction of the selection that made the adaptation. Darwinian adaptations and other individual traits may be currently adaptive, maladaptive or neutral. One relatively recent meaning of adaptation is inconsistent with the Darwinian conception of adaptation. The inconsistent meaning characterizes much research on humans and non-human species in behavioural ecology. Its focus is on equating Darwinian adaptation with current adaptiveness. Current adaptiveness is not an actual scientific prediction of a hypothesis about Darwinian adaptation. Some aspects of the discussion in the evolutionary literature surrounding the current adaptiveness view of adaptation are evaluated. Contrary to claims by some who advocate current adaptiveness, the environment of evolutionary adaptedness of humans and other organisms is scientifically knowable through discovery of the functional design of Darwinian adaptations.

1997 Characterizing human psychological adaptations. Wiley, Chichester (Ciba Foundation Symposium 208) p 4–22

The word 'adaptation' has several meanings in biology (Medawar & Medawar 1983). It is sometimes used to refer to changes within the lifetime of an individual that give it better fit to an environmental problem. Callusing from friction is one such adaptation; others are antibodies against a parasite and greater lifetime mate number and sexual advertisement in women who are reared and live in social environments of reduced paternal investment (Thornhill & Gangestad 1996). Adaptations of this kind are environmentally caused and genetically caused — all features of the individual are, and with equal input from both causes (Daly & Wilson 1983) — but they are not the result of a genetic difference between those with the adaptation and those without. That is, the state variation (with and without) is not heritable (i.e. it doesn't show what biologists and animal and plant breeders call heritability) but each state can be inherited. For example, all humans have the underlying physiological machinery to create a callus upon encounter with the evolutionary ancestral stimulus of skin friction. This machinery is inherited in that parents and offspring both have it. When parents and offspring both experience friction on the same part of the body, callusing and callus location are inherited, even though neither are heritable. When mothers and

4

daughters both have impoverished upbringings and live in such environments as adults, the inheritance of less restricted female sexuality is seen. The psychological machinery/adaptation that reads the potential for male investment is always inherited.

The respective machinery of condition-dependent callus formation, condition-dependent sexuality of women and condition-dependent antibody formation is adaptation in two other mutually dependent senses, one in terms of outcome and one in terms of the process that generated the outcome. The species-typical machinery (sex specific in the sexual example) is the long-term outcome of evolution by Darwinian selection and thus is an adaptation. Selection is differential reproduction of individuals as a consequence of fitness variation among them. As Williams (1966) cogently argued, fitness is most usefully defined as an individual's reproductive success due to its fit to/design for an environmental problem. Number of offspring produced by an individual *per se* is not the currency of fitness, nor is differential individual survival. Differential performance of individuals frequently and inevitably to some degree in each generation of every population arises from chance, not design for/fit to a problem, and biologists refer to this as drift, or as genetic drift when performance variation is heritable. When variation in design is heritable, selection increases in frequency those designs that confer the highest fitness, i.e. the best phenotypic solution to the ecological problem. Darwinian adaptations adapt organisms to their environment and the historical process going on during the creation of a Darwinian adaptation is also referred to as adaptation.

Darwinian adaptation

All usages of adaptation I have discussed are interrelated. They all refer to an organism's design for environmental problems. Darwin's view of adaptation, a phenotypic trait with purposeful/functional design created by past selection, underlies all these usages. A functionally designed feature of an organism, a Darwinian adaptation, is one that performs a purpose 'with sufficient precision, economy, efficiency, etc. to rule out pure chance as an adequate explanation' (Williams 1966). Functional design rules out phylogenetic legacy (see below), drift, incidental effect and mutation as explanations for the existence of an individual's feature, and favours selection as an explanation because only selection can make phenotypic design. Teleonomy (Pittendrigh 1958), a subdiscipline of evolutionary biology, is based on the recognition, elucidation and analysis of functional design in living systems. If selection created a trait, that trait will be functionally designed for the ecological problem generating the selection. Thus, each Darwinian adaptation contains in its functional design the data of the cause—the selective force—that created it. These data are both necessary and sufficient to demonstrate scientifically the historical environmental problem that was causal in creating the adaptation. The concept of Darwinian adaptation is central in biology because Darwinian adaptations are the biologist's sole source of information about the forces of selection that were actually effective in designing organisms during the evolutionary history of life.

Consider any hypothesized selective force that may have acted in the history of life. It can only be demonstrated to have acted effectively when it produces an evolved outcome—adaptation to deal with that selection. For example, the hypotheses that male human ancestors adaptively coerced matings and that selection for anti-rape was effective on females in human evolutionary history would be supported by evidence, in the first case, that men have a feature(s) functionally designed for rape, and, in the second, that women have anti-rape adaptations. This is the only evidence that supports these hypotheses (Thornhill & Thornhill 1992, Thornhill 1996).

Darwin's method of historical science applied to adaptation was simply to hypothesize about the selection that made an adaptation, and then test predictions derived from the hypothesis (e.g. Ghiselin 1969). The predictions are about what must be true of the design of an adaptation if the hypothetical historical selective force was, in fact, causal (also see Williams 1966, Dawkins 1986). The Darwinian conception of adaptation is part of what Williams (1992) calls the view of 'organisms as historical documents'. Only a small subset of an organism's features are its Darwinian adaptations. Their functional designs document historical selection. The rest are incidental effects of adaptations (Gould & Lewontin's [1979] spandrels), phylogenetic legacies and phenotypic effects due to either drift or deleterious mutation. A potential spandrel is human rape. If so, this behaviour must arise from psychological Darwinian adaptation to an environmental problem other than rape (Thornhill & Thornhill 1992). Phylogenetic legacy refers to features that record evolutionary phenotypic compromises and constraints from which Darwinian adaptation may be moulded (Williams 1992). Because selection is blind to future needs and gradually modifies pre-existing phenotypic features rather than starting from phenotypic scratch, phylogenetic legacies abound. Well-understood examples in vertebrates are such features as the inversion of the retina and the crossing of the digestive and respiratory systems. These legacies themselves have no functional design, but they are very much a part of the organism as a historical document. They are part of the picture of phylogenetic descent during the evolutionary history of life. Deleterious genes in mutation–selection balance are seen in all species, but at low frequencies because of negative selection. If a trait is more abundant than the mutation–selection balance can account for, say it has a frequency of 2% in the population, it was probably positively selected, at least in some environments or times. When such a trait is shown to have functional design for offsetting an ecological problem, it then goes on the list of Darwinian adaptations, even though it may be relatively rare in the population. Drift can only cause the evolution of traits without significant cost to their bearer because selection will act to determine a trait's fate when cost exists.

It is not uncommon in the literature for a trait's evolutionary origin to be confounded with its selective history. Origin and selective history deal with different historical causes. The human chin appears to have originated as an incidental effect of Darwinian adaptation of the human mandibles, but evidence indicates that sexual selection then acted on male and female chins in the human evolutionary lineage because the adult chin seems functionally designed to signal health and thereby evolutionary historical mate value (see Thornhill & Gangestad 1996). Teleonomy coupled with phylogenetic analysis can illuminate the original phenotypic states from which Darwinian adaptation subsequently evolved by selection.

Current selection

Relatively recently, there has been a change of view about the meaning of adaptation in the minds of some evolutionary researchers, particularly many behavioural ecologists. Reeve & Sherman (1993) espouse this view when they define adaptation as any trait that is currently adaptive, i.e. linked to current reproductive success (offspring number), preferably lifetime reproductive success. This is the same as advocating that an adaptation is a trait under current selection because if current trait variation is correlated with current variation in reproductive success, selection is going on, either directly on the trait in question or on a correlated trait.

It is difficult to measure from studies of current selection the relevant parameters for understanding evolutionary maintenance of a Darwinian adaptation or other trait. Sherman & Reeve (1997) argue that if a trait is currently associated with reproductive success in a certain context, then it is likely that the trait is maintained by selection in that same context. Selection is variable in its occurrence, intensity and direction in space and time (also see below), and thus it is unlikely that a study of current selection in part of a species range over a brief time will provide evidence for the evolutionary maintenance of a trait. Maintenance can only be understood in relation to historical evolutionary forces that can be determined by Darwin's method of historical science.

Some comparative biologists have claimed that an appropriate sample size in comparative studies is the number of times the adaptation evolved from some prior state and not the number of species that show the adaptation (Harvey & Pagel 1991). Williams (1992) in criticizing this claim argued that if an adaptation in a species is currently under stabilizing selection, it should count as an independent datum in the correlation across species between ecology and adaptation. However, neither the form of current selection on a trait nor the presence or absence of selection on a trait is relevant to whether a trait is a Darwinian adaptation. Instead one must look for functional design. If multiple species of monkeys in the same genus live in conditions of high sperm competition (multi-mate groups and females mating with multiple males) and all have large testicles, large ejaculates and high sperm number, the N is species number and not $N = 1$ (i.e. for the genus). The machinery of adaptation in these independent gene pools evolved once from prior primate features, but it must have been maintained by selection processes having acted independently on the gene pools. Past selection is the only game in town for explaining the persistence of Darwinian adaptation in a phylogenetic line, even if the adaptation occurs in all species of a taxon. Phylogenetic inertia, although said to be a prime mover of social evolution (Wilson 1975), is not an ultimate cause; one must look to historical selection to explain the persistence of adaptation among species within a taxonomic group.

Wilson (1975) argued that there are two prime movers of social evolution: phylogenetic inertia and ecological pressures. The latter is Darwinian selection. The former pertains to factors affecting resistance to or lability for evolutionary change. A range of diverse factors is involved: genetic and phenotypic variability; prior Darwinian adaptation and incidental effects (also called preadaptations); and trade-offs among interacting adaptations (e.g. predator defence vs. colourful courtship display). Many would call these and related factors potential evolutionary constraints because some can be circumvented by subsequent selection and some apparently cannot (see Williams 1992 for discussion). Error always arises when evolutionists use constraintist or phylogenetic inertia hypotheses to explain Darwinian adaptation. For example, Wilson used haplodiploidy to explain the restriction of eusociality to *Hymenoptera*, with the exception of termites. Of course, now we understand that eusociality is seen outside these taxa in the diplodiploid mole rats and aphids. Also, we see it in temporary form in the forfeiting of offspring production by nepotistic helpers in many birds and some mammals. The general point here is that phylogenetic inertia is not a prime mover of evolution comparable to the actual agents of evolution (drift, gene flow, selection and mutation). Evolutionary agents cause evolutionary change. Phylogenetic constraints do not cause evolution in the same sense that evolutionary agents do. The constraints affect what the agents have to work with, but they are no more causal in the explanation of an adaptation's functional design than the existence of air or genes; without both, adaptations could not have evolved. Evolution by selection is both necessary and sufficient to explain the functional design of an adaptation.

Some advocate that current selection is the most useful or only relevant way to study adaptation (e.g. Wade 1987) or that it is just as valid a way to answer questions about selective background of an adaptation as teleonomy (e.g. Sherman & Reeve 1997). As discussed in detail by Williams (1966, 1992), Thornhill (1990, 1991), Symons (1992), Tooby & Cosmides (1990), Irons (1997) and others, current selection cannot identify Darwinian adaptation. Evolutionary novel environments, that is those that lack the ancestral circumstance(s) (have modern human contraception, novel resource levels, etc.) which created the evolutionarily historical adaptive relation between a trait and its focal environmental problem, can eliminate the causal relation between an adaptation and current reproductive success.

Furthermore, evolutionarily novel traits and incidental effects of adaptation can be currently adaptive. Burley et al (1982) have shown that artificial red leg bands on male zebra finches result in high male reproductive success because of female preference. The female preference for the leg bands is an incidental effect of female adaptation for preferring males with red natural ornamentation. Actually this incidental effect may lower female reproductive success because sons' attractiveness stemming from the band is not inherited from the father.

The view that evolutionary theory makes the prediction that Darwinian adaptation will be found to be currently adaptive is erroneous. Because of evolutionary novelty, negative reproductive success results can't falsify that view. Similarly, positive reproductive success results for a trait cannot identify it as a Darwinian adaptation. A scientific prediction is one that must be true if the hypothesis generating it is true. Current reproductive success expectations are not actual scientific predictions of hypotheses about Darwinian adaptation.

Furthermore, it remains an open question whether current adaptiveness can be measured meaningfully. Even if current adaptiveness is studied in a truly natural, evolutionary historical environment, there remains the major measurement problem associated with variable current selection. As Williams (1992) put it, 'It would be rash to assume that any brief and local biological study can yield reliable data on long-term directions and intensities of natural selection.' I would add that Williams' conclusion applies equally to not so brief and non-localized studies. It is difficult to measure current adaptiveness in nature, except of course in a way that may allow rejecting the null hypothesis of no relation between reproductive success and a trait in a given study. Those interested in measuring current adaptiveness may benefit from using meta-analytical techniques (Arnqvist & Wooster 1995) to analyse many studies of the same species in different parts of its range and times. Not only does meta-analysis allow statistically accurate summary conclusions from multiple studies (e.g. that overall in studies of 40 human societies there is a statistically significant positive relationship between men's physical attractiveness and the number of their surviving offspring), but it can also robustly identify where among the studies the relation does not hold if this is the case (e.g. where modern medicine eliminates a significant amount of variance in offspring viability). Of course, the meta-analytical study of current selection on a trait cannot address the question of Darwinian adaptation.

Data on current adaptiveness or any other data related to current evolution in a population (e.g. heritability, gene frequencies) are not actual evidence bearing on hypotheses about long-term historical evolution. Yet microevolutionary data can be persuasive to evolutionists. Take the laboratory experiments by Rice (1996) on *Drosophila* in which he stopped females from co-evolving antagonistically with males and demonstrated that males in about 40 generations of artificial selection show better designs than wild-type males for circumventing female adaptation which appears to protect female interest surrounding mating. These results were held out by some evolutionists as the first strong evidence that selection on males and females in the context of sexual conflict is real (see Chapman & Partridge 1996).

Rice's elegant experimental results serve merely as a hypothesis for what may have occurred in the long-term evolutionary process in nature. Short-term evolutionary effects may not be extended into long-term evolution because of changes in future generations in the correlations between traits, heritability and unforeseen counter-selection. Also, artificial selection, in the lab or elsewhere, may be significantly different than selection in nature (e.g. Orr & Coyne 1992). Finally, the genetics of the experimental system were highly contrived, although appropriately chosen for the purpose of the experiment.

Strong evidence for effective selection having occurred in co-evolutionary arms races between the sexes in long-term evolution in nature actually exists. There are multiple examples of rape adaptation in male animals (review in Thornhill & Sauer 1991), and female water striders have anti-rape adaptation (Arnqvist & Rowe 1995). There is some evidence for anti-rape adaptation in women as well (Thornhill 1996). In the striders there is clear evidence of adaptation for rape in males and adaptation to prevent rape in females. These adaptations in the strider are strong evidence that selection in the context of sexual conflict has been effective on both sexes in long-term evolutionary history in nature.

The environment of evolutionary adaptedness

Sherman & Reeve (loc. cit.) argue that measuring fitnesses in the environment of evolutionary adaptedness (EEA; see Symons 1992) is impossible; but this is exactly what teleonomy does (Sherman & Reeve 1997). Darwinian adaptations are caused ultimately by evolution via selection in the past; that is, the functional design of an adaptation demonstrates scientifically that that functional design promoted higher reproductive success in the species' EEA than alternative designs. A species' EEA is the set of environmental features which generated the selection that made the species' adaptations. The EEA is stamped in the functional design of adaptation.

There is much discussion in publications on evolutionary methods about what is seen as the great problem of identifying the EEA of humans. (The methodological issues are no different in considering the EEA of any species.) Hrdy (1981) wrote an entire book on her proposal that non-human primates contain the data on the nature of the human EEA and the data on evolved human nature. She stated:

> The purpose of this book . . . is to suggest a few plausible hypotheses about the evolution of woman that are more in line with the data . . . Since we cannot travel back in time to see that history in the making, we must turn to those surrogates we have, other living primates, and study them.

It is scientifically incorrect to infer the selection that made women's adaptations from the adaptations of female non-human species. Symons (1982) has provided a detailed critique of Hrdy's method. In short, the only way to discover the selective forces that shaped women's sexuality is to characterize the functional design of women's sexual adaptations. The data on the sexuality of existing non-human primates don't provide any evidence for how human evolutionary ancestors conducted their sex lives. These data, when they comprise a demonstration of Darwinian adaptation, are evidence for the selection that worked in the evolutionary history of the primate involved. The human EEA is instantiated only in the functional designs of human Darwinian adaptations. Hrdy's view that we can't travel back in time to see human history in the making misses Darwin's adaptation concept and William's view of the organism as a historical document. The functional design of an adaptation is the record of the salient, long-term environmental problem involved in the creation of the functional design. Thus, we can actually scientifically go back in time and discover the creative selection pressures that were effective in human evolutionary history.

How do hunter–gatherers fit into the issue of the mystery of the EEA? As Betzig (1997) points out, living human foragers differ from one another, often greatly, in foraging behaviour, mating system, etc. She feels that this makes it unclear where one should begin for a reconstruction of the selection pressures that have shaped *Homo sapiens*. But it is completely clear. We start and finish with the functional design of human Darwinian adaptation. It is likely that the kinds of adaptations underlying the cultural variation discussed by Betzig (1997), as well as Kacelnik & Krebs (1997), arise from a universal human nature, or two human natures, one for males and one for females, depending upon the particular adaptation. In humans it has proven useful to look for species-typical (often sex-specific) psychological adaptations, the evolved information-processing mechanisms of the human mind, that underlie the vast behavioural variation across and within cultures (Symons 1987, Tooby & Cosmides 1990, Brown 1991, Thornhill & Gangestad 1996).

This is not to say that using Darwin's method is simple. The skill involves identifying Darwinian adaptation and elucidating its functional design. Although many adaptations are species typical because of past selection fixing adaptive features throughout a species, this is certainly not a criterion of Darwinian adaptation. There are numerous well-known cases of alternative evolved functional designs within species. The most common alternative functional designs within species involve sexually dimorphic adaptations. Also, the teleonomist anticipates that Darwinian adaptation will exhibit special-purpose functional design, not general-purpose design (see Symons 1987). Because environmental problems generating selection are specialized problems (e.g. spatial orientation, coercing matings and finding a mate with good prospects for successful reproduction), selection is expected to favour problem-specific designs. This is the theoretical reason for expecting special-purpose adaptation. Empirically, there is no example of a general-purpose adaptation. For evidence in humans, look in *Gray's Anatomy*. For other species, pick up an insect or plant physiology book, for example. The same goes for what is known of psychological adaptation in humans and non-human animals. Special-purpose functional design is the rule, which has profound implications for how teleonomists approach the study of adaptation.

In conclusion, misunderstandings of historical science have generated claims that teleonomy has inherent problems that limit its utility and claims that study of current selection or microevolution in general are ways to understand the salient features of life's evolutionary history. Detailing the reproductive successes and failures of individual insects, birds and people is interesting, exciting, challenging, rewarding intellectually and a worthy life endeavour for a biologist. But those who wish to know the evolutionary history of their study species must obtain data of the outcomes or phenotypic stamps of that history.

Acknowledgements

Martin Daly, Don Symons and Margo Wilson provided useful criticisms of this manuscript.

References

Arnqvist G, Rowe L 1995 Sexual conflict and arms races between the sexes: a morphological adaptation for control of mating in a female insect. Proc R Soc Lond Ser B 261:123–127

Arnqvist G, Wooster D 1995 Meta-analysis: synthesizing research findings in ecology and evolution. Trends Ecol Evol 10:236–240

Betzig L 1997 Not whether to count babies, but which. In: Crawford C, Krebs D (eds) Evolution and human behavior: ideas, issues, and applications. Lawrence Erlbaum Associates Inc., Hillsdale, NJ, in press

Brown DE 1991 Human universals. McGraw-Hill, New York

Burley N, Krantzberg G, Radman P 1982 Influence of colour-banding on the conspecific preferences of zebra finches. Anim Behav 30:444–455

Chapman T, Partridge L 1996 Sexual conflict as fuel for evolution. Nature 381:189–190

Daly M, Wilson MI 1983 Sexual development and differentiation. In: Daly M, Wilson MI (eds) Sex, evolution and behavior, 2nd edn. Duxbury Press, Boston, MA, p 243–277

Dawkins R 1986 The blind watchmaker. WW Norton & Co, New York

Ghiselin MT 1969 The triumph of the Darwinian method. University of California Press, Berkeley, CA

Gould SJ, Lewontin RC 1979 The spandrels of San Marco and the panglossian paradigm: a critique of the adaptationist program. Proc R Soc Lond Ser B 205:581–598

Harvey PH, Pagel MD 1991 The comparative method in evolutionary biology. Oxford University Press, New York

Hrdy SB 1981 The woman that never evolved. Harvard University Press, Cambridge

Irons WG 1997 Adaptively relevant environments versus the environment of evolutionary adaptativeness. Evol Anthropol, in press

Kacelnik A, Krebs JR 1997 Field studies in sociobiology: what are they for? In: Betzig L (ed) Human nature: a critical reader. Oxford University Press, Oxford, p 21–35

Medawar PB, Medawar JS 1983 Aristotle to zoos: a philosophical dictionary of biology. Harvard University Press, Cambridge, MA

Orr HA, Coyne JA 1992 The genetics of adaptation: a reassessment. Am Nat 140:725–742

Pittendrigh CS 1958 Adaptation, natural selection, and behavior. In: Roe A, Simpson GG (eds) Behaviour and evolution. Yale University Press, New Haven, CT, p 390–416

Reeve HK, Sherman PW 1993 Adaptation and the goals of evolutionary research. Q Rev Biol 68:1–32

Rice WR 1996 Sexually antagonistic male adaptation triggered by experimental arrest of female evolution. Nature 381:232–234

Sherman PW, Reeve HK 1997 Forward and backward: alternative approaches to studying human social evolution. In: Betzig L (ed) Human nature: a critical reader. Oxford University Press, Oxford, p 147–158

Symons D 1982 Another woman that never existed. Q Rev Biol 57:297–300

Symons D 1987 If we're all Darwinians, what's the fuss about? In: Crawford C, Smith M, Krebs D (eds) Sociobiology and psychology: ideas, issues and applications. Lawrence Erlbaum Associates Inc., Hillsdale, NJ, p 121–146

Symons D 1992 On the use and misuse of Darwinism in the study of human behavior. In: Barkow J, Cosmides L, Tooby J (eds) The adapted mind: evolutionary psychology and the generation of culture. Cambridge University Press, Cambridge, p 137–159

Thornhill R 1990 The study of adaptation. In: Bekoff M, Jamieson D (eds) Interpretation and explanation in the study of behavior, vol II. Westview, Boulder, CO, p 31–62

Thornhill R 1991 Teleonomy and the study of sexual selection. Acta XX Congressus Internationalis Ornithologici III:1361–1366

Thornhill R 1997 Rape-victim psychological pain revisited. In: L Betzig (ed) Human nature: a critical reader. Oxford University Press, Oxford, p 239–240

Thornhill R, Gangestad S 1996 The evolution of human sexuality. Trends Ecol Evol 11:98–102

Thornhill R, Sauer KP 1991 The notal organ of the scorpionfly (*Panorpa vulgaris*): an adaptation to coerce mating duration. Behav Ecol 2:156–164

Thornhill R, Thornhill N 1992 The evolutionary psychology of men's coercive sexuality. Behav Brain Sci 15:363–375

Tooby J, Cosmides L 1990 On the universality of human nature and the uniqueness of the individual: the role of genetics and adaptation. J Pers 58:17–67

Wade MS 1987 Measuring sexual selection. In: Bradbury JW, Andersson MB (eds) Sexual selection: testing the alternatives. Wiley, New York, p 197–207

Williams GC 1966 Adaptation and natural selection. Princeton University Press, Princeton, NJ

Williams GC 1992 Natural selection: domains, levels, and challenges. Oxford University Press, Oxford

Wilson EO 1975 Sociobiology: the new synthesis. Harvard University Press, Cambridge, MA

DISCUSSION

Mealey: How long would it take, bearing in mind that constraints and selection pressures are changing over time, for data to be lost (as opposed to being gained) as an adaptation becomes more fixed and less variable?

Thornhill: Evolutionary science cannot predict future evolution because the genetic parameters can change and there can be counter-selection. Therefore, we cannot foresee what will happen in the future.

Mealey: But I wasn't referring to the future, rather to the present versus the past, i.e. some of the human data may have been lost, given the dramatic environmental and selection changes that we have experienced.

Thornhill: The data are not lost if one thinks in terms of the way in which the organism has been redesigned by selection in recent evolutionary history. Indeed, in this case the data are gained rather than lost.

Maynard Smith: The data are lost in *Homo erectus* concerning adaptation to an earlier environment, and you implied that we should therefore look at *Homo sapiens* rather than *H. erectus*.

Thornhill: If one is interested in the selective history of *Homo erectus*, then one should look at the adaptations of *Homo erectus*; and if one is interested in the selective history of *Homo sapiens*, then one should look at the adaptations of *Homo sapiens*.

Dawkins: This cannot be correct because your whole thrust has been that we should be looking at the history of the forces that have led to the species that we are now looking at. When one looks at the genes of a modern animal, one is looking at a description of ancestral environments. If one could read the language, one would be able to read the ancestral environments that led to the development of the modern animal. In that sense, the environment of *H. erectus* is also written into modern *H. sapiens*.

Tooby: When you say 'written' do you also mean 'overwritten'?

Dawkins: There will presumably be a bias in favour of recency and in favour of the places that the ancestors have lived in for a long time. For example, if our ancestors took a brief dip into the sea then there would be a brief description of that, but mostly the environments of longer duration would be written in. It is not possible to just draw a line around one particular species, say *H. sapiens*, just because it is the species that you're working on. Your historical bias towards the long term requires that you study those earlier times.

Thornhill: When I look at the adaptations of *H. sapiens*, I want to know the selective pressures in the lineage that led to *H. sapiens*, and that's a different question to that of the adaptations of *H. erectus*. However, I do agree that in the genes (or rather I prefer to use the term 'in the adaptations') of those species there are data about the kinds of environments that were historically significant in terms of the reproductive success of individuals, and that these are stamped in each species-specific adaptation.

Kacelnik: I would like to mention a somewhat trivial semantic problem. In the phrase 'an adaptation can be currently maladaptive' the word 'maladaptive' could be taken to mean a trait that was previously an adaptation but is currently not an adaptation, namely it doesn't promote fitness gains. The problem with reading history by looking at current adaptations is that the state of the genome can only tell the set of all possible histories that could lead to that particular state, and the data on the actual histories have been lost. That is, the current genome could have been reached via many different pathways that cannot be reconstructed by looking at the genome today, and therefore the analysis of current fitness consequences cannot be used to identify past adaptations.

Miller: I'm concerned about a possible circularity between the analysis of adaptations and the analysis of the environment of evolutionary adaptiveness (EEA). By your argument, if, in 100 years, every human could read, cognitive psychologists might claim that reading is an adaptation because they could elucidate the functional design features of reading and they could simply read off what the selection pressures were for this adaptation. This is an extremely weak strategy for demonstrating to psychologists that evolutionary thinking is useful. Much of the heuristic value of evolutionary psychology comes from people's tacit understanding of what the EEA must have looked like, and it doesn't come from studying the adaptations at all. When we rule out entire classes of possible adaptations; it is not because they don't have certain design features it is because they don't match what we think was happening during human evolution.

Thornhill: The only way you could be sure that you had identified an EEA feature that mattered in terms of generating reproductive success differentials is to find evidence for its existence in the design of the adaptation. We can think about EEA scenarios to guide our reasoning, but the proof that predator A or social circumstance B, for example, was present in human evolutionary history and generated net positive selection is in the identification of human adaptations for dealing with predator A or social circumstance B.

Cosmides: Geoffrey Miller's argument presumes that universality is the overriding criterion for claiming that an aspect of the phenotype is an adaptation. But universality is neither necessary (ovaries are adaptations, but they are not universal) nor sufficient (everyone held underwater for long enough drowns, but drowning is not an adaptation). To claim that an aspect of the phenotype is an adaptation, one needs to show that it is well designed for solving an adaptive problem (a problem that recurred over many generations and whose solution affected reproduction) and that its presence is not better explained by appeal to physical law or as a by-product of some other adaptation.

There would be no danger of mistaking reading for an adaptation even if, in 100 years, everyone could read. Despite its universality, there would still be no evidence that there are any cognitive mechanisms that are *well designed* for acquiring the ability to read. It is certainly possible that, in the long-term future, our species might evolve adaptations for reading. To show that this had happened, what kind of evidence would future scientists look for? They would look for evidence that the human cognitive architecture contains mechanisms well designed for learning how to read; that their properties are not better explained as features designed for speaking (or some other function); that, given the appropriate ontogenetic triggers, these reliably develop in all intact humans across the normal range of environmental variation; and that their development manifests a panhuman ontogenetic time table.

Pinker: As Darwin said, humans have an instinctive tendency to speak, as we see in the babble of our young children, but no child shows an instinctive tendency to write. By studying the psychology of the individual child one can obtain evidence that the development of language (but not reading) is an adaptation. For example, children develop language without being taught: if children without a language are grouped together a form of language emerges but, in contrast, a writing system does not emerge spontaneously when children interact, or even when adults interact for thousands of years—writing systems have been invented only a few times. It is difficult for children to learn to read unless there is an explicit programme designed to teach them.

Daly: I am aware that you're not saying this, but I would watch out for being taken to imply, because of your emphasis on children, that the criterion of adaptation is appearance early in development.

Cronin: You said that it is difficult for children to learn to read, but what would happen if we devised more effective schemes for teaching children to read?

Pinker: If children showed evidence of developing the ability to read spontaneously on exposure to print, and if, having been exposed to part of the alphabet system, were able to guess the rest, then one would guess that some of the information in the alphabetic system came from the brain of the child.

Tooby: Geoffrey Miller brought up the problem of circularity. Almost all interesting scientific claims, when stated in their most abstract form, sound circular. For example, in the equation $F = ma$, how do you measure any one of the variables? You use the other two to measure the third. So force is measured by how much it accelerates a given quantity of mass, and mass is measured by how much an object is accelerated by a given amount of force. These variables are interdefined, or 'circular', which means they stand in a clearly defined, logical and non-arbitrary relationship to each other. This gives Newtonian mechanics, and other such scientific systems their power. What leads to the justification and widespread adoption of such theories is that when they are applied to different sets of observations, they produce an endlessly rich network of mutually consistent observations and inferences. That is, any one use of mass and acceleration to measure force is tautologous, but the mutual consistency of large sets of measurements of different interacting objects using Newtonian mechanics was not a foregone conclusion, and shows the utility of the theory. This 'circularity', or conceptual and methodological interdependence, is not a weakness but a strength for evolutionary biology as well. I endorse the suggestion that if the goal was to discover human selective history then it would be necessary to look at present adaptations. However, what makes science particularly interesting is that it is possible to tie together different data sets from different origins which demonstrate mutual consistency, so that independent descriptions of the environments in which humans evolved can be obtained. These descriptions of ancestral environments can be used to formulate testable hypotheses about design features of specific adaptations in the modern human psychological architecture—and allow the discovery of psychological machinery that would not have otherwise been discovered. By the same token, it is entirely valid to use some features of modern adaptations to make reliable inferences—or 'observations'—about the specific features of the ancestral world that are necessary to explain the evolution of modern designs. These determinations about otherwise unobservable features of the ancestral world can then be used to formulate new and testable hypotheses about yet other cognitive adaptations to look for in modern humans.

So while it may sound circular to use modern adaptations to discover and characterize ancestral environments and selection pressures, while also using characterizations of ancestral environments to discover and characterize modern adaptations, it is not in fact circular. This is because in each case it is different converging and interlocking data sets and inferences that are being used to go in each direction. For example, facts about modern human and ape sperm competition are being used to generate reliable knowledge about an ancestral hominid mating system that had to be in place to give rise to, and explain, these modern adaptations. The determination, through these observations, of what type of ancestral mating system hominids had allows one to make other inferences about, and to test for the existence of design features of sexual jealousy mechanisms that no one would otherwise have thought to test for.

Bouchard: I would like to discuss microevolutionary experiments because I believe they carry too much weight. They are often carried out in a restricted range, they often reflect odd, extreme selective processes, and yet they are sometimes used as the basis of the rules by which one looks at human evolution. On the other hand, there is nothing wrong with looking at the neural circuit and looking for the genes that shape that circuit because, when we have a handle on some of these, it will be possible to look back and see whether they represent something we didn't think existed in the past environment.

Thornhill: Finding the neural genes would be useful in the sense of increased knowledge, but it is not the correct approach for characterizing adaptations.

Bouchard: I disagree. It can be fundamental to the enterprise in the sense that it may elucidate features of something in the past that were not immediately obvious. Our characterization of the EEA is coarse and it needs to be refined. Also, the picture of the past environment obtained from analysing complex adaptive structures may be completely different to that obtained from looking at the genes themselves.

Thornhill: But the genes *per se* will not tell you anything about history. Knowledge of the history of effective selection can only be obtained from the functional designs of adaptations because effective selection is the ultimate cause of the designs. Adaptations, and not genes *per se*, are the ontogenetic products of gene–environment interactions.

Sherry: Randy Thornhill, you mentioned that the criterion for an adaptation is not current adaptiveness, and not necessarily a feature of the nervous system or genome, but is instead a functional design feature. However, you haven't mentioned anything explicitly about the criteria for good functional design.

Thornhill: Certain adaptations I have worked on show functional design in a way that falsifies certain hypotheses about their design and strongly supports only one functional hypothesis. That is, the functional design of the adaptation meets a priori design specifications of one hypothesis about the selection responsible for it, but not the predictions of alternative hypotheses about selection pressures that theoretically could have created the design. For example, my studies have shown that in certain scorpionflies the clamp organ on the back of the male's abdomen is a rape organ.

Sherry: But what criteria did you use to convince yourself that it is a rape organ?

Thornhill: It's a matter of argument from design. You argue what the function is from the ability of the design to solve the particular ecological problem. The clamp is designed specifically for forcing mating on unwilling females. It is not designed for sex recognition, species' recognition, preventing the mating male's mate from being taken over by another male or any other function that might theoretically play a role (digestion, excretion, flight, etc.).

Kacelnik: Half of this story seems to be missing. Imagine an animal living on the ground which has skin covering its eyes. You may say that this is an adaptation to something in its past, and you may hypothesize that this animal has previously lived in caves, but unless you find either fossils of that animal in caves, so that you can reconstruct the actual history or make claims about the relative cost of these adaptations today, you can't say that it is or it isn't an adaptation.

Dawkins: Randy Thornhill's fundamental logic depends on whether the adaptive function leaps out when we simply study the form.

Thornhill: Yes, I would say the adaptive function leaps out in the sense that you can predict and understand the fine detail of the trait's design for meeting an ecological problem.

Daly: There are a few iconic examples of elegant design, such as wing design in different birds. A falcon who pursues other birds aerially has a wing with proportions that can be assessed aerodynamically for their efficiency in this particular kind of flight, and that are different from the wings of another bird which has a different flight pattern. These are the best examples of a formal analysis of good design. However, when we try to apply words such as 'mechanism' in psychology, are we not claiming a degree of precision without delivering it?

Dawkins: You're talking like an engineer. If engineers looked at a wing they wouldn't say it vaguely looked as though it would be good for flight, but that all the component features are quantitatively what they would have predicted.

Maynard Smith: I believe that this sort of analysis is possible for behavioural adaptations. What worries me is that I can imagine a situation in which something looks functionally well-designed for doing X, but is not an evolved adaptation for doing X; it has just been learnt because it is a sensible way of doing X. For example, suppose that we have evolved a human universal to classify living organisms, and suppose it is true that all human societies that have been studied have this natural history-classifying behaviour. It would then be an adaptive behaviour, and we would all have a genetic predisposition to exhibit this behaviour. Alternatively, we could learn to classify in this way because animals do in fact belong to natural kinds. How does one decide whether the behaviour is a learnt adaptation? It is possible that it is something you learn using a general-purpose computer that isn't particularly concerned with natural history, but is simply confronted with a world in which there are natural kinds. Alternatively, it could be a pre-programmed behaviour.

Thornhill: I don't see these ideas as being alternatives. All behaviours have environmental ontogenetic causes. In some cases we call these causes learning. All learning requires psychological adaptation for learning and is thus preprogrammed.

Daly: But an interesting question is whether you can demonstrate adaptation for natural history categorization over and above mental adaptation for categorization. Do we possess cognitive procedures that were designed by selection specifically for dividing the world into natural categories?

Cosmides: To show that there are cognitive adaptations that are functionally specialized for identifying species and their taxonomic relationships, one needs to (1) define the relevant selection pressures, i.e. what adaptive problems could have been solved better given the ability to correctly identify species and their taxonomic relationships; (2) use this definition to identify what principles of categorization would have been especially appropriate for solving this problem; (3) demonstrate that people do in fact employ these principles when identifying and making inferences about species and their taxonomic relationships; (4) demonstrate that these principles are *different* to those employed when one is categorizing other kinds of entities (e.g. human-made artefacts) — i.e. that the categorization of species is not a by-product of cognitive adaptations designed for categorizing entities from some other adaptive domain; and (5) show that these principles include features that cannot be explained as the expression of some more general-purpose or content-free categorization mechanism (i.e. one designed to operate across many different content domains).

One would *not* need to show that species categories are 'unlearned'. Categorization procedures are designed to operate on data derived from the world; presumably, these procedures require a certain amount of experiential input to build whatever categories they are going to build. Equally, those who propose that species are categorized via a general-purpose categorization mechanism need to meet certain standards of evidence. First and foremost, they need to avoid vagueness: they need to state what the design features of this proposed mechanism are. Otherwise, the hypothesis that a general-purpose mechanism is responsible is empirically empty. After there is a specific proposal on the table, then a learnability theorist (such as Steven Pinker) can determine whether a general-purpose mechanism with the hypothesized properties can, at least in principle, build the categories that people ordinarily create, given the information available to them (and do this without simultaneously building categories that no human would ever come up with). Many general-purpose theories fail this basic test of theoretical adequacy.

Miller: It seems as though the most universal aspects of the world that create the deepest adaptations are the universals that are the most difficult to notice. What implications does this have, bearing in mind that we're trying perform a functional analysis of the design of adaptations? For a long time learning theorists didn't even notice that there was a problem with generalization. What strategies can we use to try to notice these features that are so universal but often get overlooked?

Thornhill: Adaptations are often overlooked, whether we're talking about fish adaptations, insect adaptations or human mental adaptations. With regard to universality, there's nothing in adaptationism that says that an adaptation is defined as a trait that is seen in all members of a species. Functional design is the criterion for recognizing adaptation. Even if a functionally designed trait is present in only 10% of the population it is an adaptation.

Dawkins: It's not by definition.

Thornhill: But if it shows functional design, then neither drift, mutation nor incidental effect are responsible for the existence of the trait.

Daly: It's a probability model. Some sort of null model of the likelihood of this degree of functional design emerging by other processes is required. There are hypotheses other than selection that, in principle, are logical.

I would now like to ask Dan Sperber to comment on the subject of categorization and the evidence there is for and against the existence of specialized modules.

Sperber: Randy Thornhill has argued that, just by looking at current adaptations, it is possible to read something about the past environment. Similarly, Richard Dawkins has suggested that there may be a description of the past environment in the genes. This is arguable only up to a point, because the 'description' of the past environment you retrieve in this way is coarse. As Alex Kacelnik has pointed out, many different pathways in the past could have led to the present state of affairs.

One important aspect you cannot 'read' just by looking at the present design of organisms is what exactly their apparent adaptations were adaptations to. In order to identify the function of a specific adaptation (and, in fact, to confirm that you are dealing with an adaptation rather than with a side-effect of an adaptation), you need information about the past environment. This information may be so trivial and obvious that you rely on it as mere background knowledge, without realizing that you are using it at all but even then questions will be raised if this background knowledge is not addressed. Imagine, for instance, that you identify in a species a stable behavioural disposition that, at first sight, you attribute to a fear-of-snakes psychological mechanism—an adaptation easy enough to describe and explain. If you are correct then you are indeed reading in the present design of the organism a feature of its ancestral environment: it must have contained snakes (and you may want to look for confirming evidence).

At this stage your description, however plausible, might be false: suppose there were no snakes in the ancestral environment, but other dangerous animals sufficiently snake-like for the mechanism you describe to be an adaptation to their presence. Even if your description of the mechanism as a fear-of-snake one is correct it is still rather coarse, and you may want to refine it. To begin with, fear of snakes in general, as opposed to fear of venomous snakes, is unlikely to be an adaptation at all. Reacting with fear to the presence of non-venomous snakes is best viewed as a side-effect. So, you must assume that there were not just non-venomous snakes, but also venomous snakes, in the environment, and that the psychological mechanism was an adaptation to the risks these venomous snakes presented. So far, you can still read the past from the present, so to speak, but you are not through yet. The adaptation did not evolve as an adaptation to venomous snakes in general, but to actual snakes of one or several specific species represented in the environment. The function of the adaptation must have been to lower the risks caused by these specific snakes. You cannot read this specific function from the current design of the organism. You need, for this, to take into account independent information about the past environment, or else you must leave question marks in your account.

The kind of difficulty that the fear-of-snakes example illustrates extends to core issues in evolutionary psychology. Take the question of how humans categorize biological kinds. A number of us have argued that there is a genetically determined disposition to classify organisms (as opposed to other objects in the world) in a quite specific manner (see for instance Atran 1994, Keil 1994, Sperber 1994). Humans everywhere use a taxonomic form of classification with no overlap of categories to classify biological kinds, and they assume, at least implicitly, that each category (dog, snake, oak, mushroom, etc.) is characterized by an underlying 'nature' or 'essence'. The psychologist Frank Keil has been one of the most effective advocates of this view (e.g. Keil 1994). However, Keil doubts that this psychological mechanism, which he has helped identify, is genetically dedicated to the task of biological classification to which it happens to be put. What Keil does — as would any psychologist — is to look at the current design, and he sees nothing there which speaks of animals or plants. Keil therefore argues that humans have a non-dedicated, non-specialized 'mode of construal' that they happen to apply to animals and plants. They do so because the biological domain happens to provide the best possible inputs for this mode of construal. I believe that Keil is, strictly speaking, correct in arguing that the function cannot be read from the design. From an evolutionary point of view, however, it is straightforward to consider the hypothesis that, if a specific psychological mode of construal evolved, it is likely to have been as an adaptation to the problems raised by some specific cognitive domain. Although the function cannot be read just from the design, it can be reasonably surmised from knowledge of the design taken together with knowledge of the ancestral environment. In this particular case, the relevant environmental knowledge is utterly trivial: there were plants and animals, and the fitness of our ancestors depended to a large extent on their ability to identify them and to infer hidden properties from these identifications. The taxonomic and essentialist mode of construal is best viewed as an adaptation to this particular challenge. It is not inconceivable, on the other hand, that more subtle aspects of the way humans classify biological kinds can be explained by looking in detail at the character of this challenge.

Thornhill: The first problem there is to categorize the adaptation properly. If you're talking about a psychological adaptation, such as fear of snakes, you must ensure that you're focusing on the information that the hypothetical piece of mind uses. Particular movements in a particular size range may cause a fear of snakes, and several components of information may be processed. When you know all this information then you can say that you had fully categorized this piece of mind.

Tooby: A formal theory of adaptation, where you're looking at the probability of goodness of fit, has to include the fact that the adaptation has to be to something, and so any claim about an adaptation involves a formal claim about past environments. You can't have one without the other. They are both necessary for making any adaptationist argument.

Kacelnik: With respect to the point on specificity of representations and mental architecture, if you start with a system with a completely unspecified architecture and you make it more dedicated to perform certain tasks, there comes a point where you have to accept that the system is in some sense 'representational'. We seem to be discussing the possibility of whether specific representations exist in the brain, because everybody accepts that some degree of specificity actually exists. People seem to be interested in problems at the two ends of a piece of string. Some are looking at those aspects involved with learning about the world, i.e. those that no one can be born knowing, and others start from the problem that it is not possible to learn without having certain a priori specifications. The interesting battleground is on the details of where we accept the specificity of dedication of the human mind. It's an empirical question, and not an abstract one that we can solve by arguing about the abstract notion of adaptation.

References

Atran S 1994 Core domains versus scientific theories; evidence from Itza-Maya folkbiology. In: Hirschfeld LA, Gelman SA (eds) Mapping the mind: domain specificity in cognition and culture. Cambridge University Press, Cambridge, p 316–340

Keil F 1994 The birth and nurturance of concepts by domains: the origins of concepts of living things. In: Hirschfeld LA, Gelman SA (eds) Mapping the mind: domain specificity in cognition and culture. Cambridge University Press, Cambridge, p 234–254

Sperber D 1994 The modularity of thought and the epidemiology of representations. In: Hirschfeld LA, Gelman SA (eds) Mapping the mind: domain specificity in cognition and culture. Cambridge University Press, Cambridge, p 39–67

The genetic basis of human scientific knowledge

Roger N. Shepard

Department of Psychology, Building 420, Stanford University, Stanford, CA 94305-2130, USA

Abstract. The ecologically most significant respect in which humankind now dominates all other terrestrial species is in its scientific understanding and technological manipulation of the world. What psychological adaptation underlies this seemingly discontinuous development? There is reason to believe that natural selection has endowed the perceptual/representational systems not only of humans but also of other perceptually and cognitively advanced animals with an implicit knowledge of pervasive and enduring properties of the world. Perhaps especially in the human species, natural selection has, in addition, favoured a heightened degree of voluntary access to the representational machinery embodying this implicit wisdom, thus facilitating the realistic mental simulation of possible actions in the world before taking the risk of carrying them out physically. This, together with the emergence of an unprecedented motivation toward understanding, seems to have enabled some human individuals to use 'thought experiments' to convert more and more of the implicit knowledge that we all share into a self-consistent set of explicit scientific laws. Although knowledge of the world must ultimately come from the world, as empiricists claim, it can in this way come through one's genes as well as through one's own direct perceptual interactions with the world.

1997 Characterizing human psychological adaptations. Wiley, Chichester (Ciba Foundation Symposium 208) p 23–38

Through natural selection, humankind, like other animal species, has evolved not only physical but also psychological adaptations to its particular ecological niche. Yet, even among mammals, species differ widely in the breadth of the niche to which they have evolved their adaptations. The extreme of narrow specialization is illustrated by the panda, which survives and reproduces only in a particular habitat of bamboo, on which it has come exclusively to feed. The opposite extreme of adaptability is represented by *Homo sapiens*, an omnivorous, tool-developing and expansionist species that has established communities from equatorial to arctic regions, from sea-level islands to high altitude mountains, and from dripping rain forests to arid deserts. Further developing a technology previously confined to the making of clothing, shelters, stone tools, containers, hunting weapons and fires, this species has extended its dominion deep into the earth and the sea, and high into the upper atmosphere. Now it even contemplates the colonization of the moon and other planets. At the same time, it is rapidly transforming its original terrestrial environment on such a global scale as to threaten the survival of vast numbers of more specialized and less adaptable species, such as the panda.

It has been suggested that *H. sapiens* is now claiming for itself a new kind of niche — the 'technological niche' or perhaps, more generally, the 'cognitive niche'. What, then, is the origin and character of this uniquely human adaptation, which, even if it arose in one particular environment, has led to this unprecedented ability to understand, to manipulate and to dominate an ever-expanding realm?

In focusing on the uniquely human understanding and mastery of the world, I propose to consider: (a) how it might be understood as arising through natural selection; (b) in what respects it appears to be continuous with the pre-scientific capabilities of ancestral species; and (c) what exactly has made possible the apparently discontinuous jump to human technology and science — as pre-eminently exemplified by the emergence of mathematics, physics and, of course, the Darwinian theory of evolution.

In the process I shall argue that the still widely held version of empiricism that insists that any knowledge which an individual possesses about the world can only have come through that individual's own sensory interactions with the world is wholly untenable. The remarkable behavioural capabilities of other animals and the development, in humans, of modern science depend alike, I claim, on a deep, if implicit, wisdom about the world that is our genetic legacy from countless aeons of ancestral interactions with the world.

To understand the potential reach of the 'cognitive niche', we must recognize that every niche, in addition to any properties that are more or less peculiar to that particular niche, possesses properties that are shared with many other niches. Indeed, all natural niches that are hospitable to biological life as we know it share some general, perhaps universal, properties. Here are three examples: (a) on a biologically relevant scale, space is three-dimensional and Euclidean; (b) material bodies can move or be moved about in this space, but cannot occupy the same location at the same time and do not materialize out of, or vanish into, empty space; and (c) at least on the planetary surface, moreover, intervals of relative light and warmth regularly alternate with intervals of relative darkness and coolness (and, on this particular planet, do so with a 24 h period).

Selection pressures in any particular niche that favour individuals with perceptual/ cognitive/manipulative capabilities are also likely to favour individuals that have internalized these pervasive and enduring properties of the world. Each individual that begins its life without the benefit of such internalized knowledge would have to acquire that wisdom from scratch, through trial and possibly fatal error. We are largely unaware of the implicit, internalized wisdom that underlies our own ability to make our way in the world because natural selection has ensured that such wisdom serves us swiftly and automatically at a largely unconscious level.

Capacity for learning as an evolved adaptation

A restrictive type of empiricism that fails to recognize the existence of such a genetically transmitted wisdom tends to be accompanied by another prevalent preconception. This is the (often implicit) equivalence assumed between intelligence and aptitude for learning. Such an equivalence is implied, for example, in the common dismissal of much of the behavioural capacities of 'lower' animals — however complex and adaptive they may be — as arising from 'mere' instinct. Such an equation fails to notice that an all-knowing agent (paradigmatically, an infinitely intelligent God) would have neither need nor use for a capacity for learning. To admit, of any agent, that it could learn something is to admit (contrary to the hypothesis) that there is something that it does not already know.

Clearly, any principles governing an individual's learning are not themselves learned, and so must have been shaped by natural selection in the world. Indeed, it can be shown that there can be no principles of learning that are effective universally, absolutely, and in the abstract (e.g. Schaffer 1994, Shepard 1987, Wolpert 1992). Any candidate principles of learning are effective only relative to a particular class of concrete possible worlds (with particular regularities and correlations). However effective a proposed set of learning principles may be for some set of possible worlds, there will always be an equally large set of possible worlds for which those same principles will lead the agent to make the wrong inferences and generalizations.

Why, then, is a capacity for learning so ubiquitous in animal species? And what accounts for its evident effectiveness in the case of the human species? The answer suggests itself: our evolutionary adaptation to the world (a) is still far from having internalized all there is to be known about that world; nevertheless (b) it has internalized enough about that world to afford us some basis for learning and generalization in that world. Similarly, for the discovery of scientific laws: just as learning is impossible in the absence of already internalized principles of learning attuned to our particular world, the inductive formulation of explicit general principles of the world is impossible in the absence of already internalized principles of induction that reflect existing regularities of that world.

Representation of spatial transformations as an evolved adaptation

We should not be surprised that our representational systems have also internalized the geometry of Euclidean three-dimensional space in which we and all of our ancestors have evolved. But how can we confirm that we have indeed internalized the geometry of this particular space? The internalization of the circadian cycle was established when animals were raised in artificial isolation from the terrestrial 24 h variation in illumination and temperature, and found to maintain, nevertheless, their characteristic 24 h activity cycle. But the very universality of the three-dimensionality of our world precludes our taking an individual out of this world to see whether it would continue to perceive and to think three-dimensionally. We can, nevertheless,

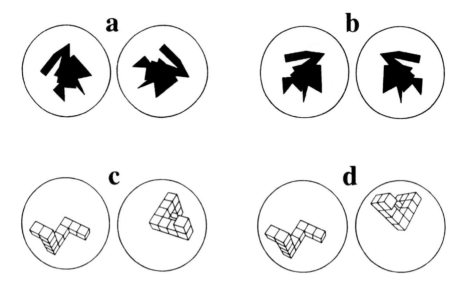

FIG. 1.   Pairs of shapes that when alternately displayed give rise to different types of apparent motion: (a) identical planar objects giving rise to a rigid rotational motion in the plane; (b) enantiomorphic planar objects giving rise to a rigid rotation in space; (c) identical solid objects giving rise to a rigid screw displacement in space; and (d) enantiomorphic solid objects giving rise to non-rigid motion in space. (Reproduced from Shepard 1994. Copyright 1994 by the Psychonomic Society with permission.)

investigate whether an individual, although remaining in three-dimensional space physically, is able to take an object out of that space mentally, when only such a mental move achieves compliance with another deeply internalized principle, such as the principle of object conservation.

For such a test, we can make use of the perceptual phenomenon of apparent motion, which is induced in an observer by alternately presenting two identically shaped objects in different positions in space. That we experience a single object moving back and forth, rather than two visual stimuli going on and off separately is, I suggest, the manifestation of the internalized principle of object conservation. It is simply more probable in our world that an enduring object suddenly moved from one position to another than that one object ceased to exist and, at exactly the same instant, a separate object materialized nearby. Moreover, our perceptual system seeks to represent the transformation as one that preserves as much as possible of the structure of the object transformed and that is kinematically as simple as possible and, hence, that can be simulated as quickly as possible (Shepard 1994).

Suppose identical two-dimensional shapes, like the Cooper polygons shown in Fig. 1a, are alternately presented in orientationally different positions in their common two-dimensional plane. In accordance with Euler's theorem of kinematic geometry, there is always a rigid rotation about a fixed point in the plane that will rigidly carry the object back and forth between the two positions. It is precisely this simplest motion that observers experience.

But what happens if the two alternately presented shapes are mirror images, as shown in Fig. 1b? These cannot be transformed into each other by any rigid motion within the plane. A rigid motion is still experienced. But it is experienced as a rotation out of the plane, through three-dimensional space. Evidently we are just as capable of representing a rigid motion in three-dimensional space as in the two-dimensional plane. But only the motion in three-dimensional space can attain the conservation of shape that is probable in the world.

Suppose we now step everything up one dimension. When identical three-dimensional shapes, like the Shepard–Metzler objects shown in Fig. 1c, are alternately presented in their common three-dimensional space, again a single such object is experienced as rigidly undergoing a rotational motion in three-dimensional space, just as in the case illustrated in Fig. 1b. Despite the absence of any external motion, we experience the simplest apparent motion in compliance with Chasles's theorem — the generalization to three dimensions of Euler's theorem. For any two positions of an object in space, there is a screw displacement that will carry the object back and forth between the two positions. Although this theorem was not formally established until the last century, we now have experimental evidence indicating that it has long been deeply embodied within the perceptual/representational systems of ourselves and (I believe) of other visually advanced animals.

But what happens if the two alternately presented three-dimensional shapes are not identical but enantiomorphs, like a right and left hand, as shown in Fig. 1d? Such shapes cannot be transformed into each other by any rigid motion within their common three-dimensional space. A rigid motion is still mathematically possible — but, again, only by breaking out into a higher-dimensional space. In failing to avail ourselves of this four-dimensional possibility, human viewers always experience a motion that is confined to three-dimensional space and hence, for enantiomorphic shapes, non-rigid. For the particular objects illustrated here, one of the arms of the object typically appears to rotate independently, as if connected by a swivel joint (Shepard 1994).

These phenomena of apparent motion are thus consonant with the Kantian idea that we are constituted to represent objects and events only in Euclidean space of three (or fewer) dimensions. In our three-dimensional world there may simply not have been sufficient selective pressures for the evolution of the more extensive neuronal machinery needed for the concrete representation of higher-dimensional spaces or the additional rigid transformations they afford.

Self-initiated mental transformations

But if we share with other species the ability to perceive and to represent transformations in three-dimensional space, what else is required for the discovery and explicit formulation of scientific laws? Explicit formulation, clearly, requires the full power of language, which, among terrestrial species, may be unique to humans (although we don't yet know what powers of communication are possessed by cetaceans). Discovery, however, requires something quite different I believe — namely, the ability mentally to simulate possible events and operations in the world or, in short, to perform thought experiments.

Einstein reported that he came to his discoveries not through thinking in words or mathematical symbols (Wertheimer 1945) but through 'visualizing ... effects, consequences and possibilities' (Holton 1972) by means of 'more or less clear images which can be voluntarily reproduced and combined' (Hadamard 1945). An individual may have perceptual/representational machinery that swiftly and automatically constructs the simplest connecting transformation between two externally presented glimpses of an object in three-dimensional space, as demonstrated by apparent motion. But such machinery, having initially evolved in the service of perception, need not be accessible voluntarily, from within, to simulate such transformations in the absence of external stimulation. Indeed, there is reason to believe that this is a much more recent and, as yet, relatively tenuously developed capability even in humans.

My students and I have pursued the systematic study of just such voluntary mental transformations in our work on 'mental rotation'. This line of work itself began with a thought experiment (described in Shepard & Cooper 1982). The results from the first actual experiment that we then carried out (Shepard & Metzler 1971) provided quantitative confirmation of the results suggested by the thought experiment. The time required to say whether two perspective images portrayed the same three-dimensional shape (Fig. 2a) increased linearly with the angular difference between the orientations in which those shapes were portrayed, and with essentially the same slope whether the orientations differed by a rigid rotation in the picture plane or by a rotation in three-dimensional depth (Fig. 2b).

My co-workers and I have since obtained evidence that in carrying out an imagined spatial transformation, intermediate spatial positions of the object are internally represented, in succession, over a particular path of transformation. Moreover, that path tends to be the simplest one according to the principles of kinematic geometry (Shepard 1994, Shepard & Cooper 1982).

Thought experiments in the discovery of scientific laws

The mere accumulation of observational facts does not lead to Newtonian mechanics, electrodynamics, relativity theory, quantum mechanics or the theory of evolution. As Steven Weinberg noted, 'It is glaringly obvious that Einstein did not develop general relativity by poring over astronomical data' (Weinberg 1992). In a book in preparation, based on my 1994 William James Lectures at Harvard, I argue that physical laws are discoverable and may actually have been discovered through experiments carried out in thought rather than in the physical world. Here, I can do no more than present one example, to give some feel for how compellingly a thought experiment can reveal a seemingly empirical fact about the physical world.

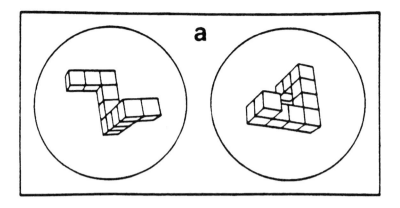

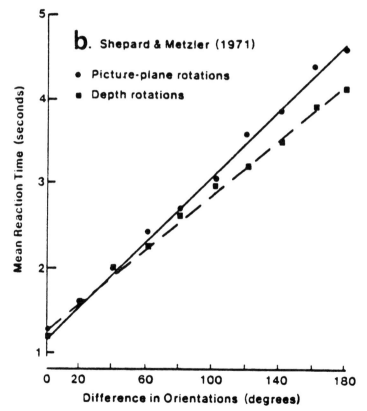

FIG. 2. (a) A pair of three-dimensional objects that can be seen to be identical in shape by imagining one rotated into congruence with the other. (b) The time required to carry out this 'mental rotation' as a function of the angular difference in portrayed orientations (either in the picture plane or, as illustrated, in depth). (Adapted from Shepard & Metzler 1971 with permission. Copyright 1971 by the American Association for the Advancement of Science.)

My example is a slight variant on the thought experiment actually described by one of the masters of the thought experiment, Galileo Galilei. In this variant, I think of Galileo imagining himself at the top of the leaning tower of Pisa with three identical bricks. By symmetry, Galileo could only suppose that these three mutually interchangeable bricks, on being individually hefted, would manifest exactly the same downward force and, on being released together over the edge, would reach the ground at the same instant.

'But wait!' Galileo might suddenly have thought, 'It would take no more than a piece of string or dab of glue to make two of these three bricks into the equivalent of a single larger brick.' True, on hefting, this larger brick would manifest a doubled downward force. But surely the mere addition of a virtually weightless bit of string or glue could not cause the two bricks, which would otherwise have fallen side-by-side with the third brick, now to fall twice as fast as Aristotle had supposed!

How can the effectiveness of such thought experiments be reconciled with the prevailing empiricist dogma—that knowledge about the world can only come from observations and experiments on the world? The explanation must be, in part, that thought experiments tap into innate knowledge about the world internalized through natural selection in the world. If only to shake ourselves out of our standard empiricist ways of thinking, we might benefit from considering the extent to which the following pair of propositions might have some validity:

(1) every real experiment is preceded by a corresponding thought experiment; and
(2) every thought experiment has been preceded by countless real experiments consisting of (a) the formation of new designs through mutation and recombination and (b) the selection of the most effectively proliferating designs that together, across innumerable ancestral generations, have constructed the mind that is now performing the thought experiments.

The reason that relatively few members of our species—pre-eminently the Galileos, Newtons, and Einsteins—have successfully used thought experiments to establish valid scientific laws is that the internalized knowledge is in an implicit, unverbalized and, hence, unsharable form. Its conversion to an explicit, sharable set of consistent principles through thought experiments is an evolutionarily recent and tenuously established development. A consideration of what distinguishes conclusive from inconclusive or even misleading thought experiments indicates, moreover, that the former make use of logical and mathematical truths. For example, symmetry principles, as in the example of the interchangeable bricks, appear to play an especially effective role.

I have thus come to question the long-standing distinction between logical and empirical truths. There are now good reasons to suspect that the line between the a priori and the 'empirical' has been too sharply drawn. It is even possible, as some theoretical physicists and cosmologists have recently speculated, that the ultimate laws of physics may constitute the only mathematically self-consistent set and, hence, could be discovered by thought alone, if we were but smart enough. (To me, anyway, the alternative—that the ultimate laws are purely arbitrary—is less than fully satisfying.)

Conclusion

Perhaps psychological science need not limit itself to the description of empirical regularities observed in the behaviours of the particular, more or less accidental collection of humans or other animals accessible to our direct study on planet earth. Possibly we can aspire to a science of mind that by virtue of the evolutionary internalization of universal regularities in the world, partakes of some of the mathematical elegance and generality of theories of that world.

Acknowledgements

Preparation of this article was supported by National Science Foundation Grant DBS-9021648. The article has benefited from suggestions made by John Tooby.

References

Hadamard J 1945 The psychology of invention in the mathematical field. Princeton University Press, Princeton, NJ

Holton G 1972 On trying to understand scientific genius. Am Scholar 41:95–110

Schaffer C 1994 A conservation law for generalization performance. In: Cohen WW, Hirsh H (eds) Proceedings of the eleventh international conference on machine learning at Rutgers University, New Brunswick, NJ, July 10–13 1994. Morgan Kaufmann, Palo Alto, CA, p 259–265

Shepard RN 1987 Toward a universal law of generalization for psychological science. Science 237:1317–1323

Shepard RN 1994 Perceptual–cognitive universals as reflections of the world. Psychonom Bull Rev 1:2–28

Shepard RN, Cooper LA 1982 Mental images and their transformations. MIT Press/Bradford Books, Cambridge, MA

Shepard RN, Metzler J 1971 Mental rotation of three-dimensional objects. Science 171:701–703

Weinberg S 1992 Dreams of a final theory: the scientist's search for the ultimate laws of nature. Pantheon, New York

Wertheimer M 1945 Productive thinking. Harper, New York

Wolpert DH 1992 On the connection between in-sample testing and generalization error. Complex Syst 6:47–94

DISCUSSION

Dawkins: If we ask the question why, for example, quantum theory is so impossible for anyone to understand, is it simply because we have been naturally selected in a world of large objects that behave in the way that we think is reasonable? Or is there really something weird about quantum theory? One could imagine a thought experiment in which a cognitive being had been naturally selected but its body was so small that it was influenced by quantum events, then its intuitions as to what were just normal would be different from ours.

Shepard: It is interesting to speculate that life may have evolved in a totally different environment from ours. I'm really just talking about life as I know it, but I would agree with your first point that the reason it is so difficult for us to understand quantum theory is that it deals with a scale of space/time/energy that is beyond what we have been evolutionarily prepared to deal with.

Kacelnik: Did you have anyone in mind when you criticized people for attributing everything to learning? Because not even Skinner would deny that reinforcement principles were acquired and inherited through evolutionary processes.

Shepard: Perhaps not, but there are people who have different views on this subject. Many people take the position that most of what's important in human affairs is learned within a particular cultural context.

Cosmides: Almost all psychologists assume that there has to be some structure to the mind. However, the question is, is all the knowledge that you obtain about the structure of the world driven by content-free learning mechanisms operating only on sensory stimuli as experienced by the organism via its perceptual system or is there knowledge about the world that is embodied in the structure of the learning mechanisms themselves? When you put the question that way, you are asking how domain specific various kinds of mental mechanisms are. Domain specificity makes most psychologists uncomfortable. I would estimate that 90% of psychologists would argue that most learning is accomplished by domain-general learning mechanisms.

Kacelnik: If the claim is that we come into the world completely unbiased between all kinds of perceptions, then this is unsustainable, and not claimed even by Skinner. However, once we elaborate the specific details of whether we respond to this or that, then we can have an informed discussion.

Cosmides: I disagree with how you are characterizing Skinner's claims. The equipotentiality assumption — that the brain can pair any stimulus equally easily with any response — has been the mainstay of behaviourism since Watson. Equipotentiality is not correct. There is contentful structure to the minds of animals.

Kacelnik: But there are so many textbooks from about 20–25 years ago saying that equipotentiality is unsustainable.

Shepard: The problem with Skinnerian conditioning is that if you're going to have a general theory of learning then you have also got to have a general theory of generalization. Because one never confronts the identical situation twice, every response that we make is, in a sense, a leap in the dark. We can only behave reasonably on the basis of principles that allow us to generalize from past situations to the present situation. Learning theorists never properly appreciate this. Suppose the pigeon in a Skinner box is presented with a stimulus and it receives a reinforcement for pecking a key (e.g. see Skinner 1953). In order to understand the effect of that single reinforced trial we would have to specify how the probabilities change for every possible stimulus that we could present. No one can claim to have a way of doing this. We all come into the world with an innate similarity metric. Even young infants can be shown to have this. If a stimulus is presented to an infant several times and then a different stimulus is presented, then depending on how different the second stimulus is from the first one, the infant will show dishabituation, i.e. the infant will demonstrate surprise. In order to have a complete theory of learning you have to have a complete theory of the innate metric of similarity that has evolved in the service of generalization (Shepard 1987). Machine-learning theorists have made progress in the characterization of structures that are necessary for learning in a general, abstract way, without requiring a detailed knowledge of neural circuitry (Schaffer 1994, Wolpert 1992).

Nisbett: Spelke and others have managed to convince people that the notion of equipotentiality is quite impossible and that there is, for example, a pre-wired theory of mechanics (Spelke et al 1995). However, the battleground has now shifted substantially to, for example, whether or not there is a wired-in theory of mind. In my opinion there is.

Sperber: The notion that there may be domain specific, genetically determined mental mechanisms is still met with great resistance. For instance, Plunkett & Elman (1997) argue that there may be 'architectural innateness' in the organization of the human mind, but not what they call 'representational innateness'. They are willing to accept that there may be predetermined computational procedures, but not predetermined content to these procedures. However, it may not be that easy to distinguish architectural from representational innateness. If one thinks, as many of us here do, that some elements of the mental architecture are narrowly specialized for handling specific features of the environment, then these elements of the architecture 'are about' — or represent, in a quite ordinary sense — the environmental features they handle.

Daly: Can you clarify the distinction between content and procedure, because if you say that we have evolved complex procedures with implicit knowledge of what the world we live in is like, then doesn't that represent content?

Sperber: Consider the mental procedures involved, for instance, in the categorization of plants and animals. If these procedures are just domain-general feature extractors, that also work equally well for the categorization of stones, clouds, artefacts and so on, then these procedures are contentless. If, on the other hand, as I believe, these procedures are specialized for the categorization of living kinds, they can be said to represent the living kinds domain. In other terms, some notion of what living kinds may be is already present in the very procedures we specifically apply to living kinds.

Gigerenzer: Where does the content blindness of many theories come from? There is an old, beautiful dream (belonging to Leibniz, but not his alone) that all reasoning and decision making could be reduced to one universal calculus. Such a rational calculus would put an end to scholarly bickering: if a dispute arose, the contending parties could settle it quickly and peacefully by sitting down and calculating. For some time, the Enlightenment probabilists believed that the mathematical theory of probability had made this dream a reality. The essence of the dream is that sound reasoning should follow a general-purpose syntactical procedure, detached from the specific content of the problem and the pragmatics of the situation. Although most mathematicians gave up this dream by around 1840, many psychologists still hold on to it. Examples abound, from Piaget's formal operations to Rips' mental logics and to Johnson-Laird's mental models. The ideal is that content should play as little role in sound reasoning as the black and white swans do in logical exercises. There are two current variants of this dream: the rational and the irrational. The rational variant assumes that such a general-purpose calculus exists and is embodied in the human mind; the irrational variant accepts the same definition of sound reasoning but assumes that the human mind cannot stand up to this ideal, rather it is filled with cognitive illusions. Skinner propounded the related ideal of 'stimulus equipotentiality' in his behaviourism, assuming that the content of the stimulus would not matter for the laws of operant conditions; but John Garcia and others falsified this assumption experimentally. Yet it somehow survived the cognitive revolution. There are few theories of reasoning that allow the content (and the pragmatics) to play a decisive role, such as described by Cosmides & Tooby (1992). The fact that many researchers still hold on to the old beautiful dream is one explanation for the strong separation between structure and content we have in theories of reasoning and decision making.

Bouchard: Roger Shepard always presents this information in such a compelling way, and it appears that every mind carries out the process in the same way. However, there are dramatic individual differences. His figures have been instantiated in the Vandenburg test, and there are enormous differences in performance. For example, I have observed striking sex differences in the capacity to carry out these rotations: females seem to have more difficulty with the 180° rotations. I would like to ask Roger if he has observed the same result.

Shepard: Our strategy was to obtain extensive data from a few subjects, i.e. we took a more psychophysical approach and weren't looking at individual differences. However, in our experience there weren't any noticeable differences between the sexes. I would like to mention the important point that the wisdom about the world is deeply internalized and is not accessible to conscious introspection or control. Therefore, we see it operating most effectively when we present a stimulus or situation that has some relation to what our ancestors experienced. When we study perceptual phenomena, or even apparent motion which is quasi-perceptual, we find that everyone tends to experience the same motion. I strongly believe that animals and young infants would respond the same way to these apparent motion displays. However, mental rotation is a completely different kind of task that requires some kind of internal conscious control. It's not an automatic response and there is probably more variation between subjects. This is why the successful thought experiments in physics are largely carried out by a few exceptional individuals such as Galileo, Newton and Einstein. Another example is the paper-and-pencil test of intuitive physics that Michael McClosky and his co-workers at Johns Hopkins

University have studied, which gives the impression that people are rather stupid, but of course paper and pencils are an extremely recent addition in evolutionary history (McClosky 1983). However, if you present the same problems in a way that directly engages our innate perceptual machinery, people come across as being much smarter.

Cronin: Tom Bouchard pointed out that there are individual differences and sex differences in performing the rotation task, but this does not in the least undermine the claim that our perception of three-dimensionality is a human universal. Although there are differences in ability to rotate figures, those differences do not contravene three-dimensional perception. So, for example, women don't try to flip the figures into a fourth dimension in order to rotate them.

Bouchard: No, but there are evolutionary forces at work that maintain sex differences, although we don't necessarily know what their driving forces are.

Cronin: But they are not forces that contravene the universality claim. In spite of differences, every human being incorporates the same information about the world's three-dimensionality.

Wilson: If there is a sex difference in mental rotation, which there seems to be, and if we can identify the underlying mechanisms, then we might expect to observe variation within both men and women as a function of fetal androgenization or exposure to other mechanisms that generate sex differences. Variation in sexually differentiated developmental processes might impact on the distribution of mental rotation abilities in both sexes in adulthood.

Pinker: The stimuli that Roger Shepard used are difficult for people because of the hypotheses he wanted to test, and they were included in the Vandenburg test precisely because they differentiate between individuals. Roger mentioned that if you go from a mental rotation paradigm, where you have to mentally flip one object onto the other, to an apparent motion paradigm, where the two views of the object alternate and the question is whether people see a single object rotating back and forth, then probably 100% of women interpolate the trajectory and see the object rocking. With simpler stimuli, where the corresponding parts of two views are marked more saliently, most women can also do the mental rotation. An even simpler example of a mental rotation that everyone can do comes from the old joke where there are two carpenters hammering in nails, and one of them throws aside every other nail that he picks out of the box. The second carpenter asks what he is doing, and the reply is 'they're defective—the heads are on the wrong side and the pointy end is on the wrong side'. The other carpenter then says 'you idiot, those are for the other side of the house'. I am sure that 100% of women get that joke. It's not that women lack a mental rotator, it's more likely that women on average have a smaller capacity to co-ordinate the sequence of rotations that you have to do to get every single one of those parts into place.

Bouchard: That's exactly my argument. We all want to perform experiments that are representative of the environment in which evolution took place, but the examples are generally much simpler. When the task becomes more difficult, and probably more representative of what happens in the real world, one might observe larger differences.

Shepard: I would argue against that. I would argue that when you present stimuli or tasks that more closely resemble what our ancestors had to deal with in the real world, you observe more uniformity of behaviour.

Dawkins: Pigeons have been exposed to the same rotational constancy experiments with different results, but presumably they live in the same three-dimensional world.

Kacelnik: When identifying patterns on the ground from above they face two-dimensional problems with rotations.

Dawkins: Then it is interesting to explain why pigeons give different results.

Cosmides: There's a large difference in the types of stimuli. Many adaptations will be engaged by certain kinds of psychophysical cues, such as motion cues, in the environment. The more motion cues are introduced, the more uniformity is observed. In the case of pigeons, they are usually shown one static image and then another, i.e. a series of static stimuli. Therefore, what is actually being tested is their ability to imagine or simulate object motion, and not necessarily their ability to make inferences about objects in the real world.

Shepard: We should try the apparent motion experiments in pigeons. In the mental rotation experiments we instruct the subjects as to what to do; pigeons might not be able to understand the idea of what they are to do in mental rotation experiments.

Hauser: I would like to ask Roger Shepard about the relationship between domain-general versus domain-specific learning and thought experiments. Are there certain domains of knowledge in which thought experiments are more constrained?

Shepard: Yes.

Hauser: Why?

Shepard: Because, for example, it's difficult to do thought experiments on quantum mechanics: we haven't internalized the knowledge about the quantum world that would enable us to do this.

Hauser: Does the range of experiences that we've had, therefore, account for all the differences in the range of thought experiments?

Shepard: Yes, but not necessarily the experience we've had as individuals, but the kinds of problems that the human ancestral line has had to deal with.

Hauser: I'm not sure that these experiences are the only constraint that influences the kind of thought experiment one has.

Cosmides: There are probably some mechanisms that lead us astray as scientists. For example, the 'theory of mind' mechanism generates explanations of behaviour as 'caused' by internal mental states such as 'beliefs' and 'desires' (*Behaviour:* Why did Mary go to the water fountain? *Explanation:* Because she wants water [*desire*] and she *believes* she can get it at the fountain). Such explanations strike most people as accurate and complete accounts of human behaviour. But as a scientific theory, belief/desire explanations are woefully incomplete (and, in many cases, incorrect). Nevertheless, they are intuitively compelling — so much so that it is difficult to see why any further explanations are necessary (especially ones invoking complex mental machinery). There are many phenomena — seeing, falling in love, finding something beautiful — caused by complex computation mechanisms, operating automatically and outside conscious awareness, that structure our thought so powerfully that the phenomena they produce seem inevitable. Yet they are not: different machinery would cause different phenomena (dung flies are attracted to dung; we are repelled by it). Because we are unaware of the cognitive process that cause us to see or fall in love or explain behaviour in terms of beliefs and desires, it is difficult to realize that these cognitive processes exist at all. They provide intuitively compelling ways of understanding the world, which often mislead psychologists (and others, of course) into thinking that no further explanations are necessary. Just as our internal theories of the physical world are misleading when one wants to understand subatomic particles, our internal theories of behaviour are misleading as scientific accounts of the mind.

Tooby: It seems to me that these foundational ways of thinking, such as the theory of mind mechanisms, intuitive object mechanics, the teleology module and so on, shape how we do science at the basic level. In fact, I don't think it could be otherwise. What we accept as understanding something is the final translation of observations and higher level concepts into the terms that our various evolved computational devices can accept. This suggests an entire alternative approach to the history of science and the philosophy of science. For example, observations did not compel Democritus and the Stoics to develop atomic theory. The theory that all things are composed of tiny rigid solid-shaped objects, and that everything happens by direct mechanical causality is simply the application of our evolved rigid mechanics module to everything. When modern science emerged during the Renaissance, it did so again through the attempt to submit everything to the mechanics module, making mechanical properties such as shape, location, speed, contact and collisions relevant, making other observable properties, such as colour, irrelevant and other quite real phenomena, such as action at a distance, unacceptable. Over the ensuing centuries, this equation of science with intuitive mechanics made the products of other modules, such as teleology, vitalism and mentalism, the products of other modules, seem increasingly inconsistent with a scientific, that is, a mechanical world view. The activation and clash of our various evolved inference engines is just as visible in how we do science today, even in how disciplines get divided up. So psychologists, behavioural ecologists, economists, anthropologists, sociologists and so on, all study humans. But anthropologists and economists conceive of explanations not in terms of the design of machinery that produces behaviour and decisions, but instead implicitly rely on the theory of mind mechanism, with its associated belief–desire psychology. So the standard anthropologist considers it an explanation of why the Masai behave in a certain way to identify the values they have and the beliefs they hold. They don't even notice what, say, a cognitive scientist would find completely missing from such an explanation: a mechanistic description of the computational devices in the brain that give beliefs and values — to the extent they really exist — causal reality. So, cognitive scientists interpret and reason, sometimes metaphorically, with the tools of mechanistic causality, in contrast with other fields that interpret and reason through other modules, such as the theory of mind module. Equally, the fact that we all come equipped with an evolved theory of mind module that is automatically triggered when we think about people makes a science of human behaviour difficult because the theory of mind module continually pre-empts or trumps the development of more complete, more mechanistic descriptions of all the machinery in the mind. Just as our intuitive mechanics module usefully, but falsely, informs us that the world is composed of solid slabs of homogeneous material, making quantum mechanics difficult to understand, so our theory of mind module usefully, but falsely, informs us that the beliefs and desires are the best explanatory concepts for humans. For this reason, psychologists, neuroscientists, economists — indeed everyone — keep sliding back down into folk theories, such as learning theory, that are only slightly formalized versions of the false but useful concepts our evolved psychologies provide us with.

Nesse: It's important to note that it's odd that 50 years after much of this evidence was thought of as fairly solid we're still arguing about it. In *Behavioral and Brain Sciences* about two years ago there was an article by Davey, arguing that there was no preparedness in the propensity to phobias of certain objects as compared to other objects (Davey 1995, Nesse & Abelson 1995). Sue Mineka et al (1984) and others have demonstrated conclusively that there is preparedness. It's fascinating that the debate is vastly oversimplified, and that people take sides and feel compelled to defend them.

References

Cosmides L, Tooby J 1992 Cognitive adaptations for social exchange. In: Barkow JH, Cosmides L, Tooby J (eds) The adapted mind: evolutionary psychology and the generation of culture. Oxford University Press, Oxford, p 163–228

Davey GCL 1995 Preparedness and phobias: specific evolved associations or a general expectancy bias? Behav Brain Sci 18:289–325

Nesse RM, Abelson JL 1995 Natural selection and fear regulation mechanisms. Behav Brain Sci 18:309–310

McClosky M 1983 Intuitive physics. Sci Am 248:122–130

Mineka S, Davidson M, Cook M, Keir R 1984 Observational conditioning of snake fear in rhesus monkeys. J Abnorm Psychol 93:355–372

Plunkett K, Elman JL 1997 Exercises in rethinking innateness: a handbook for connectionist simulations. Bradford Books, Cambridge, MA

Schaffer C 1994 A conservation law for generalization performance. In: Cohen WW, Hirsh H (eds) Proceedings of the eleventh international conference on machine learning at Rutgers University, New Brunswick, NJ, July 10–13 1994. Morgan Kaufmann, Palo Alto, CA, p 259–265

Shepard RN 1987 Toward a universal law of generalization for psychological science. Science 237:1317–1323

Spelke ES, Phillips A, Woodward AL 1995 Infants' knowledge of object motion and human action. In: Sperber D, Premack D, Premack AJ (eds) Causal cognition: a multidisciplinary debate. Clarendon Press, Oxford, p 44–78

Skinner BF 1953 Science and human behaviour. Macmillan, New York

Wolpert DH 1992 On the connection between in-sample testing and generalization error. Complex Syst 6:47–94

Evolutionary conflicts and adapted psychologies

Anders Pape Møller

Laboratoire d'Ecologie, CNRS URA 258, Université Pierre et Marie Curie, Bât. A, 7ème étage, 7 quai St. Bernard, Case 237, F-75252 Paris Cedex 5, France

Abstract. Animal information processing and decision making are often considered to be adaptations that allow individuals to behave optimally under particular ecological conditions. Numerous examples demonstrate how cues from the biotic and abiotic environments affect the ways in which animals process information and make decisions. Information gained from interactions with living organisms is the most complex because individuals have to respond to heterospecifics or conspecifics which may decide on what to do depending on the behaviour of a focal individual. Evolutionary conflicts of interest include: (i) interactions between hosts and parasites, predators and prey, and between competitors; (ii) sperm competition interactions between females, male mates and male non-mates; and (iii) interactions between mate-searching females and their potential mates. Brains may evolve particularly rapidly under the influence of evolutionary conflicts and they may enhance the importance of adapted psychologies in these contexts.

1997 Characterizing human psychological adaptations. Wiley, Chichester (Ciba Foundation Symposium 208) p 39–50

A brief introduction to signalling theory

Animals must continuously respond to challenges from the environment (both abiotic and biotic aspects) because resources are limiting and the environment can generally be considered to be stressful. Stress in this context represents a challenge to any biological system that potentially has long-lasting and debilitating effects. Stress resistance is energetically costly because resources that could otherwise be used for growth, reproduction and survival have to be used for maintenance (Hoffmann & Parsons 1991). The important point is that both abiotic and biotic factors will have a negative impact on any organism. The following arguments will be based entirely on biotic interactions because the abiotic environment poses fewer and often more predictable challenges than the biotic environment.

Competition for limiting resources and the negative impact of conspecifics and heterospecifics on fitness maximization of any organism are some of the main motors of adaptive evolution. Individuals that are better able to cope with the continuous challenge from the biotic environment will on average leave more offspring and contribute disproportionately to future generations. If conspecifics or heterospecifics are preventing or reducing the ability of individuals to achieve this goal, this opens up the possibility that additional resources may be acquired at the expense of other individuals by means of communication. The means by which such excess resources are acquired by some individuals are less important as long as there are no long-term damaging effects on future interactions. Communication is based on the production of signals that have to be interpreted by receivers, which may respond depending on how the signal has been perceived. Signals may be of a number of different kinds, and the communication and the response of receivers may depend on the extent of common interests of the two parties (Maynard Smith & Harper 1995). Signs are simple, uncostly signals used in interactions between two individuals that share a common interest. This situation may be relatively uncommon because individuals, even closely related ones such as a mother and her offspring, may rarely have fully congruent interests. A possible example is road signs in human societies; they are simple, relatively cheap signals that convey reliable information about the names of particular places and perhaps the distances to such places.

A second kind of signal is a reliable or honest signal that is costly to produce. Such signals have sometimes been called handicaps (Zahavi 1975). Imagine the situation when a male intends to signal to prospective mates that he should be chosen. The fitness of males is usually limited by access to females, whereas female reproductive success may depend much less on the number of males with whom a particular female has mated. However, some males may be resistant to parasites, while others are susceptible to a particular parasite that currently affects the population in question. If the resistance has a genetic basis a female choosing a resistant male would benefit by rearing resistant offspring. How should the female decide which male to chose? A global solution is to choose the male that produces the most extravagant display (the largest antlers, the loudest calls, the most complicated movements and the brightest colours). The reason for this simple solution is that only males in prime condition are able to signal at the highest possible level; males in poor condition will simply be too handicapped by their poor condition during pregnancy and birth, their poor adult condition and their high parasite load. In other words, it is the differential cost of a given level of signalling that prevents males of inferior quality from faking the signal and acquiring a disproportionate share of the females (Grafen 1990). In accordance with this idea females tend to prefer males with phenotypes at the upper extreme: larger, brighter and louder. Such a signal will not only be reliable for females, but also for conspecific males and for heterospecifics, including predators.

A third kind of reliable signal is phenotypic symmetry. Almost all organisms are radially or bilaterally symmetrical, and symmetry enhances performance for mobile organisms. It is difficult to produce a perfectly symmetric phenotype, particularly for an exaggerated signal, and symmetry thus reveals the ability of an individual to cope with the environment during its ontogeny (Møller & Swaddle 1997).

Antagonistic interactions as generators of adapted psychologies

If signallers are limited in their success by other individuals, an evolutionary conflict of interest will arise between the interacting parties. A categorization of conflicts of interest is presented below.

Within species:

(1)   Parent and offspring.
(2)   Male and female mates.
(3)   Male mates and non-mates.
(4)   Male non-mates and females.
(5)   Female sexual competition.
(6)   All individuals that share a common resource such as mates and food.

Among species:

(1)   Competitor and competitor.
(2)   Parasite and host.
(3)   Predator and prey.

Conflicts of interest are the driving force behind co-evolution: the continuous evolutionary process whereby individuals of one species or one sex evolve adaptations to exploit individuals of another species or sex. In turn, individuals of this other species or sex will evolve new adaptations to exploit the first, which in turn will evolve new adaptations (Thompson 1994). Typical examples of such co-evolutionary processes include the conflicts of interest between the sexes of a species, between competitors of two or more species, between predators and their prey, and between parasites and their hosts. For example, predators may co-evolve with their prey since the ability of prey to evade capture will ultimately drive the predator extinct. Running speed in predators is therefore likely to co-evolve with running speed in prey, and the evolution of crypsis is likely to co-evolve with the evolution of visual acuity that manages to break crypsis. Similarly, brood parasitic cuckoos rely on hosts for successful reproduction, but the host can evolve an ability to discriminate against the offspring of the parasite and reject such offspring. This will select for better mimicry among the parasite and better ability of the host to discriminate against the parasite.

Deviant phenotypes are produced in every generation due to continuous deterioration of the environment and the debilitating effects of most mutations. Such deviants are likely to fall prey of predators and become hosts of parasites (Møller & Swaddle 1997). Obviously, the selection pressure on the predator or the parasite is much stronger than the selection pressure on the prey or the host because all predators must eat prey to survive, and all parasites must parasitize a host. The opposite is not the case. This is the so-called life–dinner principle (Dawkins & Krebs 1979).

Interactions between parties with conflicting evolutionary interests also give rise to interactions on an ecological time-scale. Even though predators and prey are not obviously benefiting from exchanging signals, there is increasing evidence for predator–prey signals conveying reliable information (Caro 1995). If only a single prey item is being consumed, the predator may decide which prey individual to choose on the basis of the behaviour of all the prey. Stotting and other kinds of extreme prey behaviours may allow the majority of the prey individuals to spend a relatively small amount of energy on signalling without causing the predator to engage in long-lasting, unsuccessful pursuits of prey. Again, honest signalling is a key to understanding the interaction. Similarly, asymmetry may be an important feature of these interactions between predators and prey (Møller & Swaddle 1997). An individual prey with a symmetric colour pattern will appear to be facing a predator when detecting it, but this will not be the case for an individual with an asymmetric colour pattern. If asymmetry reflects poor environmental conditions during ontogeny, or a poor genetic constitution, the predator is likely to capture the asymmetric prey upon approach. The predator will thereby learn that asymmetric prey are easy to capture because of their poor condition.

Continuous interactions between parties differing in evolutionary interests are likely to generate abilities in both interactors to mind-read the other party. Any individual with a superior ability of such activities is likely to become over-represented in future generations. Any individual of any organism is likely to be engaged in a number of different evolutionary conflicts of interest, and the adapted psychology of a particular species is likely to be tuned in on the interactor that contributes the most to differences in fitness. Natural selection will have tuned the psychology onto the most important features of the environment. A brief list of some examples of the action of adapted psychologies is provided in Table 1.

Sexual selection as a generator of adapted psychologies

A particularly important arena for evolutionary conflicts of interest is conflicts between the sexes of a single species. Imagine the situation in mammals where the reproductive tract of a female immediately after copulation is invaded by millions of leukocytes that destroy a large fraction of all sperm (Birkhead et al 1993). The remaining sperm must swim in a chemically hostile environment to reach the site of fertilization, and few if any sperm will even manage to go that far. The reason for the evolution of such a hostility of the female reproductive system to sperm is that females are selecting among sperm provided by a single male, but also among sperm provided by different males. If we could track the evolutionary history of this phenomenon, initially we might have seen females that allowed relatively more sperm to reach the site of fertilization. Females that sometimes had their eggs fertilized by poor quality sperm would have offspring of poor quality or no offspring at all, and they would be selected against. The hostility

TABLE 1 A selective list of examples of activities performed by animals with adapted psychologies arising from evolutionary conflicts of interest on an ecological time-scale

Conflict	Challenge	Reaction	Reference
Between the sexes	Individuals should maximize fitness benefits from attractiveness	When phenotypes are manipulated, males adjust their sexual behaviour to their new level of attractiveness	Møller (1988a)
	Individuals should minimize investment in offspring relative to their partner	When male phenotypes are manipulated, attractive males reduce their investment	Møller (1994)
	Males should guard their mates to prevent other males from getting access to their female	When male access to a female is changed by temporary removal of a male mate, neighbouring males with a fertile female, but not males with a non-fertile female, increase the level of their mate guarding	Møller (1987)
Between parasite and host	Hosts should avoid being parasitized by cuckoos, but cuckoos should prevent hosts from defecting	When magpie hosts reject an egg of a great spotted cuckoo, the cuckoo retaliates by destroying the clutch of the host, which then alters its behaviour towards the cuckoo in future attempts	Soler et al (1995)
Between competitors	Socially dominant conspecific and heterospecific individuals prevent subdominants from access to food	Subdominants give alarm calls (usually only used when a predator is present) when access to food is prevented and these calls make dominant individuals seek shelter. If the food source is dispersed, and food is readily accessible, no alarm calls are given	Møller (1988b)

of the female reproductive tract and the ability of male ejaculates to cope with such hostility are therefore believed to have co-evolved to greater levels of choosiness in females and greater abilities of males to cope with the female tract. Males and females of the same species may differ in their interests, and it is difficult to imagine how one party may evolve rapidly without co-evolution by the other. Artificial arrest of the co-evolutionary process, as has been performed in *Drosophila* fruit flies, results in deleterious mating effects in individuals of the sex that does not evolve (Rice 1996).

A particularly important conflict of interest throughout the animal and plant kingdoms arises from sperm and pollen competition because females may not have interests in common with male mates that do not have interests in common with male non-mates (Birkhead & Møller 1997). Sperm competition has been shown to be particularly prevalent in socially monogamous species with relatively intense sexual selection, such as most birds and humans. There are several reasons for this. First, sexual selection for male genetic quality may be maintained over evolutionary time in such species because the reduction in genetic variance caused by fixation of alleles due to selection is balanced by mutational input. This is not the case in mating systems such as polygyny or lekking where selection is more intense. Individuals of the choosy sex (usually females) may therefore obtain considerable genetic benefits from their mate choice in socially monogamous species. Second, sexual selection gives rise to the evolution of sexual ornamentation, which is costly to produce and maintain and does not enhance survival. The costs of the signal are only balanced by the benefits in terms of mating success. Sexual ornamentation will be particularly costly, and therefore particularly reliable, in socially monogamous mating systems because males often play some or even a major role in reproductive activities other than the act of copulation. Females of socially monogamous species are thus likely to be more readily able to tell males of differing quality apart. Third, social monogamy differs from other mating systems by imposing a serious constraint on female mate choice. Only a single female is able to acquire the most attractive male, whereas many females may share a polygynous male. This opens up the widespread pursuit of sperm competition because females mated to less attractive males are able to secondarily adjust their mate choice by copulating with one of the most attractive males and letting this male sire some or all of their offspring (Møller 1992).

Relatively intense sexual selection in socially monogamous mating systems is likely to result in the evolution of adapted psychologies because both sexes may lose from defection by the other party. This will particularly be the case when both males and females provide extensive parental care, which will be wasted by the individual providing expensive care for unrelated offspring. The whereabouts of the partner and the continuous signalling of bonding and fidelity, while being confronted with conspecifics of different qualities, will result in a burst of evolution of different kinds of signals and behaviour to convince, control or impress the partner.

Obviously, this machinery will be exploited in mating contexts between a female and a male non-mate. Since males in most species will be limited in their reproductive success by access to females, male psychologies will have evolved to persuade females that they will get what that want, whereas female psychologies will have evolved to test male quality. I would argue that this is the breeding ground for extreme abilities of mind-reading. Anything close to a human would never have evolved in an extremely polygynous lineage because mind-reading is likely to be a more important determinant of fitness, when sexual conflicts of interest are ubiquitous.

Co-evolution of signals and brains

If evolutionary conflicts of interest are an important generator of signal evolution and adapted psychologies, it is easy to suggest that these abilities will co-evolve with brain structure and size. Sexually dimorphic structures exist in the brains of a wide variety of organisms, and such differences affect learning abilities important for mate competition (Jacobs 1996). For example, repertoire size in different bird species is directly related to the relative size of the higher vocal centre of the brain (DeVoogd et al 1993). Larger repertoires are more efficient at attracting females in several bird species. It is not known to what extent females have similarly evolved larger capacities for handling information on male quality as determined from the size of their repertoires. I suggest that feedback between signal evolution and brain evolution may be particularly important in the context of sexual selection. It is likely that sexual selection in socially monogamous species with challenging evolutionary conflicts of interests between the sexes will have experienced co-evolutionary changes in the size or the structure of the brain. Such changes will have facilitated the evolution of adaptive psychologies that promote further co-evolution.

If these ideas are generally true, then I would predict that: (i) brains should be larger and/or have evolved specialized abilities to cope with problems of sexual evolutionary conflict in species with relatively moderate intensities of sexual selection, such as socially monogamous birds and humans; (ii) aspects of adapted psychologies should be more well developed in the same species; and (iii) the relative importance of sexual and social evolutionary conflict could be determined from investigating the specializations mentioned above in moderately sexually selected species such as sexually dichromatic monogamous birds and in highly social species such as colonially breeding birds.

Acknowledgements

R. Thornhill provided constructive criticism. This study was supported by a grant from the Danish Natural Science Research Council.

References

Birkhead TR, Møller AP 1997 Sperm competition and sexual selection. Academic Press, London

Birkhead TR, Sutherland W J, Møller AP 1993 Why do females make it so difficult for males to fertilize their eggs? J Theor Biol 161:51–60

Caro TM 1995 Pursuit-deterrence revisited. Trends Ecol Evol 10:500–503

Dawkins R, Krebs JR 1979 Arms races between and within species. Proc R Soc Lond Ser B 205:489–511

DeVoogd T J, Krebs JR, Healy SD, Purvis A 1993 Relations between song repertoire size and the volume of brain nuclei related to song: comparative evolutionary analyses amongst oscine birds. Proc R Soc Lond Ser B 254:75–82

Grafen A 1990 Biological signals as handicaps. J Theor Biol 144:517–546

Hoffmann AA, Parsons PA 1991 Evolutionary genetics and environmental stress. Oxford University Press, Oxford

Jacobs LF 1996 Sexual selection and the brain. Trends Ecol Evol 11:82–86

Maynard Smith J, Harper DGC 1995 Animal signals: models and terminology. J Theor Biol 177:305–311

Møller AP 1987 Mate guarding in the swallow Hirundo rustica: an experimental study. Behav Ecol Sociobiol 21:119–123

Møller AP 1988a Female choice selects for male sexual tail ornaments in the monogamous swallow. Nature 322:640–642

Møller AP 1988b False alarm calls as a means of resource usurpation in the great tit Parus major. Ethology 79:25–30

Møller AP 1992 Frequency of female copulations with multiple males and sexual selection. Am Nat 139:1089–1101

Møller AP 1994 Symmetrical male sexual ornaments, paternal care, and offspring quality. Behav Ecol 5:188–194

Møller AP, Swaddle JP 1997 Asymmetry, developmental stability and evolution. Oxford University Press, Oxford

Rice WR 1996 Sexually antagonistic male adaptation triggered by experimental arrest of female evolution. Nature 381:232–234

Soler M, Soler J J, Martinez JG, Møller AP 1995 Magpie host manipulation by great spotted cuckoos: evidence for an avian mafia? Evolution 49:770–775

Thompson JN 1994 The coevolutionary process. Chicago University Press, Chicago, IL

Zahavi A 1975 Mate selection-a selection for a handicap. J Theor Biol 53:205–214

DISCUSSION

Hauser: In relation to your idea about social systems what kind of pressures would cause an evolutionary change from an organism that can mind-read in the sense of predicting behaviour to an organism that can mind-read in the sense of predicting mental states? In monogamous systems animals are in close contact and thus an individual's experiences tend to be shared immediately with its partner. In contrast, the fission–fusion system of the chimpanzee is characterized by individuals leaving one social group, gaining unique experiences on their own or with another social group, and then joining up with either their old group or some subset of it.

Møller: It's extremely difficult to answer this question because we don't have good clues about the mind-reading capabilities of any organism.

Cosmides: We do have a certain amount of information about mind-reading in humans. One of the most interesting things is that humans, at least by the time they are four years old, understand that another person can have a belief about the world that's false, i.e. they understand that the person's behaviour is based on his or her belief as opposed to the way the world actually is. This suggests that humans don't have to rely on contingencies alone to predict behaviour. Therefore the question can be rephrased to ask: to what extent do the kinds of interactions in monogamous systems lead to the ability to model the belief states of others?

Daly: I would like to bring up the issue of whether animals behave differently when they're being watched, i.e. whether they have any notion that some of their important interactants are aware of their behaviour. Are any of these birds sensitive to audiences?

Møller: There are a number of studies that are addressing this. Peter Marler and his co-workers have looked at this in chickens (Gyger et al 1986).

Daly: In the context of extra-pair copulations, some people have argued that animals act as though they are concerned with who sees this happening.

Møller: That's true, but these are only anecdotal reports.

Nesse: Lee Dugatkin's studies on guppies are relevant to this. He has shown that the brightly coloured so-called 'bold' guppies go up to the predator and observe it, whereas the 'dull' ones don't because they're more timid. However, as soon as the female is removed, there are no behavioural differences between the two types of guppies (Godin & Dugatkin 1996).

Tooby: Conflicts of interest will drive the elaboration of adaptations for mind-reading: one organism reaps the benefits of developing mind-reading, but this results in the selection of strategies to block mind reading in the partner. Therefore, it's not clear from an a priori stance at what level of complexity the adaptations in the two parties will stabilize, in systems where fitness interests strongly conflict. It is possible that ecological or other circumstances will allow one party to be quite good at blocking cues that would allow the other to infer mental states, and therefore the adaptations for mind-reading among antagonists won't evolve very far in such a species. On the other hand, in situations where there are benefits from mutual co-operation, the evolutionary elaboration of mind-reading adaptations might proceed towards extreme complexity due to the positive feedback inherent in such a situation. For some species, the ability to co-ordinate behaviour might lead to large fitness pay-offs for both parties, which could select for each party to evolve machinery to help the other read its mind, and for each party to evolve machinery to be able to draw richer and more precise inferences from the cues that the other emits. In situations of social monogamy there are conflicts of interest, but there are also harmonies of interest. Therefore, monogamy may drive certain types of elaboration of mind-reading adaptations because of the necessity of complex co-operation. Humans supply a large array of cues when they want to have their minds read, and yet when they don't want their intentions to be apparent they can become considerably more opaque.

Møller: In my opinion, conflicts of interest do not drive co-evolution, rather different interests are the driving force behind co-evolution.

Tooby: As the Red Queen principle suggests, co-evolutionary races often do not get anywhere because of the antagonistic nature of the interactions, which leads to mutually destabilizing change but no progress. However, co-operative co-evolution can result in the development of relatively elaborate interlocking behaviours and structures because each progressive step is made even more advantageous by subsequent selection on adaptations used in the complementary role. For example, human language is a spectacular elaboration of a mind-reading ability, depending on immensely sophisticated complementary adaptations in the sender and receiver. In language the sender has the complete freedom not to send information if it is against his/her interest to do so. Moreover, conflicts of interest are everywhere in the natural world, while sophisticated mind-reading abilities appear to be relatively rare. Humans are unusual not in their conflicts of interest, which are standard, but rather in the frequency with which they engage in extended activities in which the participants have strong common interests. Indeed, consider the fact that humans, unlike our great ape relatives, maintain high contrast scleras into adulthood, allowing others to perceive eye direction at greater distances, and also consider the leakiness of so-called involuntary human emotional expressions. The fact that these automatically provide potentially damaging information to surrounding individuals implies a fairly high degree of co-operation among ancestral humans, compared to other organisms.

Møller: I don't agree that antagonistic co-evolutionary systems don't go anywhere. For example, if you look at the way cuckoo offspring interact with their hosts, you observe that there are continuous co-evolutionary changes in these lineages.

Rogers: So why do male sage grouse maintain their elaborate plumage if female sage grouse don't care what the males look like?

Møller: I agree, this is puzzling. However, there are studies in different species that suggest this is the case; for example, the study of Pruett-Jones (1990) in New Guinea.

Daly: There is also the issue of what correlational methods can show in the absence of experiments. Many people have had difficulty showing that females choose males in lek-breeding species if they rely on natural variation and don't manipulate male traits.

Møller: But the observation that these correlations are still present in monogamous species is puzzling. The reason why people believe that this is the case is that one male will perform most of the copulations at a lek during the season, but there is little evidence of any phenotypic correlations with mating success. It's also interesting in this context that the females in monogamous relationships are exposed to a much larger range of phenotypic variance. In lekking species there is little scope for choice because the males look much the same.

Rogers: But the absence of variation sounds like the signature of strong selection, and what's imposing the selection if it's not some kind of choice?

Møller: What I imagine has happened is that under these extreme skews in male mating success, the qualities of males that females have been interested in have been selected out. The selection pressures must have been strong to result in the disappearance of these 'quality' features.

Kacelnik: Can you elaborate on the experiments with the polygamous long-tailed widowbirds?

Møller: There have been about 15 studies of these birds. Half of them show no relationship between male mating success and their phenotype, particularly in highly polygynous species. It's interesting that situations where there is strong selection pressure and where male success varies the most are the situations where there is little, if any, evidence for discrimination based on phenotypic traits.

Cosmides: I would like to mention a puzzling observation related to co-evolution and deception. Paul Ekman (1992) has shown that people are not good at detecting deception when they're listening to somebody talk and looking at their face. What's interesting about this is that there are cues in the stimuli, i.e. in the micro-movements of different muscles, such that if you know what they are you can tell who is lying and who is not. Therefore, it is not the absence of cues that results in poor deception detection. Interestingly, the only group of people who were good at detecting deception in Ekman's experiments were secret-service police. Secret-service police (who protect the President) have to *prevent* something from happening, as opposed to regular police who have to investigate situations *after* they have happened.

Cronin: We've been exposed to these cues for two million years but can't detect them, and yet within an hour Paul Ekman can teach people to pick them up.

Miller: Leda Cosmides' point illustrates that human communication is not under strong selection to be truthful or honest. People don't seem to care much about whether what somebody is saying is true.

Pinker: No! People are obsessed with whether what somebody is saying is true. Remember that these experiments create an artificial situation in which a person is brought into an unfamiliar circumstance and has to guess, out of context, whether a stranger is lying. In real life, where you know someone long enough to know their repertoire of ordinary twitches, flinches and reactions, it would probably be much easier to detect deception.

Mealey: If the base rates of lying are low enough, then the costs of cheater detection in those circumstances will include not only the costs of evolving and maintaining the ability to detect cheaters, which aren't the most expensive, but also the costs of picking up false positives. If you're falsely detecting lies then you may later be exposed to the revenge of potential co-operators, which will be an enormous cost. This is probably why we don't invest more in cheater detection because we prefer to avoid all those false positives.

Kacelnik: Under the logical signal-detection theory, you would want to have the most accurate information possible and then set your response taking into account that information. If false alarms are dangerous then you would want to know whether the person is lying, but then you might decide that if there is a 30% probability, for example, that s/he is lying, you will still treat it as if s/he is telling the truth. Therefore, there is no explanation as to why you would want to fool yourself if you are not using those cues which are there.

Pinker: We can't tell from the data whether subjects were or were not setting a high signal threshold for deciding that the actor was lying. The experiment does not allow one to do the proper signal-detection analysis.

Nisbett: One can think of this issue in terms of arms races: one person detects another person's cheating but that person prevents detection by increased powers of deception, resulting in a steady-state situation.

Daly: I would like to know more about the ancestral social environments in which these selection pressures arose, because only then could we make a plausible argument about the ecological milieu within which our inabilities and talents in this regard evolved. The long-term reputational consequences of lying about anything anyone cares about have to be built into models of costs over and above the short-term consequences and the value of truth in immediate decision making.

Sherry: As with other cognitive systems, there may be subsequent effects on behaviour that are separate from the ability to articulate consciously that a particular individual had just lied.

Rogers: If this is really an arms race between deceit and detection, one would expect it to proceed quickly, and there may even be different signals of deceit in different places.

Møller: There are a number of examples of brood parasites that lay eggs in the nests of hosts. However, in some populations there are variations in the response of the hosts ranging from complete acceptance to non-acceptance.

Rogers: Are you implying that this is an arms race where there is geographic variation?

Møller: Yes, and one way in which antagonistic interactions between these species can be maintained is if there is some kind of meta-population structure in which there are localized populations of cuckoos and hosts. If particular populations of hosts start to discriminate against the cuckoo, then migration to other naïve populations can maintain the cuckoo–host interaction.

References

Ekman P 1992 Telling lies. Norton, New York
Godin J-G J, Dugatkin LA 1996 Female mating preference for bold males in the guppy, *Poecilia reticulata*. Proc Natl Acad Sci USA 93:10262–10267
Gyger M, Karakash S J, Marler P 1986 Avian alarm calling — is there an audience effect? Anim Behav 34:1570–1572
Pruett-Jones MA, Pruett-Jones SG 1990 Sexual selection through female choice in *Lawes parotia*, a lek-mating bird of paradise. Evolution 44:486–501

Normative and descriptive models of decision making: time discounting and risk sensitivity

Alex Kacelnik

Department of Zoology, South Parks Road, Oxford University, Oxford OX1 3PS, UK

Abstract. The task of evolutionary psychologists is to produce precise predictions about psychological mechanisms using adaptationist thinking. This can be done combining normative models derived from evolutionary hypotheses with descriptive regularities across species found by experimental psychologists and behavioural ecologists. I discuss two examples. In temporal discounting, a normative model (exponential) fails while a descriptive one (hyperbolic) fits both human and non-human data. In non-humans hyperbolic discounting coincides with rate of gain maximization in repetitive choices. Humans may discount hyperbolically in non-repetitive choices because they treat them as a repetitive rate-maximizing problem. In risk sensitivity, a theory derived from fitness considerations produces inconclusive results in non-humans, but succeeds in predicting human risk proneness and risk aversion for both the amount and delay of reward in a computer game. Strikingly, and in contrast with the existing literature, risk aversion for delay occurs as predicted. The predictions of risk aversion for delay may fail in many animal experiments because the manipulations of the utility function are not appropriate. In temporal discounting animal experiments help the interpretation of human results, while in risk sensitivity studies human results help the analysis of non-human data.

1997 Characterizing human psychological adaptations. Wiley, Chichester (Ciba Foundation Symposium 208) p 51–70

The human brain is the result of our evolutionary history, and this history has been driven by natural selection. This is not controversial. Nor is it controversial that human psychology, both in terms of mental experience and of overt behaviour, is under the control of the brain and in consequence is itself the result of natural selection. What is less obvious and a matter of intense interest at present is the extent to which applying evolutionary thinking makes a difference to the understanding of human psychology here and now. A broad survey of different attitudes to these questions and my perspective in the matter can be found elsewhere (Betzig 1997, Kacelnik & Krebs 1997).

My view is that the best way to argue either way is to examine concrete research rather than engaging in abstract arguments on whether people maximize fitness always or only sometimes, or whether some psychological trait is an adaptation (it gave heritable advantages in the past) or is still adaptive (it is genetically advantageous today). It is moot to discuss whether 'in principle' humans may be fitness-maximizing entities under all circumstances, when cultural heterogeneity is considered. No animal species is, and it would be curious for those of us that favour putting humans in their place as yet another result of unguided evolution to claim that natural selection has treated humans as its 'chosen species' by endowing *Homo sapiens* with evolutionary long-sightedness. The opposite view (that humans don't care about fitness) is also indefensible, as it flies in the face of both folk (but accurate) psychology (why do people love their children?) and of anthropological evidence showing that in many pre-demographic revolution cultures social and reproductive success are strongly correlated.

The concrete task then is to examine active research programmes and enquire as to what extent being furnished by good evolutionary theory makes a difference to scientific progress in the field of human psychology. This judgement is not always easy to make. For instance, Steven Pinker's work (Pinker 1997, this volume) provides evidence that human language abilities sit comfortably in the general picture of human behavioural evolution, but to date I have seen no evidence that language research itself has been dramatically affected by adaptationist thinking. This will probably happen soon, but for the time being people who are sceptical about the adaptationist programme could argue that most of what we know about the topic could be — and was — achieved without any help from Darwinism[1]. In contrast, the study of criminal and abusive behaviour has uncovered a strong relation between the social statistics and the expectations derived from an adaptationist analysis (Daly & Wilson 1996). In this case genetic relatedness became a variable of interest because relatedness is crucial in understanding natural selection, and indeed had been overlooked in the past. Hopefully, a critical case-by-case enquiry will tell us if the adaptationist perspective is just a luxury or a fundamental pillar of psychological sciences that is here to stay.

To follow my own recipe as best I can, I will present a selective review of two areas of research where adaptationist and descriptive theorizing converge. I chose these examples because in this area the combination of normative and descriptive modelling is not accessory but central. Both examples, temporal discounting and risk sensitivity, address ubiquitous issues in human and non-human behaviour, but are perhaps much less familiar to non-specialized audiences than other universals involving social exchanges, such as communication, mating or aggression.

---

[1]Noam Chomsky, referring to the evolution of language, (Chomsky 1988) expressed this scepticism boldly: 'Evolutionary theory is informative about many things, but it has little to say, as of now, of questions of this nature.'

Case 1: time discounting

Time discounting is the drop in subjective value of events as a function of the delay expected until their occurrence. The reason why time discounting is important is not difficult to see. Decision making in all human (and many non-human) affairs involves choosing between actions that take time to produce consequences. Environmental policy often requires trading consequences with a differing time horizon: the immediate loss of prosperity resulting from banning forest cutting against the delayed, long-term loss caused by losing the atmospheric activity of that forest. Personal saving requires curtailing present gratification to allow for future well-being. To make these choices people must, consciously or unconsciously, combine the magnitude of both immediate and future consequences and the time delay until each of them, and this requires a time-discounting criterion.

One problem with normative modelling of time discounting[2] is that it is by no means obvious if there is only one way to be 'rational' about discounting. For socially meaningful decisions, such as environmental policy, psychological knowledge cannot replace the need for ethical, philosophical and economic judgement. For choosing a criterion of personal saving people must follow the advice of their own psychology to consider the way they feel about postponing gratification and external (non-psychological) advice about the theoretical expectation of economic performance for the capital set aside. Whatever the final criterion we may want to impose, our judgement should be informed by understanding how individuals and societies do discount the future and what (in evolutionary sense) are they trying to achieve by doing it that way.

Normative models of discounting have been suggested by economists (Samuelson 1937) and biologists (Kagel et al 1986). Using suitable simplifying assumptions, the most common normative models yield discounted functions with an exponential form. The assumptions are: (a) that there is a constant probability of loss (or a constant interest rate) of the reward per unit of waiting time; and (b) that the subject's behaviour is fully tuned to it. The derivation is easiest to see using money gains as an example.

Say that the current interest rate is x, and the choice is between two sums of money, one to be received immediately and another to be received after a fixed number T of capitalization time units. The issue is to identify the sums of money with equal value taking into account the time gap. After one time unit a starting sum S will be worth $S(1+x)$, after two time units it will be $[S(1+x)](1+x)$ or $S(1+x)^2$, and after T time units it will be $S(1+x)^T$, Thus, because of the capital growth, the delayed sum that equals the value of the immediate sum is:

---

[2]'Normative' here is equivalent to making predictions on the basis of achieving a specified goal or maximizing some quantity.

$$V = S\,(1+x)^T.\tag{1}$$

Equation 1 describes gains accrued by the immediately received sum during the delay, but the argument can be reversed by thinking of the delayed reward losing value because of some chance of it being lost during the delay. If x is the expected loss per unit time, then the value of a delayed sum when evaluated at the beginning of the delay is:

$$V = A\,(1-x)^T\tag{2}$$

where V is the present value of a delayed reward of magnitude A and $0 < x < 1$. A mathematically equivalent, but more conventional form of this expression is

$$V = A\,e^{(-kT)}\tag{3}$$

where $k = -\ln(1-x)$. Equation 3 thus gives the present value of a delayed reward for any combination of its amount and its delay time. The constant k is not normally established a priori but used as a free parameter, chosen according to the best fit to the subjects' choices.

So much for the normative prediction. The trouble is, this is not the function that real animals (human or non-human) follow. The empirical database rejecting exponential discounting is rich, but one clear example is provided by the research of Cropper et al (1991).

They phoned a large number of subjects and asked the following question:

> 'Without new programmes, 100 people will die this year from pollution and 200 people will die 50 years from now. The government has to choose between two programmes that cost the same, but there is only enough money for one. Programme A will save 100 lives now. Programme B will save 200 lives 50 years from now. Which programme will you choose?'

Different subjects were asked using different values for the number of future lives and the time difference. The curve dividing the preference responses for the two programmes in a plot of number of future lives saved versus delay of programme B gives the discounting function. Its shape is significantly different from the exponential function in Equation 3. To rescue exponential discounting, the discounting factor k has to be steeper for short delays than for longer ones. However, although to take socially important decisions (or making electoral promises) we may be satisfied that Equation 3 does not describe what society wants collectively, for psychological analysis these results are not conclusive because we need to know how individuals, not groups, discount. Non-exponential social discounting as found in these surveys does not exclude exponential individual discounting because even if each individual followed an exponential law, the population preferences could be non-exponential due to individual differences in discount factor.

Intra-individual discounting has been assessed by Myerson and Green (Myerson & Green 1995, 1996). They asked each subject to make many choices between cards taken from two piles. Cards on the left when turned showed a number indicating a nominal amount of money, different from card to card, to be received immediately. Cards on the right showed either of two numbers ($1000 or $10 000, in separate treatments) indicating a nominal amount to be paid after a delay. Notice that the Cropper et al survey varied the delayed outcome and the Myerson and Green one varied the immediate outcome. The purpose of having two treatments differing in the delayed sum is to examine if the discounting factor is independent of this sum (as suggested by Equation 3) or not. The delays were one week, one month, six months, and then one, three, five, 10 or 25 years. The exponential equation failed to predict individual results, which followed a similar law to that of the social data in the survey of Cropper et al. There is a different equation which is good at describing the results. This equation also has a free parameter, k, but it accommodates the observed rapid discounting of short delays and slower discounting of long delays. The equation is:

$$V = \frac{A}{1 + kT} \tag{4}$$

where A is the size of the delayed rewards. Equation 4 is well known in psychological studies of discounting and self-control (Mazur 1987, Logue 1988, Rachlin 1989), and it is usually called hyperbolic discounting. The addition of a unit constant to the denominator principally serves to make the function behave sensibly when delay tends to zero. Additional key references to the sources of Equation 4 can be found in Green & Myerson (1996) and Kacelnik & Bateson (1996). To compare the two equations graphically, Myerson and Green's results are shown in Fig. 1, averaged across individuals.

An important difference between Equations 3 and 4 is that the former fails to account for the phenomenon known as preference reversal. Preference reversal refers to the observation that, for a constant difference in delay between an earlier smaller reward and a larger more distant one, large rewards may be preferred when both rewards are distant, but there is a strong tendency to prefer the least delayed reward regardless of its smaller size when both of them have greater immediacy. Figure 2 shows how hyperbolic discounting accounts for this observation.

Equation 4 *per se* is not strongly explanatory because it did not emerge from an a priori analysis but purely from its power to describe data efficiently. In contrast, because of the strong appeal of the a priori argument favouring exponential discounting, several re-elaborations have been made to rescue the rationale that led to it. For instance, it has been shown (Green & Myerson 1993) that since the discount parameter varies with amount of reward (larger rewards are discounted less steeply; Raineri & Rachlin 1993) both exponential and hyperbolic discounting could account for the observed preference reversals. This, of course, leaves out the issue of why, according to the normative rationale, rewards of different size should be

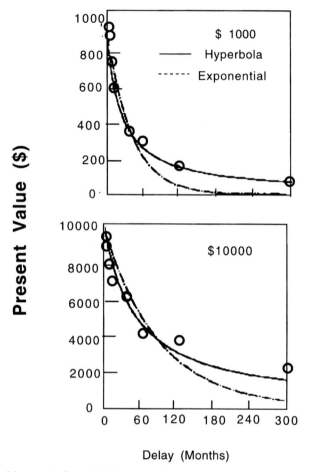

FIG. 1.   Myerson & Green (1995) asked subjects to choose between an immediate amount and a delayed amount of money. The figure shows the immediate amount that has equal value to delayed amounts of either $1000 (top) or $10 000 (bottom). The lines show the best fit according to two models of discounting (from Myerson & Green 1995).

discounted differently. On the other hand, hyperbolic discounting *per se* does not exclude the rationale based on probability of loss, because hyperbolic discounting can be interpreted in terms of probability of receiving the delayed rewards, given that one assumes that this probability is proportional to $1/(1+kT)$ (Green & Myerson 1996). This, in turn, begs the question of why (normatively speaking), the probability of loss should behave that way. Another strategy is to suggest that individual behaviour is not tuned to a single rate of discounting but to a distribution of possible processes as statistically encountered in nature (Y. Iwasa, personal communication 1992). In summary, hyperbolic discounting as an empirical fact does

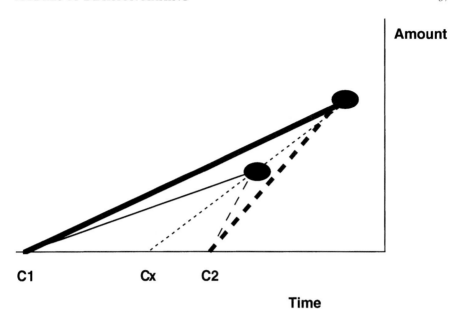

FIG. 2.   The figure depicts the problem of choosing between two potential rewards that differ in size and time of delivery. The subject makes the choice at times C1 or C2 using the equation 4 $[V = A/(1 + T)]$ as its currency. In this case T is the interval between the choice and the reward. Ignoring the unity in the denominator (it makes no difference here), V equals the slope of the line passing through the choice point in the time axis and the reward location as a function of its size and time of delivery. Thick lines identify the large reward, thin lines the small reward, solid lines C1 and dashed lines C2. At C1 the thick line is steeper than the thin one, indicating that the large reward should be preferred, whereas at C2 the opposite occurs even though the time gap between rewards is constant. Cx indicates the choice time at which indifference is expected, as it follows from the fact that the dotted line passes through both rewards. This phenomenon is called preference reversal and is a robust empirical finding.

not exclude the logic leading to the normative model based on probability of interruption or loss, but forces the rejection of the simplest and perhaps most appealing derivation. An interesting twist, however, is that it is possible to make a normative case for Equation 4, as I will illustrate by reference to non-human data.

Non-human discounting has been studied by offering animals (most commonly pigeons or rats) choices between stimuli (such as coloured pecking keys) which lead to characteristic food rewards after characteristic delays. As with humans, experimental animals trade amount for time: delayed rewards are discounted. The obtained discounting function is also hyperbolic in the sense of Equation 4 (Mazur 1987).

Although the effect of expecting a nominal delay in human data and of repeatedly experiencing a real delay in non-human experiments is mathematically similar, the difference between the procedures should not be ignored. The repeated procedure required to train animals to choose among arbitrary stimuli associated with different delays or amounts of reward necessarily exposes the subjects to the frequency of losses, which can only be guessed by the human subjects who are asked to choose after thinking about (rather than experiencing) the consequences of each option. In fact, in most animal studies the frequency of loss of reward during the delay is zero, so that, given that they show steep discounting, the application of the normative argument that led to Equation 3 requires the assumption that the subjects are incapable of learning this absence of interruptions. Since independent research shows that the same species do have the ability to learn the difference between continuous and partial reinforcement, this defence is weak.

A different normative argument can be made for non-human experiments precisely because they all use repeated choices. If a subject makes one choice after another between the same alternatives, and it chooses consistently one option that produces a fixed delay and a fixed reward, then it experiences a rate of gain over time given by the following equation:

$$\text{rate} = \frac{A}{t + T} \tag{5}$$

where T is the delay between the choice and the reward and t is the sum of all the other times in each cycle.

Equation 5 is almost identical to Equation 4, but it has no free parameter and, instead of adding 1 to the denominator it adds t, which has an interpretable meaning. In natural foraging t may relate to inter-prey searching times or inter-patch travel time, and in the laboratory to inter-trial intervals. This equation has been extensively used as the basis of normative modelling in classical optimal foraging theory (see reviews in Stephens & Krebs 1986, Krebs & Kacelnik 1991). The rate maximization approach is substantially different from the rationale that led to Equation 3 because it does not include any stochastic process that could cause losses of the reward. A recent study illustrates a direct test of this equation. Bateson & Kacelnik (1996) used a titration procedure in which starlings chose between two coloured pecking keys. One colour (the standard) gave a fixed amount of food S after a fixed delay D. The value of S varied between treatments. The other colour (adjustable) changed its delay (but not its amount) against the subjects' preferences: when the bird preferred the standard, the adjustable delay in future trials became shorter, when it preferred the adjustable, the adjustable delay became longer. The centre of oscillation of the adjustable option delay reflects the point where the subject is equally likely to choose between the alternatives. In this experiment t was considerably greater than one time unit. To fit the data, however, Bateson and Kacelnik found that t must be assumed to be small. Figure 3 shows Bateson and Kacelnik's results.

A discussion of the possible reasons for the differential effect of the time between choice and rewards (T) and the other times in the cycle (t) would take us away from

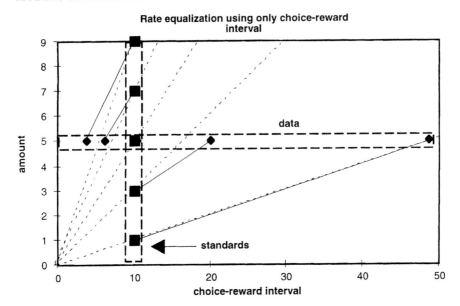

FIG. 3.   Bateson & Kacelnik (1996) trained starlings to choose repeatedly between two keys. One (the standard) gave (in different treatments) one, three, five, seven or nine units of food after 10 seconds. The other (adjustable) always gave five units of food, but the delay varied until the subject was indifferent between the two options. The solid lines join the standard of each treatment (squares) with the mean results (diamonds). Equalization of the ratio of amount over delay between options requires that the diamonds sit on top of the broken lines passing through each standard.

the focus of this chapter and has been discussed elsewhere (see Bateson & Kacelnik 1996, 1997, Benson & Stephens 1996, Kacelnik & Bateson 1996 for access to earlier literature). Here I will leave aside this problem and accept that a deterministic, rate-maximizing argument with some warts (non-equivalent effect of different times in the cycle) gives a qualitative and quantitative fit of the discounting of delayed rewards obtained in non-humans tested under repeated choices. This result does not use any free parameter to approximate individual data.

Since the successful equation is virtually the same as that fitting human discounting in non-repetitive problems, it is tempting to suggest that humans may show hyperbolic discounting because, like other species, their discounting mechanism evolved to achieve rate maximization under repetitive choices (a similar argument was made by Rachlin & Raineri 1992).

In summary, we are replacing one normative argument by another. We do this because one of them describes the data better, not because of evidence that its assumptions are closer to the evolutionary scenario. Neither of the models is well tuned to the experimental conditions. The rationale for exponential discounting is under-defined because we know nothing of the hazard rate that should be used to discount nominal rewards nor can the subject know it. The rate-maximizing rationale is quantitatively strong because it lacks free parameters, but it does not apply to the human tests because the latter are not repetitive but basically one-off choices.

Our present argument is that the discounting process used for one-off events seems to obey a law that evolved as an adaptation to cope with repetitive events. This can only be a temporary stand while we test further predictions that follow from this interpretation.

Case 2: risk sensitivity

My second example addresses another fundamental problem in decision theory, that of choosing between actions with uncertain outcomes. As in the previous example I start with a normative analysis. Appropriate sources to follow these ideas in detail have been listed elsewhere (Kacelnik & Bateson 1996; some useful readings are Holloway 1979, Caraco 1980, Stephens 1981, Houston & McNamara 1982, Stephens & Krebs 1986, McNamara & Houston 1987, 1992, Houston 1991, Reboreda & Kacelnik 1991). Here I am more concerned with getting the concept across than with identifying the original source of each explanatory device.

Consider the relative value of two actions, one (I shall call it uncertain or variable) that yields multiple possible outcomes with specified probabilities and another (I shall call it fixed) that yields a predictable outcome equal to the average of those of the first action. Which option should be preferred, the uncertain or the fixed one? I have used outcome as equivalent to pay-off in the currency measured by the experimenter, such as amount of food gathered by an animal, the number of points scored in a game, the number of nestlings fledging from a nest or the number of copulations achieved by a mating polygamous male. Value, or utility, is normally thought of in a more fundamental currency, perhaps hidden to direct observation, such as the fitness gain or the subjective gratification resulting from that outcome. It is values that should be compared to predict choices. Say, for instance, that the different outcomes are different amounts of food to be obtained by a foraging animal as a consequence of choosing different prey types. Then pay-off will be the expected amount of food, but value depends on how food gains translate into fitness gains. This introduces complications because the relation between pay-off and value need not be a stable function. Fitness gains as a function of food gain can be state dependent: if the animal is well fed, little fitness is gained by gaining a unit of extra food, whereas if the subject is near a threshold level of reserves required to achieve a goal such as reproduction, migration or even overnight survival, then a small increase in feeding pay-off can yield a large increase in fitness. Put simply, the answer of which of the two options has greater value depends on the shape of the value versus pay-off function. If value versus pay-off follows an accelerated relation, then the variable alternative has greater expected value than the fixed one because of the higher marginal gains due to

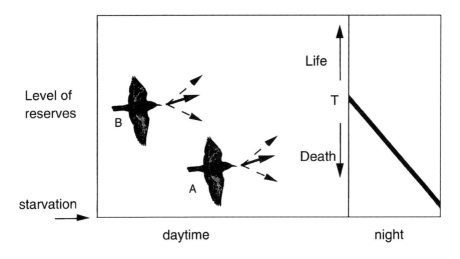

FIG. 4. A day in the life of a risk-sensitive bird. The 'state' of the bird is given by its stored reserves (vertical position) and the time of day (horizontal position). Since there is no foraging during the night, and reserves are used for metabolism, to survive the day cycle the bird needs to reach a threshold level (T) of reserves at dusk. Here there are two foraging options which yield the same outcomes at all times of the day. One (fixed) changes the state of the subjects as shown by the thick solid arrow. The alternative (variable) has either of the effects shown in thin broken lines with equal probability. The subject in state A would die if it chose consistently the fixed option but may survive if it chooses the variable option. The subject in position B would survive if it chooses the fixed option but it may die if it chooses the variable one. Subject A should be risk prone and subject B risk averse.

the events yielding higher pay-offs. The opposite is true when this function is decelerated. It is easy to see this using an example with a discrete rather than a continuous function.

Consider a small bird in a cold and short winter day choosing between two foraging sites that have equal expected pay-off, but differ in their variance. Fitness here is given by a binary outcome, achieving or not achieving the accumulation of enough reserves to survive the long, cold night. Figure 4 presents this idea pictorially.

According to this interpretation, we should be able to manipulate experimentally the animal's preferences. If we put a laboratory animal in a situation where the average rate of gains expected from two alternative options with equal mean pay-off is below the minimum needed to survive, it should prefer the uncertain option (or be 'risk prone'). If the average gains are sufficient, it should prefer the fixed option (or be 'risk averse'). This logic applies whatever the reason of uncertainty. For instance, in foraging tasks

TABLE 1   Risk sensitivity studies in which non-human animals chose between fixed and variable reward sources and the value function was manipulated. The table shows the number of publications that used the amount of reward or delay to reward to introduce uncertainty, classified by whether they obtained the predicted shift in preference, an ambiguous result or no effect at all (Kacelnik & Bateson 1996)

Variability	Predicted shift	Some effect	No effect
Amount of reward	8	6	6
Delay of reward	0	2	4

the uncertain option can be unpredictable in its amount of reward and fixed in its delay to reward or fixed in amount and variable in delay.

Once again, the theory is sharp and clean, but the evidence is not. In a recent review of experimental non-human studies, for preferences between a fixed outcome (an amount of food or water after a time delay) and an uncertain outcome (in either amount or delay), Kacelnik & Bateson (1996) found an inconclusive picture. This is summarized in Table 1.

Inconclusive results are a frequent problem raised by testing normative theories. Since the theories are derived from first principles, it is not appropriate to judge the internal logical consistency of the theory using empirical results. For this kind of theory it is not obvious what the meaning of an empirical rejection or confirmation is. It would be wrong to believe that because there exist cases of both positive and negative results, the predictions did not follow from the assumptions in all cases; they do. What needs to be re-examined is if the assumptions were really met in the experiments. Given that the theory is internally consistent, it must be true that at least for the negative cases (and perhaps then also for the positive ones) the assumptions are inappropriate. This does not mean that there is no way to abandon a normative theory. If the negative cases proliferate, that line of theorizing will simply be considered unproductive.

Meanwhile, the negative examples can have enormous heuristic value, as they sharpen our judgement and promote new research. A number of insights have been derived from failures of risk sensitivity theory as a predictive tool. Contrasting the theory and the experimental results suggests that the issues of amount and delay variability are more different than it was first assumed, and good theoretical reasons have been discovered, why this should be so. McNamara & Houston (1987) have pointed out, for instance, that while amount variance concerns only the state of the subject at the time of the next choice, delay variability also affects the number of choices and consequently the degrees of freedom left to the subject at the time of the forthcoming choice. This makes the possibility of predicting what an animal should prefer more remote and may explain a number of failures, but it does not give a full explanation of some observed regularities, especially that there has never so far been a successful case of a switch from risk proneness to risk aversion in experiments

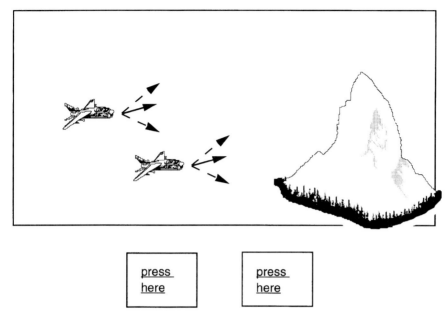

FIG. 5.   A computer game that mimics the problem described in Fig. 4. With no action, the plane approaches the mountain while losing height steadily. The subject plays by pressing either of the two buttons. One of them has the outcome shown in the thick solid arrow while the other has either of the two outcomes shown with broken arrows. Only one plane is normally present in the screen. Points are scored by flying over the mountain. One group was paid an amount of money that increased with points scored following an accelerating function while another followed a decelerating function (A. Kacelnik, W. James & I. Todd, unpublished data 1996).

manipulating delay variability. A careful re-analysis of each of the available experiments shows that in most cases the weakest links are the assumptions about the function relating value to pay-off, namely the value function.

One way to test the usefulness of the theory is to conduct an experiment where the value function can be more directly controlled, and this can be more easily achieved using humans as subjects. To illustrate this point I will now describe an experiment (A. Kacelnik, W. James & I. Todd, unpublished data 1996) using a computer game fashioned to mimic the situation described in Fig. 4. The game is explained pictorially in Fig. 5.

The subjects (Oxford undergraduates) were divided in two groups according to their (manipulated) value function. The rules for point scoring were identical between groups but the money paid as a function of total point scoring differed. Subjects in group A (accelerating value function) were told that if they scored between zero and 10 points they would receive £0.50, between 10 and 20 points £2.00, and more than 20 points £12.50. For the same scoring categories, subjects in

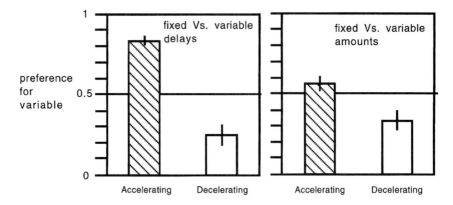

FIG. 6.   Results of the experiment described in Fig. 5. Subjects in the accelerating group were risk prone whether variability was in amount or delay, whereas those in the decelerating group were risk averse in both treatments. All results differ from 50% with significance of at least 5% (A. Kacelnik, W. James & I. Todd, unpublished data 1996).

group D (decelerating value function) were told they would receive £3.50, £5.50 and £6.00. The assumption made is that while money within this range has a linear relation to utility (and perhaps survivorship) for impoverished students, point scoring was only seen as a means to gain money, so that they were expected to show a flexible strategy for point scoring to implement monetary maximization.

Within each group subjects played two games a number of times, each round differing in the dimension for which the variable alternative had uncertain outcome: amount (the height gained after each choice) or time (the delay between the subject's choice and the flying plane gaining a fixed height).

Figure 6 shows the results of the experiment. The main conclusion is that the switch of preference between the two groups was manifest in both amount and delay. Group A was risk prone for both amount and delay variability and group D the other way round, as predicted by risk sensitivity theory. The delay results are specially striking because this is the first report of a switch in risk preference for delay.

This success of risk sensitivity as a descriptive theory does not of course imply that the theory accounts well for results in non-human animals, but it suggests that the suspicion that the main source of failure of the theory lies in wrong assumptions about the fitness versus pay-off function is well founded.

Conclusions

My goal in this brief and selective review was to advocate that a normative approach can play an important role in guiding research in human psychology, but that for normative models to be useful they have to be tested and often rejected by rigorous confrontation with real data. This implies a permanent exchange with descriptive approaches. Additional sub-themes are that important decision-making processes in humans are isomorphic with problems faced by other species, justifying a consistent theoretical approach across species, but that there are no reasons to expect that humans or non-human animals will act consistently as fitness maximizers in all conceivable circumstances. The psychologist's and ethologist's task is to uncover the psychological mechanisms, flexible or inflexible, that members of the species inherited. Theoretical speculation about fitness consequences of different mechanisms is yet another helpful tool in accomplishing this task.

I used two examples, time discounting and risk sensitivity. In both cases sharp normative predictions were made from explicit assumptions about function, and experimentation was carried out to contrast the normative predictions with decision-making processes in both human and non-human subjects. In both cases some aspects of the models were supported and others were not, and much progress was made and is still being made by these two roads of exchange: between normative and descriptive theorizing and between human and non-human experimentation.

I finished the article with a positive result: risk sensitivity theory, a normative framework developed in the context of fitness maximization in stochastic environments, succeeded in predicting human subjects' choices under an artificial situation with people working to score points that they could exchange for money in a non-linear way. Although this last experiment proves that normative risk sensitivity theory can predict flexible human responses to uncertainty, the experiment does not prove the evolutionary credentials of the theory because the experiment was not based on fitness manipulations, and thus the link between this form of optimality and evolutionary theory remains speculative. My message is that evolutionary psychology can produce unique contributions and can follow a rigorous and systematic path integrating normative theorizing and data-driven descriptions.

Neither animals nor humans are likely to be driven directly by the maximization of fitness, but we may understand the psychological mechanisms that do control their behaviour by asking about the fitness consequences of different courses of action. Concrete research progress, not abstract discussion, will determine the relative development of different approaches. With regard to the long-term perspectives of evolutionary psychology, I am cautiously optimistic. I believe that in many areas of psychological research entirely new questions are being asked and new answers obtained under the influence of normative models based on adaptationist thinking, but that the case is not yet fully made, and I am suspicious of those who write with messianic conviction: we should be scientists struggling to learn if we are right or wrong, not salespeople struggling to convince.

Acknowledgements

This work is supported by the Wellcome Trust (Grant 046101). I am grateful to my colleagues William James and Ian Todd for letting me present our unpublished results and to Miguel

Rodriguez-Girones, Steven Pinker, Peter Cotton, Melissa Bateson, James Fotheringham and Fausto Brito e Abreu for useful commentaries.

References

Bateson M, Kacelnik A 1996 Rate currencies and the foraging starling: the fallacy of the averages revisited. Behav Ecol 7:341–352

Bateson M, Kacelnik A 1997 Risk-sensitivity or preference for variance? Anim Behav, in press

Benson KE, Stephens DW 1996 Interruptions, trade-offs and temporal discounting. Am Zool 36:506–517

Betzig L 1997 Human nature: a critical reader. Oxford University Press, Oxford

Caraco T 1980 On foraging time allocation in a stochastic environment. Ecology 61:119–128

Caraco T 1981 Energy budgets, risk and foraging preferences in dark-eyed juncos (*Junco hyemalis*). Behav Ecol Sociobiol 8:213–217

Chomsky N 1988 Language and problems of knowledge: the Managua Lectures. MIT Press, Cambridge, MA

Cropper ML, Aydede SK, Portney PR 1991 Discounting human lives. Am J Agric Econom 73:1410–1415

Daly M, Wilson MI 1996 Violence against stepchildren. Curr Dir Psychol Sci 53:77–81

Green L, Myerson J 1993 Alternative frameworks for the analysis of self-control. Behav Philos 21:37–47

Green L, Myerson J 1996 Exponential versus hyperbolic discounting of delayed outcomes: risk and waiting time. Am Zool 36:496–505

Holloway CA 1979 Decision making under uncertainty: models and choices. Prentice Hall, London

Houston AI 1991 Risk-sensitive foraging theory and operant psychology. J Exp Anal Behav 56:585–589

Houston AI, McNamara JM 1982 A sequential approach to risk taking. Anim Behav 30:1260–1261

Kacelnik A, Bateson M 1996 Risky theories — the effects of variance on foraging decisions. Am Zool 36:402–434

Kacelnik A, Krebs JR 1997 Yanomamo dreams and starling prey loads: the logic of optimality. In: Betzig L (ed) Human nature: a critical reader. Oxford University Press, Oxford, p 21–35

Kagel JH, Green L, Caraco T 1986 When foragers discount the future: constraint or adaptation? Anim Behav 34:271–283

Krebs JR, Kacelnik A 1991 Decision making. In: Krebs JR, Davies NB (eds) Behavioural ecology: an evolutionary approach. Blackwell Scientific, Oxford, p 105–136

Logue AW 1988 Research on self-control: an integrated framework. Behav Brain Sci 11:665–709

Mazur JE 1987 An adjusting procedure for studying delayed reinforcement. In: Commons ML, Mazur JE, Nevin JA, Rachlin H (eds) Quantitative analyses of behaviour: the effect of delay and of intervening events on reinforcement value. Lawrence Erlbaum Associates Inc., Hillsdale, NJ, p 55–73

McNamara JM, Houston AI 1987 A general framework for understanding the effects of variability and interruptions on foraging behaviour. Acta Biotheoretica 36:3–22

McNamara JM, Houston AI 1992 Risk-sensitive foraging: a review of the theory. Bull Math Biol 54:355–378

Myerson J, Green L 1995 Discounting of delayed rewards: models of individual choice. J Exp Anal Behav 64:263–276

Pinker S 1997 Language as a psychological adaptation. In: Characterizing human psychological adaptations. Wiley, Chichester (Ciba Found Symp 208) p 162–180

Rachlin H 1989 Judgement, decision and choice. Freeman, New York

Rachlin H, Raineri A 1992 Irrationality, impulsiveness, and selfishness as discount reversal effects. In: Loewenstein G, Elster J (eds) Choice over time. Russell Sage Foundation, New York, p 93–118

Raineri A, Rachlin H 1993 The effect of temporal constrains on the value of money and other commodities. J Behav Decis Making 6:77–94

Reboreda JC, Kacelnik A 1991 Risk sensitivity in starlings: variability in food amount and food delay. Behav Ecol 2:301–308

Samuelson PA 1937 A note on the measurement of utility. Rev Econ Stud 4:155–161

Stephens DW 1981 The logic of risk-sensitive foraging preferences. Anim Behav 29:628–629

Stephens DW, Krebs JR 1986 Foraging theory. Princeton University Press, Princeton, NJ

DISCUSSION

Daly: I'm interested in hyperbolic discounting functions and why humans and non-human animals behave in this seemingly maladaptive way. We expend effort protecting ourselves from anticipated preference reversals as temptation approaches and from the anticipated regret of having succumbed to impulsiveness or temptation because our time horizons seem too short for our own good. Could you clarify why it's adaptive in the case of starlings?

Kacelnik: If you maximize energy over time then you have time in the denominator. The energy-over-time ratio is a hyperbolic function of time.

Daly: So it's as if the animal is deeming the delays as time invested in the acquisition of that resource.

Kacelnik: Yes. The reason that this is predicted is that if it follows the principle of lost opportunity, the time saved when choosing a less-delayed reward will be used for obtaining future rewards at the same rate.

Pinker: This would predict that you would only get hyperbolic discounting for food.

Kacelnik: No, you would get it for any reward for which you cared about the amount of reward-over-time.

Pinker: Would this also maximize the amount of reward over a lifetime, or at least over a fixed unit time?

Kacelnik: The model applies over any time horizon during which you are maximizing reward over time and does not include periods of rest. It simply tells you that throughout the time when you are working, over a fixed stable period if you maximize the reward divided by the time then what you ought to get is hyperbolic discounting. The argument doesn't take into account other complexities, such as variability at the same time as discounting. It is a complex situation because something that comes later is also more uncertain: it could be worse, as most people assume, or it could be better.

Pinker: Why is energy-over-time a different criterion than the one assumed in classical economic theory when optimal discounting strategies are analysed?

Kacelnik: If you had compound interest then you would be multiplying your capital every year by the same factor, which results in exponential growth. If you instead divided the pay-off equally over every time unit then you would have a hyperbolic line. The argument in economics is either compound interest or probability of loss. The argument in rate maximization is neither. It refers to saving time to be used for more of the same.

Hauser: Damasio (1995) studied patients with bilaterally damaged amygdalas. In card games involving potential costs in terms of monetary loss, these patients show no physiological change from baseline, whereas normal individuals do show a significant physiological response. Have you looked at tasks associated with different levels of risk within the same individual to see whether there are shifts in physiology based on these mental decisions?

Kacelnik: Not in humans, but we put animals on either a positive or negative energetic budget, in which we tested them during a week when they were losing weight versus when they were gaining weight, or when their weight was stable but they were receiving more food than they required. We compared the budget that way, but we haven't done any neurobiology studies.

Hauser: Do those functions change if you put the animals in conditions where they are vulnerable to predation?

Kacelnik: We haven't looked at predation, but all the behavioural ecology models based on discounting and predicting exponential discounting take into account the probability of loss, not only by predation but also by competitor stealing, or kleptoparasitism. This interpretation doesn't fit the data because in some of these experiments the animals undergo thousands of trials and they show exactly the same preferences each time, even though we know that they are capable of learning the probability of something disappearing.

Wilson: In these loss experiments, did the animals actually see the competitor at the time they felt the loss?

Kacelnik: This was not incorporated into the experimental design. The experiments were just designed to see if they could learn about probabilities. I would also like to mention that there are a number of ways in which the exponential discounting models can be saved. For example, if the subject knows all about the probabilities of loss but is integrating over a distribution of probabilities of loss, each producing an exponential decrease that is then integrated over a range of possible discounting factors, the integral finally results in a hyperbola. Another example is when occasionally the reward can be lost and, in such cases, the subject saves time to wait for the next reward. This results in time being in the denominator because it is again a rate equation.

Daly: It's interesting to speculate whether the subjects think they're interacting with the biotic or the abiotic world. For some of the human data on hyperbolic discounting it is possible to explain why people steeply discount future promises by arguing that what is left out of the standard economic accounts is peoples' disbelief that promised future things will really happen.

Tooby: If you are waiting for a reward and you notice that that you're not going to get it prior to when you would expect to get it, then can you go on to the next item? Surely this depends on having the information that the reward is not coming prior to when you expect to get the reward.

Kacelnik: Yes. In the cases when the reward is not forthcoming, the last account assumes that you save time by not waiting for food that will not come.

Tooby: But you wouldn't see that if the Skinner box turned on a certain schedule and you didn't know in advance which regime was operative or whether or not the reward was forthcoming.

Kacelnik: It depends on how you program the Skinner box.

Rogers: If you did a hypothetical choice experiment in which you gave the choice of $100 now or $1000 in six weeks time, and you also said that between now and whenever the choice arrives the subjects were not allowed to receive any money, then the amount divided by the time is the rate of return and it would make sense to reject the second choice because you may starve between now and then. It strikes me that we would get hyperbolic discounting under those circumstances and we wouldn't think it at all remarkable. Perhaps we have evolved a psychology that causes us to interpret these hypothetical choices in that framework because it was once appropriate to do so.

Kacelnik: I'm not saying it's not logical or reasonable. The question is what are we going to suggest next? We have obtained similar results in many different animals, including humans, rats, pigeons and starlings. We could make quantitative predictions about the consequences of the situations in which we obtained slightly different results or we could hypothesize about the minds of the people who were exposed to the problem. In my opinion this is not the best way forward. The problem is that it's difficult to do equivalent experiments in humans and animals because in order to tell the animal what the conditions are we have to expose the animal to hundreds of trials, whereas in the human experiments we give them verbal instructions and they can do the experiment only once. Therefore, in the course of training the animal to do the experiment we also train them about the probability of loss.

Møller: The species that you mentioned — rats, humans, pigeons and starlings — are all species that are demonstrating increases in population size, and fitness in increasing populations is dependent on the timing of reproduction, i.e. it pays to reproduce earlier. Therefore, you would predict that in these species it would pay to discount more than in species that have a stable or decreasing population size.

Kacelnik: Your comment touches on an important point that I have not yet mentioned, i.e. the time-scale of these functions. These ranged from up to 25–100 years in humans to 5–60 seconds in the starling. Similar mathematical descriptions apply to both, although I wouldn't stick my neck out to claim the same evolutionary forces affect these different situations. And I would also like to add, for the record, that the population of starlings is actually declining.

Sperber: Why can't you also do repetition experiments in humans?

Kacelnik: It is possible. It would take a long time to test delays of 25 years many times. It has been done with short delays, but real results have to be used rather than nominal sums of money.

Wilson: If the hypothesis is that animals pay attention to rate, and therefore that time is important, rather than an alternative hypothesis, then if you change the conditions of the animals, such as may occur in some seasonal birds like eiders that are anorexic during the breeding season, then they shouldn't show the same patterns in each of the different conditions.

Kacelnik: In a certain sense we are always unwittingly doing that kind of experiment, but it doesn't really help because when you test an animal under conditions where it doesn't care for the food you obtain noisy results. You could interpret this as a confirmation of your hypothesis, i.e. that they should not be maximizing rate when they're not hungry, but this isn't satisfactory. If we could predict the shift in the shape of the discounting factor then it may be possible to do the experiment that you suggested.

Reference

Damasio A 1995 Descartes' error. Basic Books, New York

Mate choice: from sexual cues to cognitive adaptations

Geoffrey F. Miller

Economic and Social Research Council Research Centre for Economic Learning and Social Evolution, University College London, Gower Street, London WC1E 6BT, UK

Abstract. Evolutionary psychologists have successfully combined sexual selection theory and empirical research to compile lists of sexual attractiveness cues used in human mate choice. But a list of inputs is not the same as a normative or descriptive model of a psychological adaptation. We need to shift from cataloguing sexual cues to modelling cognitive adaptations for mate choice. This theoretical chapter addresses how to make this transition in three parts. The introduction discusses four general problems with cue cataloguing as an evolutionary psychology research strategy: animals' promiscuous flexibility of cue use; cue use being marginal to cognition; cue use being marginal to the hard game-theoretical aspects of mate choice; and cue use being uninformative about the exact adaptive functions of mate choice. The middle section develops six critiques of current mate choice research: the obsession with sex differences; the over-emphasis on physical rather than behavioural cues; the assumption of weighted linear models of cue integration; the avoidance of game-theoretical problems of mutual choice and assortative mating; the neglect of co-evolution between mate choice heuristics and the cues that they select; and the failure to understand that mate choice is only worth doing if potential mates show significant genetic variance. The conclusion outlines a new normative and descriptive framework for mate choice, centred on the use of brutally efficient search heuristics that exploit the informational structure of human genotypes, phenotypes and populations to make good mate choices.

1997 Characterizing human psychological adaptations. Wiley, Chichester (Ciba Foundation Symposium 208) p 71–87

Four problems with cataloguing sexual cues

Mate choice has become the flagship domain of evolutionary psychology: it is the most well-known example of how evolutionary theory can guide the study of the human mind and human behaviour. Evolutionary psychologists, inspired by sexual selection theory (Darwin 1871, Cronin 1991, Andersson 1994), have successfully discovered and documented some of the diverse cues used in the selection of sexual partners, such as height, intelligence, walking speed, facial symmetry, sense of humour, waist-to-hip ratio, degree of genetic relatedness, full lips, political status and sexual foreplay skills (for reviews see Buss 1994, Ridley 1993, Wright 1994). Many such cues are important across a wide range of cultures and historical epochs. This universality is not surprising given that such cues show all the classic symptoms of having been sexually selected during human evolution: they distinguish humans from other apes, men from women and adults from juveniles; they have high costs and complexity to function as reliable indicators of health and developmental homeostasis; and they have exaggerated aesthetic features that play upon the intrinsic perceptual biases of our nervous systems (see Miller 1997a). These universal cues of attractiveness are the informational inputs to our psychological adaptations for mate choice.

However, imagine a software engineer's reaction if one of their programmers, charged with designing an algorithm to solve a difficult artificial intelligence problem, returned with nothing more than a list of potentially relevant input variables. The difficult part of cognitive engineering is not identifying the inputs, but knowing how to combine them in context-sensitive ways to yield adaptive behaviour. The same holds in evolutionary psychology. Describing the perceptual cues used by a psychological adaptation is just the first and often easiest step in characterizing the adaptation. I am worried that the evolutionary psychology of mate choice may get stuck at this step, producing an ever-expanding catalogue of cues demonstrated to have a statistically significant effect on attractiveness judgements in laboratory experiments, but never progressing to detailed, testable models of the cognitive adaptations that exploit these cues to make real mate choices.

One could argue that we should wait to find all the sexual cues before we try to combine them in a cognitive model of mate choice: building blocks first, architectural plans later. But this building-block approach will fail for many of the same reasons that evolutionary psychologists believe non-Darwinian psychology has failed. First, there is Brunswik's (1956) problem of 'vicarious functioning': animals are notoriously opportunistic in their use of cues. Brunswik argued that 'systematic designs' (where a single cue's value is varied and its behavioural effects are measured) are a powerful way of finding out what cues can affect behaviour in the psychology laboratory, but are an extremely weak way of finding out what cues are most informative and most often used under natural conditions, or how such cues are integrated to guide adaptive behaviour. People almost never admit indifference in mate choice, so almost any cue distinguishing two potential mates in the laboratory can reliably yield an effect on choice behaviour, whether or not the cue has much ecological validity.

Second, the longer and more diverse our cue catalogue becomes, the easier it will be for critics of evolutionary psychology to claim that general-purpose associative-learning mechanisms could account for human mate choice by correlating preprocessed cues with desired behavioural outcomes. If evolutionary psychology produces nothing more than cue catalogues, evolved adaptations will continue to be marginalized to the periphery of cognition, i.e. in low-level perception and in the motivational systems that guide associative learning. Mate choice reaches deep into the heart of cognition—judgement, decision making and reasoning—so it gives evolutionary psychology a fertile opportunity for showing why even cognition must be a set of domain-specific adaptations. But so far, that opportunity has been wasted by obsessing about sexual cues.

Third, cataloguing attractiveness cues makes it easy to avoid modelling the intricacies of mate choice as an interactive social problem of search, assessment, courtship, competition and mutual choice. The evolutionary psychology of mate choice is mostly, so far, the psychology of what happens in the first five minutes of a single virgin picking favourites from a line-up of passive strangers on a desert island. This Pleistocene tropical fantasy is pleasant for the sexually frustrated graduate student or divorced professor to contemplate precisely because it ignores the horrid game-theoretical interdependency of real-world mate choice.

Fourth, cataloguing cues makes it easy to avoid specifying the adaptive goals of mate choice in much detail. This is because attractiveness cues correlate with virtually every other aspect of an organism's phenotype (for technical reasons of developmental epistasis, physiological condition dependence and genetic correlation reinforced by assortative mating), rendering any cue a pretty good indicator of almost any underlying trait that might be worth selecting. Only when we investigate combinations of cues that indicate different trade-offs between desirable traits—such as parenting ability, current fertility, social status and immunocompetence—will we be pushed to develop more specific normative and descriptive models of mate choice. These four problems—animals' promiscuous flexibility of cue use, cue use being marginal to cognition, cue use being marginal to the hard game-theoretical aspects of mate choice and cue use being uninformative about exact adaptive functions—make cue cataloguing a rather weak method for characterizing our adaptations for mate choice. The next sections describe in more detail some pitfalls of cue cataloguing compared to cognitive modelling; then the conclusion will outline a normative and descriptive alternative.

Critiques and extensions of current research

There's more to analysing mate choice than predicting sex differences

Sex differences are easy to investigate because the contrast groups (males and females) come ready-made, predictions from sexual selection theory are often fairly simple and results attract widespread media interest. But there's much more to mate choice than sex differences. Modern sexual selection theory provides a framework for analysing one of the most important decision domains faced by all sexually reproducing animals. The theory would be as illuminating if we were hermaphrodites as it is given our two sexes. Although males and females reliably differ in some of the traits they seek and the cues they use, the basic game-theoretical problem of attracting the best mate who will accept you is similar for both sexes. I am worried that in the popular media, and even in the minds of some evolutionary psychologists, the study of mate choice has become synonymous with the study of sex differences in the relative weights given to certain cues of attractiveness.

While sexual dimorphism usually indicates sexual selection, not all sexual selection produces sexual dimorphism (see Andersson 1994). If evolutionary psychology assumes that all sexually monomorphic traits (e.g. human language, intelligence and creativity) evolved without any help from sexual selection, we prematurely rule out one of the most powerful, inventive and pervasive selection forces in nature (see Miller & Todd 1995). Also, while mate choice mechanisms may show sexual dimorphism at the level of cue perception, they may not at higher cognitive and strategic levels. Males and females face largely similar problems assessing potentially deceptive cues, integrating cues, searching through a sequence of prospects and finding the best mate who will accept them. These strategic problems will continue to be overlooked if we equate mate choice with sexual dimorphism in the traditional, rather crude way.

From physical to behavioural cues

An irony of evolutionary psychology is that many mate choice researchers have emphasized the sexual cues that are least psychological: physical attractiveness cues. Faces, breasts, buttocks, muscles, penises, symmetry, height and other morphological traits have all been subject to intense analysis, while psychological traits such as intelligence, creativity, personality, sense of humour, social skills, kindness and ideology have received mostly lip service (see Buss 1994, Ridley 1993). The reasons are twofold: the ease of experimentally manipulating stimuli that represent morphological traits, and the importance of physical attractiveness as a convenient, low-cost 'filtering cue' early in courtship. Evolutionary psychology's focus on the physical is a reasonable first step if our goal is demonstrating that there exist human universals of attractiveness, contra the claims of some cultural anthropologists and humanities scholars. But if our goal is to analyse our most important and distinctive mating strategies, selection criteria and courtship traits, then we must analyse how people make mate choices based on psychological features. The pay-offs could be significant. Whereas models of mate choice based on physical traits can only explain the evolutionary origins of our bodies, models that include psychological traits may explain the origins of our most distinctive mental capacities (see Miller 1997a,b). The co-evolution between mate choice mechanisms and the courtship behaviours that they select puts the study of mate choice at the very heart of evolutionary psychology because we are studying the core psychological adaptations that catalyse the emergence of other psychological adaptations via sexual selection.

Beyond weighted linear models of cue integration

Many mate choice researchers seem to assume that organisms register a set of cues associated with each potential mate, attach some standard weight to each cue and then add the weights together to arrive at an overall rating of attractiveness. Such 'weighted linear' models of cue integration, derived from Brunswik (1956) and revived in some recent neural network models, seem like the simplest way to start an analysis, and make it easy to interpret subjects' numerical responses on questionnaires about the relative importance of different attractiveness cues (see Buss 1994). However, the apparent cognitive simplicity of weighted linear models may be deceptive. Gigerenzer & Goldstein (1996) have analysed several alternatives to weighted linear integration that make better decisions, using less information, operating faster and better fitting some subjects' think-aloud protocols. Their 'take-the-best' heuristic, for example, would be a fast, frugal way to decide which of two prospects has higher mate value. This heuristic checks each prospect on one cue at a time, with the cues ranked in order of their ecological validity (correlation between cue value and mate value). The first cue that distinguishes the prospects would be used to make one's mate choice. If subjects using this take-the-best heuristic were asked for their 'cue weights', they would find it a meaningless question, and might report some other quantity (such as ecological validity, discrimination rate, recentness of cue use or cue ranking) on a questionnaire. Use of such heuristics may explain the puzzling finding that in many domains, subjects claim to use more cues than prove significant in *post hoc* multiple regression analyses of their decisions. This is exactly what we would expect from the take-the-best heuristic: the vast majority of decisions many be determined by the few top-ranked cues, but where those cues don't distinguish between prospects, subjects must use lower-ranked cues to decide.

Another reason for questioning weighted linear models is that the assessment costs of different features used in mate choice are so wildly disparate. Morphological features such as face and body shape can be assessed in a momentary glance, whereas resourcefulness in emergencies, parenting skills and capacity for avoiding sexual boredom can be assessed only after months of interaction. A depth-first search for all cue values in every prospect you encounter would be an idiotic way to search for a mate. Instead, people use the easily assessed physical cues as filtering devices to decide who to talk to; they use conversations to decide who to have sex with; and they use sexual relations and capacity for intimacy to decide who to have children with. What is the 'relative weight' then, of facial beauty, versus wit in conversation, versus foreplay skills? The question is meaningless, if people are integrating these cues sequentially, non-linearly and intelligently, rather than according to a weighted linear method that ignores cue assessment costs.

Moreover, this sequential cue-integration heuristic would produce behaviour that could be misinterpreted all too easily as reflecting distinctive cue weightings for short- versus long-term mating (cf. Buss 1994). People would start relationships with prospects they find physically attractive, but only continue relationships with those they find psychologically compatible. *Post hoc*, it would look as if they attached a higher 'weight' to physical attractiveness for short-term matings, and a higher 'weight' to psychological features for long-term matings. But the apparent correlation between cue weight and relationship duration could be an artefact of some cues taking longer to assess than others, with relationships ending only when someone becomes unhappy with the most recent information they learned about their partner. To investigate sequential cue integration heuristics, we need fewer questionnaire studies and single-cue experiments, and more detailed interviews (we can call them 'protocol analyses' if that helps get them published) in which we actually listen to what people say about how they confront the selective, biased, deceptive trickles of information that their would-be partners leak to them.

Mutual choice, two-sided matching and assortative mating

Although sexual harassment by males is endemic in nature, most matings seem to require mutual consent. This mutual choice constraint complicates matters because fulfilling two sets of preferences in mating is multiplicatively more difficult than fulfilling one (for discussion of this problem in sexual selection simulations, see Todd & Miller 1993). Nevertheless, evolutionary psychologists are fortunate that economists have already done some hard thinking about how mutual choice works in mating markets. This is in the literature of several hundred papers on 'two-sided matching' (see Roth & Sotomayor 1990) which, as far as I know, has never previously been cited or discussed in evolutionary psychology. A prototypical two-sided matching analysis assumes a population of men and women, where each individual has a complete and transitive set of preferences across members of the opposite sex (based on some unspecified assessment process). A 'stable matching' is defined as a pairwise assortment of men and women such that no individual would prefer to be paired to someone else who would also prefer to be paired to them. In game theory terms, a stable matching is a Nash equilibrium in the mating market. One heartening result from this literature is that at least one stable matching exists for every mating market with two sexes (Gale & Shapley 1962). Further, no stable matching exists for mating markets with one or three sexes (Roth & Sotomayor 1990).

Moreover, a simple algorithm called the 'deferred acceptance procedure' is guaranteed to find a stable matching pretty quickly (Roth & Sotomayor 1990). In this procedure, one sex proposes, and the other sex accepts or rejects. For example, each man first proposes to his most-favoured woman. Each woman rejects any suitors who are unacceptable, and each woman who receives more than one proposal rejects all but her most preferred. Any man not yet rejected is kept 'engaged'. The procedure then iterates, with any man rejected on a previous step proposing to his next most preferred woman. Engaged women can switch if a more-preferred man proposes to her. The algorithm stops after any step in which no man is rejected. Women who did not receive any acceptable proposals, and men rejected by all women acceptable to them, stay single. If preferences are strict, then the set of people who remain single is the same for all stable matchings. Surprisingly, if preferences are strict and if there are multiple stable matchings, the deferred acceptance procedure with men proposing will always find the matching that is most preferred by men and least preferred by women, while the reverse is true if women propose.

This leads to a strange corollary: although men in mating markets are supposedly competing with each other, once they realize the game-theoretical implications of mutual choice, they can recognize their common interest in devising a matching procedure that attains their most preferred stable matching. This 'male-optimal' matching makes men as happy as they could be, given the preferences women actually have, and makes women as miserable as they could be given a stable matching. Whichever sex proposes will reach its most preferred Nash equilibrium in the mating game. Moreover, I suspect that given monogamy, the preferences of the proposing sex could be shown to constitute stronger sexual selection pressures than the preferences of the other sex. These results also suggest that, if some assortative mating procedures are more efficient than others in attaining stable matchings, then group selection could favour such mating procedures, without having to overcome any individual-level selection (Miller 1994). These matching results should be more broadly appreciated by evolutionary psychologists because they identify confluences and conflicts of interest that would otherwise be overlooked.

Co-evolution of mate choice heuristics and the cues they select

Although mate choice heuristics are selected to exploit the sexual cues available in the environment, those cues themselves are heritable traits that are selected by the choice heuristics. This leads to a particularly fast, capricious and dynamic form of co-evolution (Miller & Todd 1995, Todd & Miller 1993, 1997). The diversity of sexually selected traits and mate preferences across even closely related species illustrates the speed and power of this co-evolution. Indeed, examination of secondary sexual traits, genitalia and courtship behaviour is often the only way of distinguishing between sibling species (no surprise, really, since mate preferences are what define species in the first place).

The co-evolution between sexual traits and mate preferences seems to give evolutionary psychology no logical place to start in analysing mate choice. But there are patterns to the co-evolution that run quite deep. For example, Zahavi's original 'handicap theory' was essentially a theory about how the reliability of sexual cues as viability indicators co-evolves with the mate preferences that select them (see Andersson 1994). His analysis, informed by game theory, suggested that sexual cues will typically evolve to show an intermediate degree of ecological validity (i.e. correlation with the trait they advertise) because only such intermediate degrees of reliability are evolutionarily stable. Also, analysis of some intrinsic perceptual biases that shape the evolution of sexual cues has led Ryan & Keddy-Hector (1992) to discover some aesthetic principles for courtship displays. Likewise, a better understanding of assortative mating should lead to predictions concerning typical levels of genetic linkage and phenotypic intercorrelation between sexual cues, since assortative mating tends to concentrate heritable high quality cues in certain offspring. Rather than viewing either sexual cues or mate preferences as given in our analysis of mate choice, we should view them as co-evolving traits that sometimes reach predictable, evolutionarily stable equilibria, and that other times fly off together in runaway processes that still obey certain evolutionary principles of signalling, advertisement and assortative mating.

Mate choice and genetic variance

Mate choice research reveals a curious tension within evolutionary psychology concerning within-species genetic variance. On the one hand, evolutionary psychologists downplay genetic differences within our species as superficial variants on a species-typical body plan and cognitive architecture. This is because complex adaptations are likely to pervade populations in monomorphic form, lest sexual recombination break apart the co-adapted genes that grow the adaptations (Tooby & Cosmides 1990). Such arguments have given evolutionary psychology a good pretext for distancing itself from the politically contentious research fields that study genetic variance in modern humans, such as behavioural genetics, psychometrics and Darwinian anthropology. On the other hand, mate choice is only worth doing if genetic variance is of sufficient functional importance to make the benefits of choosing well exceed the substantial search and assessment costs of mate choice (Pomiankowski 1987). The major reason for mate choice in most sexually reproducing species is that the genetic quality of your mate determines half the genetic quality of your offspring. Phenotypes are fugitive, but genes are forever. From this perspective, intraspecies genetic variance is literally the selective environment to which mate choice has adapted. So, how could there be an evolutionary psychology of mate choice, if mate choice requires a level of intraspecies genetic variance that evolutionary psychology denies could exist?

Three recent developments in sexual selection theory have illuminated this genetic variance issue. First, the gradual recognition that most mutations are harmful has led theorists to propose that this pervasive 'biased mutation' is a major reason why mate choice remains worth doing, even when adaptations have been under strong stabilizing selection (Pomiankowski et al 1993). Adaptations are continually eroded by this biased mutation, and mate choice is one of the best ways to counteract such entropy.

Second, whenever sexual selection operates like a 'winner-take-all' contest, as in polygyny, evolution favours risk-seeking behaviour — not only risky competitive behaviour as in violent conflict between males (Daly & Wilson 1988), but also genetic modifiers that maximize genetic and phenotypic variance (Pomiankowski & Møller 1995). Such modifiers explain why sexually selected traits typically show much higher coefficients of additive genetic variance than survival traits (Møller & Pomiankowski 1993). I have argued elsewhere that some of our most distinctive psychological adaptations, particularly our capacities for language, art, music, ideology and creativity, evolved largely under sexual selection (Miller 1993, 1997a,b). If so, then such capacities would be expected to show quite high genetic variance, especially in males. This may explain why some of our most complex psychological adaptations, such as intelligence measured by IQ tests, are also the most genetically variable and heritable (G. F. Miller, unpublished paper, Human Behavior and Evolution Society Annual Conference, Albuquerque, NM, 1994).

Third, sexual selection theorists are recognizing that, while random mating would break apart co-adapted gene complexes to yield monomorphic adaptations within a population, strong assortative mating can maintain relatively complex specializations in polymorphic equilibria. Indeed, speciation itself can be viewed simply as the most extreme form of assortative mating (Todd & Miller 1993). If species (i.e. self-defining reproductive communities with different mate choice and courtship adaptations) can maintain separate adaptations, then perhaps less extreme forms of assortative mating can maintain less obvious, but still significant, polymorphisms within a species. Also, assortative mating can maintain genetic variance at much higher levels than random mating, so it perpetuates its own incentives for mate choice.

Conclusion: towards a new normative and descriptive framework

Information flows successively from environment through perception, then cognition, decision making and action. But selection pressures flow the opposite way, shaping behavioural output most strongly, and trickling back to shape cognition only in so far as it guides adaptive decision making, and perception only in so far as it guides adaptive cognition (see Miller & Todd 1990). The fact that information and selection flow in opposite directions through evolving minds puts the study of perception in a curious quandary. From an information-processing viewpoint, low level perceptual mechanisms, such as those for registering cues of sexual attractiveness, look like the building blocks of cognition. The ease with which experimental psychology can investigate such mechanisms reinforces this impression. However, from a selectionist viewpoint, low level perceptual mechanisms are not the building blocks of psychological adaptation at all, but the last and most indirect products of selection pressures that have already determined an adaptive task, a set of possible behaviours, a decision-making problem and a requisite set of cognitions.

In my view, evolutionary psychology has become prematurely focused on analysing sexual cues as perceptual inputs without a sufficient normative and descriptive framework for understanding how these inputs should and could contribute to adaptive mate choices in realistic social contexts. We must remember that mate choice is fundamentally a problem of game-theoretical decision making, given sceptical prospects and hostile competitors, and not just a problem of optimal cue integration and rational social judgement. Animals encounter sexual prospects drawn from a fluid population with unknown statistical distributions of attractiveness, fertility and viability. Prospects appear in unpredictable and often irrevocable order. Some features of prospects can be assessed instantly, cheaply and reliably; others can be discovered only after long, expensive and interactive courtship. Prospects also have unknown deviations from an unknown population-typical set of mate preferences. Typically, one's own attractiveness and mate value can be inferred only indirectly. The strategies and attractions of one's competitors are even less accessible. Lost in this sea of uncertainty, deception, competition and coyness, you must try to combine your genes with the best genes that you can attract, and combine your parental effort with the most fertile and viable mate you can find.

Task complexities such as this seem overwhelming, but evolution has two secret weapons: the adaptations it constructs can ruthlessly exploit any available structure in the environment; and they can shamelessly sacrifice generality, rationality, elegance, simplicity, completeness and perfection in favour of adaptive efficiency. We must expect mate choice adaptations that take short-cuts, that use cheap and easily perceived cues first, that put cues together in the order they're available and that rely on social stereotypes whenever they're more valid than not. Our descriptive models of mate choice should draw not just on some recycled perceptual psychology, but also on the full range of information-processing heuristics wherever we can find them: judgement and decision research, social cognition, artificial intelligence, economics or whatever. We must be as inventive in developing models as evolution must have been in designing mechanisms.

We should be equally pragmatic in developing better normative models for understanding what adaptive efficiency means in mate choice, drawing not just from sexual selection theory, but also game theory, decision theory and statistics. Principally, we must develop better ways of describing the task environment in which mate choice operates: what is the sexual game being played, what are the pay-offs, what kinds of strategic decisions must be made and what information is available for making them? Mate choice mechanisms are adaptations, and adaptations are always adaptations to something: a well-specified task and a well-specified environment. If we try to model mate choice too directly, using little more than some basic sexual selection theory combined with the narrow set of empirical methods favoured by psychology journal editors, we won't be capitalizing on the full power of our Darwinian framework. This framework requires equal attention to analysing environment structure, analysing adaptive tasks and analysing adaptations themselves. Analysing environment structure doesn't just mean outlining an impressionistic reconstruction of Pleistocene social dynamics. It means detailed, quantitative analysis of human genotypes, bodies and behaviours as informational structures on which mate choice mechanisms operate. Analysing the adaptive task in mate choice doesn't just mean sketching how a despot might maximize offspring number by picking nubile slave girls. It means detailed, explicit analysis of mating games as played by ordinary humans, both ancestral and modern, with all the complexities of mutual choice, assortative mating, affordability, commitment and deception. If we face these challenges, the technical achievements of mate choice research might finally match its popularity.

Acknowledgements

This research was supported by NSF-NATO Post-Doctoral Research Fellowship RCD-9255323, the University of Sussex, the University of Nottingham, the London School of

Economics, the Max Planck Society and the Economic and Social Research Council. I thank Rosalind Arden, Helena Cronin and Peter Todd for their comments.

References

Andersson M 1994 Sexual selection. Princeton University Press, Princeton, NJ

Brunswik E 1956 Perception and the representative design of psychological experiments, 2nd edn. University of California Press, Berkeley, CA

Buss DM 1994 The evolution of desire: human mating strategies. Basic Books, New York

Cronin H 1991 The ant and the peacock: altruism and sexual selection from Darwin to today. Cambridge University Press, Cambridge, UK

Daly M, Wilson MI 1988 Homicide. Aldine de Gruyter, Hawthorne, NY

Darwin C 1871 The descent of man, and selection in relation to sex. John Murray, London

Gale D, Shapley L 1962 College admission and the stability of marriage. Am Math Monthly 69:9–15

Gigerenzer G, Goldstein D 1996 Reasoning the fast and frugal way: models of bounded rationality. Psychol Rev 103:650–669

Miller GF 1993 Evolution of the human brain through runaway sexual selection: the mind as a protean courtship device. PhD thesis, Stanford University, CA, USA

Miller GF 1994 Beyond shared fate: group-selected mechanisms for cooperation and competition in fuzzy, fluid vehicles. Behav Brain Sci 174:630–631

Miller GF 1997a A review of sexual selection and human evolution: how mate choice shaped human nature. In: Crawford C, Krebs D (eds) Handbook of evolutionary psychology. Lawrence Erlbaum Associates Inc., Hillsdale, NJ, in press

Miller GF 1997b Protean primates: the evolution of adaptive unpredictability in competition and courtship. In: Whiten A, Byrne RW (eds) Machiavellian intelligence, II. Cambridge University Press, Cambridge, UK, in press

Miller GF, Todd PM 1990 Exploring adaptive agency. I. Theory and methods for simulating the evolution of learning. In: Touretsky DS, Elman JL, Sejnowski TJ, Hinton GE (eds) Proceedings of the 1990 Connectionist Models Summer School. Morgan Kaufmann, Palo Alto, CA, p 65–80

Miller GF, Todd PM 1995 The role of mate choice in biocomputation: sexual selection as a process of search, optimization, and diversification. In: Banzaf W, Eeckman FH (eds) Evolution and biocomputation: computational models of evolution. Springer-Verlag, New York, p 169–204

Møller AP, Pomiankowski A 1993 Fluctuating asymmetry and sexual selection. Genetica 89:267–279

Pomiankowski A 1987 The costs of choice in sexual selection. J Theor Biol 128:195–218

Pomiankowski A, Møller A 1995 A resolution of the lek paradox. Proc R Soc Lond Ser B 260:21–29

Pomiankowski A, Iwasa Y, Nee S 1991 The evolution of costly mate preferences. I. Fisher and biased mutation. Evolution 456:1422–1430

Ridley M 1993 The red queen: sex and the evolution of human nature. Viking, New York

Roth AE, Sotomayor MAO 1990 Two-sided mating: a study in game-theoretic modeling and analysis. Cambridge University Press, Cambridge, UK

Ryan MJ, Keddy-Hector A 1992 Directional patterns of female mate choice and the role of sensory biases. Am Nat 139:4S–35S

Todd PM, Miller GF 1993 Parental guidance suggested: how parental imprinting evolves through sexual selection as an adaptive learning mechanism. Adapt Behav 21:5–47

Todd PM, Miller GF 1997 Biodiversity through sexual selection. In: Langton CG, Shimohara T (eds) Artificial life, vol V. MIT Press, Cambridge, MA

Tooby J, Cosmides L 1990 The past explains the present: emotional adaptations and the structure of ancestral environments. Ethol Sociobiol 114:375–424

Wright R 1994 The moral animal: evolutionary psychology and everyday life. Pantheon, New York

DISCUSSION

Cosmides: I have a comment and a question. You said that that genetic variance is the only reason for mate choice mechanisms, but in species where there is either maternal or paternal investment other, non-genetic factors can be involved. For example, the level of violence that a boy experiences early in life seems to set a threshold for violence that stays intact for his lifetime. This creates a stable difference between men in their willingness to use force, which could be a mate choice criterion for women. Another example of non-genetic variation is age of partner.

My question is that in your model on stable matching and the strategies that make males happy and women miserable, could you say what happens if women do the proposing?

Miller: If women do the proposing then all the results just reverse.

Maynard Smith: Is the list the same for every male?

Miller: No, the mathematical theorems assume that men have arbitrary preferences.

Tooby: But it wouldn't make any difference if they were all the same.

Dawkins: In that case you would get a more obvious stable response.

Miller: It's only when mate preferences diverge that you get multiple equilibria. There are some limitations but, given what I said about weighted linear integration, it is foolish to assume that arbitrary preferences occur. It's also foolish to assume that you have complete knowledge of the population size and the attributes of all individuals, and that you have instant, uniform access to all members of the population such that you can backtrack to someone who was rejected earlier, for example. Peter Todd and I are developing computer models to see whether there are some psychologically reasonable heuristics that could work without these assumptions. We could add some other reasonable assumptions; for example, the preferences are for cues that are correlated somewhat across individuals, and people have some idea of their own attractiveness in the population.

Cosmides: Why are women miserable under conditions of male optimal matching?

Miller: Under such conditions, the average extent to which a man obtains a woman who is ranked higher rather than lower is higher than under female optimal matching. One interesting observation is that this system, independent of which sex proposes, imposes stronger selection pressures than a mutual choice system. There are theoretical reasons why one sex should propose rather than both sexes, although it's not clear if there's a theorem that explains stable matching if both sexes propose.

Tooby: Why is there selection pressure to obtain a stable outcome as opposed to selection pressure on individuals or on design features to do as well as possible?

Miller: That's an evolutionarily invalid carry-over from economics, i.e. that the entire analysis is founded on finding an equilibrium. I'm not suggesting that this is a reasonable model of what happens for human or animal mate choice, rather I wanted to bring your attention to this literature because it's interesting and worth looking at.

Dawkins: One respect in which it's unrealistic is that the game you describe allows changing after the initial choice is made. In reality this choice is more likely to be a gamble, i.e. that once you have made your analytical choice you are stuck with it. I suspect that this will change the nature of the game in interesting ways.

Miller: Yes. I have also done some simulations with Peter Todd on the issue of what happens if you encounter a sequence of prospects, each of which has a value that you can perceive, but you don't know what the value of the next one might be, and therefore you have to make a choice. This is isomorphic to what's called the job search problem in economics. Economists have a theorem that you should estimate the total number of individuals or jobs encountered in a lifetime, sample the first 37% of those and then choose the next individual who is even better than the best from the initial sample. We found that this 37% rule is true if you want to maximize the likelihood of finding the best individual in the population. However, if you want to maximize the chances of getting the highest average mate value or you want to maximize the chances of obtaining someone in the top 25%, or the top 10%, of the population, then it's often sufficient to just sample the first 5–10%. Across a wide range of population sizes, sampling the first 12 and then choosing the next best is a strong heuristic that works well. This is the principal argument against economists who say that people under-sample in mate choice (Todd & Miller 1997).

Rogers: Is the cost of doing such sampling included in the model?

Miller: Yes.

Daly: This kind of analysis has also been applied to mate choice in non-human animals. I remember hearing Luther Brown and Jerry Downhower talking about mate choice in sculpins decades ago. They compared actual female decisions with the outputs of various simple decision algorithms, as well as using simulations to address how close to optimality females could get with simple rules. Gibson & Langen (1996) have recently reviewed this area.

Gangestad: I would like to mention that Randy Thornhill and I have looked at traits of asymmetry and their relation to mating success. These cues are not strictly cues, as they are small asymmetries of features such as elbows, ears and feet, and therefore they are not readily observable. The reason why we are looking at such traits is that they are markers of an underlying condition, developmental imprecision, and for the purposes of testing certain sexual selection models, we are interested in the mating fates of individuals with and without that condition. There is some evidence that males who are the most symmetrical have more mating success, yet they invest less in their relationships; for example, they put less effort into the relationship, spend less time with their partner and are less honest. These results suggest that the benefits of mating with a symmetrical male are not in terms of investment, but in terms of good genes, although it is possible that there are other kinds of benefits that we have not detected. It may first be necessary to decide what the sexual selection pressures are before one examines cue integration in detail. For instance, some of your models may not be accurate because they don't take into account genetic benefits as well as non-genetic benefits. One might expect that a mate with a higher genetic quality is more attractive initially but then is not as attentive a partner. One of your models suggested that one would leave them, but before one does so one has to think about their genetic qualities. Hence, cues of genetic quality should stay as cues throughout the process. One would also not expect there to be no consensual adultery if 'good genes' sexual selection is operating, and there are benefits from investment because females can potentially obtain the investment from one male and the genes from another male. This opens up the options that one would have to consider in determining the optimal procedure. Therefore, before one could decide what kind of procedure would be optimal, one would have to know more about the environment.

Miller: What I've attempted to do is to highlight the range of models and ideas that are worth bearing in mind when doing a particular research programme. I'm not discounting cue cataloguing. I hope that people continue to catalogue cues, but I would like to see them doing so in the context of thinking about what the cues are for in terms of mate choices in particular social contexts that have certain game-theoretical features. It is not possible to separate the sequential search problem from the cue integration problem because they are both operating at the same time.

Daly: By the phrase 'no consensual adultery', I gather that you mean 'no consensual reasortment into new mated pairs'. That does not necessarily imply anything about whether both sexes might not have incentives to engage in extra-pair copulations (EPCs).

Gangestad: Yes, that was the point I was trying to make.

Miller: This is a limited model and it assumes that matching occurs. Game theorists and economists are trying to extend the two-sided matching models to find out, for example, what happens if somebody new comes into a stable-matching situation, what happens if there are different migration rates and what happens if there are EPCs. We should follow those developments and not let the economists develop their own separate theory about mate choice.

Pinker: I don't understand your point about language and intelligence being a product of sexual selection. The criteria you use to support that claim could be used to attribute *anything* to sexual selection. You could say that the eye is a product of sexual selection because organisms with poorer vision can't see their mates and therefore can't mate with them. You could even speculate that organisms prefer mates who have better vision, and that's why vision evolved. One rejects these arguments because the value of sight for survival is so obvious, and the engineering demands of sight are so well satisfied by the demands of the eye that it is superfluous to invoke anything specifically sexual in the selection pressures that led to the eye. Now compare this to an unambiguous case of sexual selection like the peacock's tail. On engineering grounds it is badly designed to help the animal survive: it is sexually dimorphic; it is used to its full extent only in the context of showing off to females; and so on. Now if we take these criteria from the clear cases of non-sexual and sexual selection, and apply them to intelligence and language, they point to non-sexual selection. The adaptive significance of intelligence is obvious enough: it allows you to hunt animals by interpreting their tracks, to extract poisons from plants to make poison arrows and so on; the adaptive significance of language is that you can trade this information with others or share it with your kin. Also, language and intelligence are not just used when impressing mates but almost 24 hours a day all your life. So why should one think that language and intelligence are products of sexual selection, any more than that the eye is?

Miller: They are strange adaptations because they include some design features that are hallmarks of sexual selection and others that are the hallmarks of natural selection, i.e. for the ability to communicate or live socially. It's true that language and intelligence don't show much sexual dimorphism, but human language is often more grammatically complex than is necessary. I would like to convince sociolinguists to test some of these different design features. I would like to know, for example, whether the grammatical complexity of utterances shows the same kind of pattern by age and sex as my art and music data. If it does show those kinds of complexities, those kinds of patterns, then we could start to argue about the adaptive functions of language.

Tooby: Is your prediction that young males would have the most elaborate grammatical constructions?

Miller: All I'm arguing is that sexual selection is an overlooked selective force. It is a powerful way of constructing an adaptation in one species and not in other closely related species that face similar ecological problems, and it predicts certain design features other than just sexual dimorphisms that are worth looking at.

Pinker: If language were an index of health or neurological vigour, wouldn't most conversations consist of tongue-twisters and heroic, Fidel Castro-length orations? Instead language appears to be an algorithm that maps semantic structure, i.e. ideas, onto grammatical structure, which is a kind of transmittable signal. This suggests that language is an adaptation to exchange information, not to advertise health.

Miller: It would be necessary to demonstrate that certain things such as performance on IQ tests and linguistic output are especially sensitive to health status or parasite load, and that they depend on age more than the design of the eye does, for example. I am frustrated by those books on language and intelligence that do not make any detailed design feature arguments. It's worth proposing that other selection pressures might be powerful and relevant.

Hauser: Are you arguing that sexual selection plays a role in the design features of language or that sexual selection was the critical pressure in causing the origins of language? This is the kind of distinction that has emerged within current discussions of fluctuating asymmetries.

Miller: Both arguments are relevant. Some details of language can be examples of courtship displays. After listening to Randy Thornhill's presentation (Thornhill 1997, this volume), I don't want to think about which selection pressures came first. Intelligence and language are multifunctional and once they're in a rudimentary form, if a certain new selection pressure is applied then that new selection pressure may play a part.

Hauser: These arguments are similar to those of Robin Dunbar. It is clear that we use language to gossip, but we do not know whether this is what language was originally designed for.

Gigerenzer: A cognitive adaptation for solving a complex problem may be achieved by fairly simple mechanisms. In the extreme, a choice between two objects, for example between two potential mates, can be made by one-reason decision making. That is, an organism has a hierarchy of proximal cues on which it compares the two objects one-by-one, starting with the highest cue, until it finds a cue on which the two candidates differ. Then it stops searching and chooses the object with the higher value. It does not look up all the information and it does not integrate any information; that is, it is non-compensatory. This simple decision algorithm is known as 'take-the-best' (Gigerenzer & Goldstein 1996). We have studied a family of such simple, non-linear mechanisms, and there are empirical examples. For instance, female guppies seem to choose between males on the basis of phenotypic cues, such as amount of orange body colour, as well as social cues, such as the choice other females made (Dugatkin 1996). These cues appear to be ranked; imitation (the choice other females made) overrides females' genetic preference for orange body colour (unless the discrepancy in colour is extreme), and the choice seems to be made in a non-compensatory way, as does the take-the-best algorithm.

Tooby: But in the natural world there are many differences between individuals, which would lead one to predict that the chooser would always stop at the first trait.

Gigerenzer: The highest ranking trait is not always decisive for choice because, for instance, the two objects do not differ or information is lacking. Which cue or trait is decisive will vary from decision to decision, in a predictable way. The satisficing models we study do not assume that organisms cannot perceive many differences between individuals. They make, however, the bold assumption that the decision is made by one difference only: there is no trade-off between traits. Although this mechanism looks simple and stupid, it actually performs well (Gigerenzer & Goldstein 1996).

References

Dugatkin LA 1996 Interface between culturally based preferences and genetic preferences: female mate choice in *Poecilia reticulata*. Proc Natl Acad Sci USA 93:2770–2773
Gibson RM, Langen TA 1996 How do animals choose their mates? Trends Ecol Evol 11:468–470

Gigerenzer G, Goldstein DG 1996 Reasoning the fast and frugal way: models of bounded rationality. Psychol Rev 103:650–669

Thornhill R 1997 The concept of an evolved adaptation. In: Characterizing human psychological adaptations. Wiley, Chichester (Ciba Found Symp 208) p 4–22

Todd P, Miller GF 1997 Searching for the next best mate. In: Gigerenzer G, Todd P (eds) Fast and frugal cognition. Oxford University Press, Oxford, UK, in press

General discussion I

Cosmides: There have been many interesting debates in the human literature about what constitutes an adaptation. Because human behaviour is so complex, these debates have refined and crystallized many issues beyond what one finds in the non-human animal literature. I agree with Alex Kacelnik's critiques about certain evolutionarily oriented anthropologists who argue that modern humans can 'figure out' how to maximize their fitness no matter what environment they find themselves in. Alex suggested that adaptationist psychologists assume that individuals behave to maximize a motivational entity (utility), but that this utility is shaped by natural selection, and that researchers are interested in describing this utility by observing how individuals behave and in predicting it by reference to natural selection (Kacelnik 1997, this volume). This is exactly what adaptationist *psychologists* do not assume. The people he quotes as exemplifying this view define themselves *in opposition* to evolutionary psychologists. Indeed, there has been a heated debate amongst those who study humans about whether it is meaningful to just look for correspondences between various selectional theories and behaviours, or whether (as the evolutionary psychologists advocate) one should instead characterize psychological adaptations with a rigorous engineering type of approach. The field of evolutionary psychology was effectively born in opposition to this other approach. None of us intended it to be born in that way, but 'evolutionary psychology' is often used to designate those who do not think it is meaningful to look for correlations between behavioural variation and variation in reproductive success.

Daly: Alex may be being misinterpreted here. His allusions to maximization are being equated with the position of those researchers who are psychologically agnostic while trying to assess whether people are effective at maximizing fitness. But the appeal to utility is a psychological hypothesis, is it not? Utility is not isomorphic with fitness, but it's posited to be what organisms pursue more proximally.

Cosmides: In many domains there might not be any utility function of that kind. Instead there may be the execution of the adaptation.

Daly: That's true. After all, many organisms don't have a brain, let alone engage in any utility computations or mental monitoring of utility attainments, and yet they still have adaptively contingent behaviour.

Kacelnik: Suppose you think of a multidimensional utility function, for example peahens' preferences for a certain number of eye spots in the peacock's tail and a certain form of the peacock's behaviour. If you have some rationality of choice, in the sense that the choice is consistent with respect to her own choices in different circumstances, you could say that the animal is choosing as if it has a utility function.

I am not necessarily implying that the animal is experiencing such a motivational entity because I do not know what the subjective representation of this is. I am also not necessarily implying that this utility function is the one that would maximize the peahen's fitness here and now, only that it has evolved because of the combined effects of selection on the assemblage of attributes now used by peahens in choosing their mates.

Cosmides: But then you fall into the danger of being criticized for just looking for a relationship between a normative model and behaviour without worrying about evidence of 'non-optimal' behaviour. The reason why one is interested both in cases in which behaviour fits the normative model and in those in which it does not is to characterize the nature of the psychological adaptation itself. If this is done correctly then it will be possible to predict when behaviour will be adaptive and when it will not.

Kacelnik: I agree. In many cases predictions about the shape of utility functions based on evolutionary thinking are more counterintuitive, quantitative and appropriate for experimental testing than predictions formulated without any normative criteria or having no predictions at all. However, this is not always so. Some accounts of regularities in human behaviour based on normative thinking are not richly informative, or they have little predictive information value because they are similar to expectations from virtually every possible way of looking at human behaviour. For example, empirical evidence for a preference for wealthy, healthy and young partners over poor, sick and old partners is not terribly inspiring nor does it convincingly support any theory that predicts that. My grandmother could give that kind of advice if she were alive.

Cosmides: Those things that are part of our natural competences seem obvious because we have complicated evolved machinery that makes them obvious. If grandmothers everywhere think these are good things to look for in a mate, then this suggests an interesting machine is operating that is generating these kinds of preferences.

Kacelnik: My grandmother would not be trying to separate the hypotheses as we are (or should be). If we say that something is an adaptation, for example that language is a sexually selected character, we must make precise predictions of what features they should have. We can't rely on common sense for that, we have to rely on a specific prediction. If someone were to make the prediction that given a choice between a healthy and a sick partner the healthy one would be preferred, then this doesn't separate too many hypotheses.

Pinker: That's not true. The dominant view amongst intellectuals today is that people prefer healthy partners because our society puts such a value on health, youth, vigour and fitness. Part of the human psychological constitution is to innately prefer cues that correlate with health and fitness.

Gangestad: Implicit in any adaptationist claim is a cost–benefit analysis that identifies the behavioural adaptive target. The characterization of how the adaptive target is achieved at a proximate level is a separate task. Often in the field of evolutionary psychology there isn't an explicit cost–benefit analysis because we don't understand all the selection pressures. Yet, economic cost–benefit analysis can't be thrown out of the field of evolutionary psychology altogether.

Daly: One problem when people use the terms 'cost–benefit' and 'utility functions' is that what starts out as a theoretical account of the way a well-designed adaptation in some particular domain might be expected to function ends up as an inexplicit hypothesis about the structure of a psychological entity. People talk about utility in a language that seems to imply specific psychological hypotheses, but then they're likely to back away from any explicit psychologizing when asked for specifics.

Gangestad: I agree. Sometimes the distinction becomes somewhat fuzzy because, for instance, what one expects from life history expectations is that people have psychological adaptations that are sensitive to diminishing returns on some kind of investment or effort. Does that mean that they're maximizing the overall utility? Well, in some sense they are, although that's not what they're designed to do, they are designed to track diminishing returns on various sorts of effort.

Gigerenzer: I would like to challenge the rhetoric of maximizing utility. Humans and other animals make inferences about their world with limited time and knowledge. In contrast, many models in psychology, economics and optimal foraging theory treat the mind as a Laplacean demon, equipped with unlimited time, knowledge and computational capacities. Many years ago, Herbert Simon argued that organisms satisfice rather than optimize (Simon 1955). However, the notion of satisficing is often watered down to that of optimizing within the constraints of a limited budget, which takes out the very force of the satisficing approach and can make the computations even more complicated (you need to assume that the organism optimizes a more complicated equation that includes cost–benefit trade-offs). This is not how I understand satisficing. The power of satisficing is in simple step-by-step procedures with explicit stopping rules for search, and it may not even involve quantitative probabilities or utilities. Satisficing can exploit the structure of the information in an organism's environment (Gigerenzer & Goldstein 1996). I believe that human psychological adaptations should be characterized by fast and frugal mechanisms rather than by Laplacean demons.

Kacelnik: Many of these words are confusing because they are being used in different senses in the literature. In my opinion there isn't a major difference between Simon's approach and the way I would use any optimal foraging model. Simon's approach includes an optimal strategy that takes into account, for example, the cost of searching for an unconstrained best. Therefore, when you take into account the cost of searching for the absolute maximum solution for a complex problem, then it turns out that it is better, on average, to afford the probability of not reaching the very optimum, but rather reaching a decision earlier with less time wasting. One can produce a solution, such as a simple algorithm, that works well, although it generates some errors, but this wouldn't include the advantage of decreasing their decision time and it wouldn't be adapted to the restrictions of the organism itself. Therefore, Simon's satisficing is an optimality solution that sets out to define a different set of constraints. Taking a different view leads to a paradox. Say that you claim that satisficing is not fitness maximizing, and that it is a universal property of decision systems. How do you account for its universality if it is not by convergent evolution, and if you accept convergent evolution how could you eliminate natural selection as the shaping agent? If you accept that satisficing emerged through the action of natural selection, you are saying that among the available phenotypes, satisficers had maximum fitnesses and hence were the most successful maximizers. Satisficing with respect to energy or quality of partner's genes may be maximizing with respect to fitness. Here is where the confusion arises. Even though I have used the rules of thumb approach myself, I am now sceptical of inventing rules of thumb for each particular problem in an attempt to identify which one solves the problem satisfactorily. The links between maximization, satisficing and rules of thumb can perhaps be discussed using a concrete example from foraging research, i.e. the two-armed bandit (Krebs et al 1987). If you have two sources of reward, with unknown reward probabilities, and you have to decide on an optimal strategy to maximize total pay-off, you could sample each source for a long time, accumulating information until you are certain of not making an error, but this approach would take up all your time. Or you could choose one of them without any further ado, but then you may choose the poorer one and be stuck with it for a long time. Therefore, one approach is to take sampling cost into account and find out rigorously what is the best conceivable performance (we did this by stochastic dynamic programming, but it can be found by other means). This gives you the maximum achievable pay-off as a yardstick, but it doesn't tell you the psychological mechanisms employed by real subjects. For this, you may be tempted to use the rules of thumb approach. This consists of inventing simple decision rules that could guide behaviour. In a separate study Houston et al (1982) proposed a catalogue of rules that could be used in the two-armed bandit problem, and they proved by Monte Carlo simulation that several of them did reasonably well: they achieved close to optimum in a number of situations. Unfortunately, it also turned out that none of those rules reflected the psychological process that actual subjects use. I am forced to conclude that it's fine to use these rules of thumb as a guiding tool, but we should not by-pass the psychological research into the mechanisms of what people or animals actually do.

Gigerenzer: That your rules of thumb approach did not match the psychological processes of actual subjects should not be read to imply that subjects were optimizing. We need to find the appropriate tool-box subjects use for satisficing, and, I believe, we have made some progress. In our laboratory, we have found evidence that people employ fast and frugal mechanisms, and, more strikingly, that these simple mechanisms can often be as accurate as more sophisticated statistical models, such as multiple regression, in making inferences about the world. This holds for a broad range of tasks, from predicting mortality based on pollution indicators to predicting male and female attractiveness based on traits (Czerlinski et al 1997). The reason why satisficing mechanisms can often be as accurate as (or better than) weighted linear models is their power to exploit certain environments, i.e. certain structures of information. Thus, I am not convinced that the optimizing rhetoric brings us very far when there is evidence that simple psychological adaptations can make humans smart.

Daly: I'm not convinced that you are both disagreeing. You seem to both be saying that you want to understand the actual psychological processes by which organisms manage to get as close as they do to what would constitute optimal decisions if the available predictive information were all attended to and used correctly.

Maynard Smith: We do have a rather adequate theory of evolution. The problem is that the topic to which people are trying to apply it is exceedingly difficult. I've always disliked the phrase 'satisficing'. I prefer the notion that natural selection maximizes, or optimizes, and it optimizes a quite specific function, namely fitness. The problem is that it does it subject to constraints, and the difficulty we're faced with is that we don't know what the constraints are, and we're not entirely sure in what environments the optimizations took place. Nevertheless, it's important to bear in mind that the process is not just picking at random any old device that will do, but that it picks an optimization device which is subject to constraints. Unfortunately, there's no alternative in psychological research to finding out what the actual devices involved were. It may turn out that those devices are simple algorithms or that they're rather crazy algorithms because of the constraints under which the optimization took place. The term 'satisficing' offends me because it almost suggests that natural selection will put up with anything. It's not true, it puts up with only with the best it can lay its hands on, subject to what the organism can offer it.

Nisbett: There is also little evidence to suggest that satisficing actually takes place, although this is often assumed by theorists. My understanding is that when we look at consumer choices, for example, it doesn't turn out to be the case that people spend more time on more important decisions. Gerd Gigerenzer seemed to be making a sort of satisficing claim for his algorithm. How is this algorithm related to that of Herbert Simon (1955)?

Gigerenzer: We study a family of fast and frugal algorithms, not one. Like Simon's satisficing processes, these are step-by-step procedures, employing limited search and having explicit stopping rules for search. Some of the fast and frugal mechanisms, such as the take-the-best heuristic, are also related to lexicographic strategies, which have been known for several hundred years. What was not known, however, is how accurate these mechanisms can be when making inferences about unknown aspects of the real world. It's quite counterintuitive.

Rogers: Why do you claim that it isn't an optimizing strategy?

Gigerenzer: For two reasons. First, fast and frugal strategies employ psychological principles that violate classical tenets of rationality. For instance, they neither look up nor integrate all information, and some strategies can produce intransitivities. Their performance could be slightly improved by making unrealistically strong assumptions about the knowledge an individual has about its environment. Second, for many adaptive problems, no optimal strategy is known (even for chess, a well-defined game with no uncertainty in the rules). We often fabricate the fiction of optimization by making simplified assumptions about the environment. I prefer to assume simple adaptive strategies that can handle complex and uncertain environments, physical and social.

Sperber: But if you tried to optimize the cost–benefit ratio, firstly, there isn't a single solution and, secondly, you wouldn't want to optimize epistemic growth because you don't want to look at all the information. Satisficing at the epistemic level means optimizing on cost–benefit.

Kacelnik: We are disagreeing much less than it might appear, in that I have no objections to the use of the term satisficing as a description of a mechanism that the individual is using. It is possible that the best thing you can do, in a particular problematic situation, is to find something that is better than a certain threshold level and then stick to that. This is satisficing in a descriptive sense. It may well be a perfectly good mechanism, but it does not mean that it finds the optimal solution. It is possible that the best that natural selection can do, on the basis of its limited resources, is to produce subjects that use this mechanism.

Gigerenzer: The idea that a strategy should find the optimal solution is a beautiful dream. But optimal strategies are unknown for many real-world problems.

Daly: A satisficing mechanism, as I understand it, would encompass anything that a classical ethologist would have called a 'sign stimulus' or that a neurophysiologist would call a 'feature detector'. Anything an organism uses to make a binary decision could reasonably be called a satisficing mechanism, couldn't it?

Kacelnik: It's unlikely that optimization of this kind occurs in individuals in the natural environment because they are subject to information constraints and time constraints, for example. However, the way we are using optimality and maximization takes into account the constraints of limited accuracy of perception and memory, and in this situation satisficing becomes an optimality strategy.

Gigerenzer: I disagree.

Shepard: Much of this discussion seems to pre-suppose that there is just one strategy and that we're trying to attribute this to human organisms. However, it seems that humans face different sorts of problems requiring different strategies. For many problems, including many with incomplete information, we couldn't possibly enumerate all the relevant hypotheses that would have to be considered for an optimum solution. For other problems, like some that confront the visual system, essentially complete information may be available and near optimal methods of combining the optical data may have evolved. Therefore, we should keep in mind that we may deal with some problems by some form of satisficing and other problems by essentially optimal methods.

Mealey: It seems to me that the difference between optimizing and satisficing might be a minor environmental change, so that something that was optimal at one time could be only satisficing at another time in evolution, such as now. In my opinion we're unlikely to be able to distinguish between what's optimal and what's satisficing, given that the difference between them can flip-flop rapidly with minor changes in the environment.

Kacelnik: We shouldn't go back to the times when we used the term optimality to define what the animal did as being optimal or not. We should look at optimality as a research programme that simply asks, what would be the best thing to do given that the assumptions are correct? When the models fail we shouldn't conclude that the animal is not optimizing. We should conclude that something in the assumptions about the constraints or the psychological properties of the animals was, for whatever reasons, wrong. We should then ask, what is the best thing to do given that there is now an extra constraint? Finally, we may end up with something that resembles what the subject is doing. This use of repeated optimality algorithms with given sets of assumed constraints is a method, and it does not assume that the animal has an optimal strategy at any given time.

References

Czerlinski J, Goldstein D, Gigerenzer G 1997 When it pays to be a lazy thinker, submitted

Gigerenzer G, Goldstein DG 1996 Reasoning the fast and frugal way: models of bounded rationality. Psychol Rev 103:650–669

Houston AIA, Kacelnik A, McNamara JM 1982 Some learning rules for acquiring information. In: McFarland DJ (ed) Functional ontogeny. Pitman, Boston, MA, p 140–191

Kacelnik A 1997 Normative and descriptive models of decision making: time discounting and risk sensitivity. In: Characterizing human psychological adaptations. Wiley, Chichester (Ciba Found Symp 208) p 51–70

Krebs JR, Kacelnik A, Taylor P 1978 Tests of optimal sampling by foraging great tits. Nature 275:27–31

Simon H 1955 A behavioural model of rational choice. Q J Econom 69:99–118

Tinkering with minds from the past

Marc D. Hauser

Departments of Anthropology and Psychology, Program in Neuroscience, Harvard University, Cambridge, MA 02138, USA

Abstract. Cognitive scientists argue that in the absence of language, non-human animal conceptual representations are either impoverished or completely absent. One might make comparable claims about human infants, who enter the world with different conceptual representations from adults. Nonetheless, we often treat human infants like miniature adults. This is a mistake. I argue that research on human cognition, and in particular its domain-specific knowledge systems, can only succeed if it adopts a comparative perspective. To carry out this agenda, however, we require methods that can be used across species. Focusing on how numerical abilities evolved, I describe experiments on two non-human primates, representing different phylogenetic branches. Our experimental procedure — the preferential looking time technique — was designed to assess what prelinguistic human infants know about the physical and psychological world, but it is ideal for non-human animals, especially when one wishes to explore spontaneous cognitive capacities in the absence of training. Results reveal that up to a certain age, human infants and non-human primates are indistinguishable in terms of numerical competence. We must now focus on how language, together with other cognitive facilities, bring the human child to a level of numerical sophistication that exceeds non-human animals, and why non-human animal capacities stop where they do.

1997 Characterizing human psychological adaptations. Wiley, Chichester (Ciba Foundation Symposium 208) p 95–131

Mental fossils

Ten or so years ago, no one would have believed that molecular biologists could obtain DNA from long-extinct species. Thanks to the wonders of the natural world, however, some extinct species have been preserved, frozen in amber. As a result, their DNA is available, ready for extraction. This fossilized genetic material provides an extraordinary window into the phylogenetic relationships between extant and extinct species. Yes, Michael Crighton's *Jurassic Park* is not entirely fantasy.

Now consider the following Crighton-esque thought experiment. While walking on the ancient soils of east Africa, you stumble upon a human body encased in amber. Based on the size and shape of the head, and the structure of its teeth, hips and feet, you immediately deduce that it is an early hominid, probably from the

Plio-Pleistocene. But this is not your ordinary palaeoanthropological finding, devoid of the fleshy stuff. What you have uncovered is a body, completely preserved with skin, muscles, hair and perhaps even a brain inside the skull. You carefully remove the body from its amber casing. Thanks to modern technology, adding a special solution causes our amber-frozen ancestor to come to life. What an opportunity. Putting on your neuropsychologist's hat, you run your subject through a battery of cognitive tests, in the privacy of your home, away from the media hounds. You find, after years of testing, that your subject appears to have all the basic mental algorithms that one would find in a modern human adult. For example, her perception of objects is like ours: she distinguishes mechanical objects from psychological ones and uses intuitions about mental states to predict the behaviour of other members of her species. She empathizes, uses clearly recognizable facial expressions to convey her emotions and communicates by means of her vocal tract, using a system of sounds that has all the surface features of a modern human language. That is, her form of communication is nothing like the Star Trekian Klingon or closer to home, the hoots, grunts and screams of a chimpanzee. It has the familiar ring of human language.

You present your findings at a prestigious meeting, concluding that the early hominids exhibited many of the same mental capacities as extant humans. Predictably, someone in the audience stands up and says 'Well, that is all fine, and interesting but ... I still want to know where those capacities came from. What, precisely, distinguishes us and our hominid ancestors from earthworms or fish or birds or, for that matter, monkeys and apes? And what kinds of selection pressures caused the critical changes in mental capacities?'

The audience member has asked the kinds of questions that keep me up at night. In the words of Lewis Thomas, I not only want to share my late night thoughts with you, but also provide some empirical answers to the problem at hand. I will therefore sketch an argument for why questions of origins must be taken seriously, approached cautiously and removed from the level of Gedanken experiments to the real empirical trenches. I am obviously not alone in this endeavour (Baron-Cohen 1995, Cheney & Seyfarth 1990, Chomsky 1986, Cosmides & Tooby 1994a,b, Cummins & Allen 1997, Gallistel 1992, Gould & Gould 1995, Griffin 1992, Macphail 1982, 1994, Povinelli 1993, Premack 1986, Premack & Premack 1994a), but I wish to push bits of the argument to their limits and will do so in a series of stages, first conceptually and then experimentally.

Building conceptual representations

In a wonderfully playful, yet profound statement, Wittgenstein (1953) remarked that even if lions could speak we would not understand them. His fundamental point was, of course, that language places a clouded veil over the nature of our conceptual representations, and that even if lions could express their thoughts in a roar-less medium, this would do little to help our cause — we would still be no closer to their thoughts than when they let rip in lion-ese. Holding this line does, of course, have comparable implications for monkeys, apes, human babies and some brain-damaged human adults.

Wittgenstein was certainly not alone in putting forth this kind of argument, forging, as he so forcefully did, a link between language and thought. Quine (1973), for example, articulated the position that one cannot comprehend the meaning of terms within a language unless one shares with the utterer a common conceptual system for dividing up the world, its objects and events. Nagel (1974) carried this theoretical baton further, arguing that in the absence of shared perceptual inputs and conceptual representations, one can never know what it is like to be another creature, language or no language. And to take a case from linguistics, Bickerton (1990, 1995) has fleshed out the position that only creatures with language — humans, that is — have the capacity to think about and express abstract concepts, separated in time, space and even reality from current happenings in the world. Due to the structure and content of their conceptual representations, non-human animals are restricted to communicative utterances that convey information about directly perceivable objects and events in the world. And unlike human language, with its detachment from current stimuli and its generative power to create, non-human animal communication appears to be entirely controlled by stimuli in the environment (Chomsky 1966, 1986, Studdert-Kennedy 1997).

These perspectives, which continue to hold a dominant position in the cognitive sciences, lead to at least one significant hypothesis:

> Because of language, humans are expected to have (i) concepts that non-human animals lack and (ii) those concepts that they share in common with non-human animals should be theoretically more sophisticated in their power to account for both real and imagined events in the world.

In this chapter I would like to explore the topography of the hypothesis sketched by focusing on one particular conceptual domain: numerical representations. I have chosen this domain for two reasons. First, several organisms, spanning broad taxonomic groups, engage in behaviours that appear to rely on some form of numerical assessment. For example, individuals must decide whether one patch of food contains more food than another or whether a neighbouring group has more individuals, thereby providing a critical edge in competitive interactions. Although such examples may, at first glance, appear to represent mere quantity assessments (i.e. more than, bigger), further exploration may reveal more fine-grained and sophisticated systems of numerical quantification. Importantly, the behaviours discussed are typically tied to contexts that have a significant impact on an individual's fitness. Consequently, capacities for numerical assessment are of functional significance and exhibit design characteristics indicative of the blind hand of natural selection. Second, although one cannot debate the fact that language is deployed in our use of number, it is not yet clear how language directly enriches our numerical representations (Hauser & Carey 1997). Nor is it clear whether our capacity for numerical computation evolved prior to or after our capacity for language (see Corballis 1992, 1994, Bloom 1994). The line I will follow here is that comparative studies of non-human animals, who lack language, and human infants, who have yet to acquire language, provide the empirical passports into this challenging problem (see also Bloom 1994).

Giving birth to numerical concepts

Kinds of numerical processing

'Ninety-nine bottles of beer on the wall', . . . 'three-two-one, ready or not here I come', 'One potato, two potato, three potato, four . . .'. These snippets from the child's world are only possible in a world where entities of some sort can be counted, where there is a count system and a recognition mechanism that, minimally, allows for object individuation. Young children can certainly parrot such phrases, but to understand how they are deployed in either songs or games, it is necessary to understand the ordinality and cardinality of the number system, the fact that objects or events can be tagged with unique numerical labels, and that the tag applied is abstract — it is not a specific property of the object (e.g. if there are four balls each one can be a different colour and size). When do such numerical capacities emerge in the young child? What are the underlying cognitive mechanisms that support them? And do non-human animals acquire and implement, either spontaneously or through training, comparable numerical capacities? These are the questions that I now address, laying out first a suite of relevant conceptual issues that will be useful in structuring the empirical work.

Two types of mechanisms are typically invoked by cognitive researchers interested in the capacity to assess numerosities: subitizing and counting (for recent discussions of these issues, see Davis & Perusse 1988, Trick & Pylyshyn 1994, Gallistel & Gelman 1991). Subitizing represents a low level perceptual mechanism that is rapid and occurs preattentively. In contrast, counting is considered a higher level cognitive mechanism that is slower and requires greater effort, in part because it engages the attentional system. On the basis of experiments with adult humans, subitizing is recruited for small numerosities (approximately four objects, perhaps as high as six), whereas counting tends to be recruited for larger numerosities; counting can, obviously, be used for smaller numbers as when a parent illustrates to its young child that there are two marbles on the floor, and counts slowly and deliberately, pointing out each marble. As stated, then, subitizing is a relatively simple perceptual process like assigning a colour label to an object. In contrast, counting involves several processing steps, including: (i) the development of an invariant sequence of labels for items or events perceived; (ii) extraction and assignment of a label for an item or event; (iii) the ability to remember (i.e. store in memory) the last label assigned; and (iv) the capacity to understand that the final label stored in memory represents the total number of relevant objects or events perceived in an assemblage or period of time.

As many authors have noted, therefore, there is significant overlap between subitizing and counting in terms of the processing mechanisms recruited. For example, both subitizing and counting require subjects to individuate objects (see Hauser & Carey 1997 for a discussion of the relationship between object recognition and numerical systems) and maintain some representation of the objects recognized in memory. Where the two systems part company is in the domain of tagging or labelling the objects and, thus, in having a system for numerical assignment. There are currently three significant contenders for how number systems may develop in humans, and I now briefly review these.

The numeron model. In an attempt to capture the possibility that prelinguistic human infants might have a representational system for number that precedes, developmentally, the acquisition of linguistic labels for number, Gelman & Gallistel (1986) proposed the idea of 'numerons'. Numerons, like Koehler's (1950) concept of 'unnamed numbers' invoked to account for avian behaviour, were considered mental symbols that functioned like verbal labels in tagging objects or events in a one-to-one correspondence. Thus, if three balls are on the table (Fig. 1), the prelinguistic child might tag one ball with the symbol '@', the second ball with the symbol '#' and the third ball with the symbol '$'; of course we don't know how such symbols are represented in the mind, but for Gelman and Gallistel the important point is that they are not linguistic. Children, or other animals guided by such a system, would see an array of objects and use the final numeron assigned as representing the total number of objects present (e.g. three balls = '$').

The accumulator model. This representational model (Meck & Church 1983) grew out of work in classical learning theory, aimed at understanding the nature of cues used by animals to make choices during discrimination tasks and, in particular, the possibility that animals might be using timing as a mechanism. Specifically, and in direct contrast to the numeron model, the idea of an accumulator is that the brain converts numerical events into an analogue representation. Thus, when individuals are engaged in numerical assessment, a gating system opens and allows relevant information to accumulate up until and including the last item evaluated. At this point, the gate closes and a value is generated — unlike the digital output of a numeron system, the accumulator generates an analogue output. Like the numeron system, however, the analogue output is also a numerical symbol.

The object file model. Work by vision scientists in particular has provided the theoretical impetus for this model. Specifically, research into the mechanisms underlying object recognition by Treisman and others (reviewed in Kahneman et al 1992) has shown that when objects are perceived, they are tagged (bound spatiotemporally) with particular attributes that give them identity and spatial coordinates. With regard to numerical processing, the idea is simply this. When individuals scan an array of objects, they open object files for each of the relevant objects. In this schema, there are no

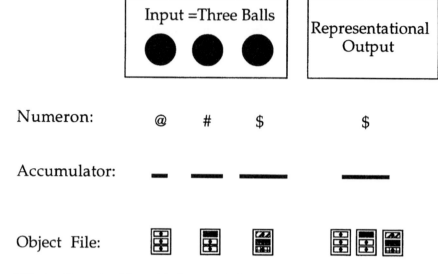

FIG. 1.   Three plausible models for numerical representation: numeron; accumulator; and object file.

numerical symbols. Rather, a set of addresses in memory are opened, each associated with a suite of relevant features.

It has been the aim of recent work on human infant numerical abilities to determine which of these models best accounts for the data. And it has been the goal of some studies of non-human animals to explore whether the representational systems underlying human numerical processing are shared with other organisms. I now turn to the empirical literature, looking first at spontaneous numerical capacities in non-human animals that arise during natural, and functionally relevant, contexts. I then turn to more controlled studies in non-human animals and humans.

Numerical abilities in the wild

Male lions commonly form coalitions in an attempt to overthrow resident males and thereby gain access to a pride . Such take-overs pose a serious threat to females because upon entry into the pride, males typically kill young infants (Bertram 1975, Packer & Pusey 1983). Male baboons form coalitions to both outcompete other males for food resources and to take control of oestrous females. And in dolphins, one group of males will form a super coalition with a second group of males in order to overthrow a third group of males, thereby gaining mating access to a receptive female.

All of these cases of social behaviour occur under natural conditions, in the absence of training, and suggest that non-human animals are sensitive to number, at some level at least. For example, subjects engaged in competitive interactions may retreat if there are simply more opponents or, alternatively, may actually count the precise number of opponents. To determine whether lions are sensitive to the number of potentially infanticidal males, researchers have conducted playback experiments comparing responses to roars from one male or female as opposed to roars from two to three males or females (Grinnell & McComb 1996, Heinsohn et al 1996, McComb et al 1993, 1994). Results showed that: (1) females were more likely to initiate evasive responses to playbacks of multiple males roaring than to playbacks of one male roaring; whereas (2) non-resident males (i.e. those not affiliated with a pride) were attracted to roars from single females and avoided roars from multiple females. From a functional perspective, these patterns of response make sense since selection should favour: (a) lionesses that can both tote up (by listening to roars) the number of potentially infanticidal males as well as the number of individuals within the pride who can defend against such attacks (e.g. young cubs clearly lack the power to defend the pride against attack from adult males); and (b) lions who can determine the number of females in a group because such assessments have a direct bearing on the probability of successfully taking over the pride.

Recently, playback experiments on inter-community aggression and numerical assessment in wild chimpanzees have been initiated (M. Wilson, M. D. Hauser & R. Wrangham, unpublished data 1996), following up on some of the same issues raised by the work on lions. Chimpanzees represent an interesting test species for exploring the functional significance of numerical assessments, for during inter-community interactions, parties differing in size may engage in lethal aggression (Goodall 1986). Consequently, and in parallel to the problem in lions, selection should favour some mechanism for determining the relative power or vulnerability of one's neighbours. Several aspects of the chimpanzee's vocal repertoire make the case of added interest. When an individual loses contact with their party, and is attempting to regain contact, they will often utter a distinctive 'hoo' vocalization. This acoustic signal is, from the listener's perspective, an unambiguous indication that a nearby competitor is alone and vulnerable. On a second level, an individual may utter a vocalization (any other call than a hoo) that is perceptually ambiguous with respect to party size or composition. For instance, if an individual produces a 'pant hoot', the chimpanzee's long-distance call, he may or may not be alone. In contrast, if an individual 'screams', then at least one other party member must be present since screams are given in response to some form of aggressive attack by a dominant group member. On a final level, one might hear a chorus of calls, representing two or more individuals. Here, although it will not be possible to ascertain total party size, it is possible that listeners can determine the number of callers from the chorused signal, thereby providing an estimate of minimum party size. It may also be possible for listeners to extract information on the identity of individuals contributing to the chorus. This perceptual ability would be highly adaptive since some age–sex classes are likely to pose a greater threat (e.g. adult males) than others (e.g. infants).

Thus far, we have only explored responses to pant hoots from foreign community males, using a single pant hoot (three to four seconds in duration) as the stimulus.

Although still in a preliminary stage, the responses have been dramatic. Upon hearing the playback, some males responded with pant hooting whereas others remained silent; females responded vocally on only one occasion. Some males and females then engaged in reassurance behaviour, hugging each other on some trials and in one instance, two adult females performed genital–genital rubbing, a behaviour commonly seen in bonobos, but rarely in common chimpanzees. Subsequently, the males (primarily the young adults) moved quickly and silently straight to the speaker, whereas the females generally stayed put or moved in the opposite direction. Given the fact that the chimpanzees were able to localize a putative intruder (i.e. speaker) 200–400 m away, their sound localization ability appears quite extraordinary relative to other primates (e.g. mangabeys; Waser 1977); human researchers tested in a comparable situation were relatively hopeless, whereas Ugandan field assistants, who were more familiar with the forest, were somewhat better (M. Wilson, unpublished data 1996)! Once the males arrived at the speaker, they tended to remain silent if the speaker was located within their home range. In contrast, they were more likely to display aggressively if the speaker was located at a community range border. We now plan to extend these observations and explore how changes in the type of vocalization, the number and sex of callers, and the relative location of the caller(s) to the receiver effects the type and strength of response.

An additional context for exploring the importance of numerical assessments in non-human animals is foraging. Group-living animals commonly confront two problems during feeding competition: (1) upon finding a food source, some animals (e.g. dominants) have the option of sharing or not; and (2) upon encountering a food source, individuals must decide whether to stay in the current food patch with their associated group or leave and join another resource patch with its associated group. In each case, there will be constraints on the decision, with an important goal being to maximize energetic returns (Stephens & Krebs 1986). Two experiments illustrate the relationship between food quantity and numerical assessments.

House sparrows typically forage in flocks, and one functional advantage of such aggregations is that individuals are more likely to evade predation. There is, however, a cost: feeding competition is increased. Elgar (1986) conducted an experiment to determine under what conditions individuals announce a food discovery by producing a characteristic call. He contrasted two situations: discoverers finding a whole piece of bread or the equivalent amount separated into small pieces. Results showed that discoverers called more often when they found pieces of bread than when they found the whole piece. This suggests that feeding competition is lower when the food item is shareable and that the advantages of group formation (e.g. predator avoidance) outweigh the costs.

In a comparable series of experiments, captive chimpanzees were observed during discoveries of food. In one experiment (Hauser & Wrangham 1987) five, 10 or 15 prunes were hidden in an enclosure. There was a positive correlation between the number of prunes found and the number of food-associated calls produced by the discoverer. This result paralleled observations by Wrangham (1977) with chimpanzees in Gombe, Tanzania, showing that individuals arriving alone at a fruiting tree were more likely to call if there was a substantial amount of food than if not. In a subsequent experiment watermelon served as bait. Here, five, 10 and 20 pieces were hidden, in addition to one large piece, equivalent in size to 20 pieces. Once again, discoverers called more when they found more food, but the large piece of watermelon elicited a level of calling that fell in between the 10- and 20-piece conditions. Here, as in the house sparrow study, it is not possible to determine whether the level of calling is indicative of the quantity of food alone, or an interaction between quantity and shareability (i.e. small pieces of food are relatively easier to share than one large piece). They do, however, provide some indication that quantity is an important factor in foraging decisions.

The discussion thus far suggests that numerical capacities are likely to have quite ancient evolutionary roots. Such capacities are, however, quite rudimentary. Moreover, they are unlikely to figure in any significant way into current debates about whether the capacity for combinatorial computation (iteration, recursion) originally evolved for numerical processing or linguistic processing (Bloom 1994, Campbell 1994, Chomsky 1957, 1966, 1988, Corballis 1992). I mention these studies to illustrate, quite simply, how functional considerations might contribute to thinking about the evolution of numerical capacities. To enter the debate about how numerical and linguistic capacities are related, both in development and evolution, we must look more closely at two different empirical arenas: the counting abilities of non-human animals tested under controlled laboratory conditions; and tests of counting and simple arithmetic in human infants.

Life after Clever Hans: laboratory tests of number in non-human animals

There once was a horse named Hans. He was a clever horse indeed, but not for the reasons originally specified. Specifically, Hans' trainer qua promoter qua manager claimed that when he wrote simple arithmetical equations on a blackboard, the horse would stomp out the correct answer. For example, when $2 + 3$ was etched on the board, Hans stomped five times. As the history of research on animal learning has taught us, it is easy to overlook the possibility that animals can pick up on cues from their trainers, thereby appearing much smarter than they actually are. Hans was no exception. It turned out that the audience members, unbeknownst to them, were bobbing their heads up and down until the correct answer had been given (e.g. five up-downs for $2 + 3$). Hans wasn't doing maths, but he had picked up on a simple solution: keep stomping until the audience stops. Quite smart.

More recent research has generally bypassed such problems and it is clear that at some level, non-human animals have the capacity to count and even perform some simple arithmetical operations following quite spectacular training regimes. They don't, it seems, have the capacity to appreciate a crucial property of number systems — discrete infinity — a capacity that allows competent users to understand that there is, literally, no end to counting. And for those interested in the evolutionary relation between language, number and generative computations, understanding discrete infinity represents the most fundamental capacity (Chomsky 1966, 1988, Corballis 1992, Bloom 1994). Since much of the literature on non-human animals has been reviewed elsewhere (Boysen & Capaldi 1993, Davis & Perusse 1988, Gallistel 1992, Hauser & Carey 1997), I restrict the following discussion to a small subset of the relevant data.

Some of the most elegant work on number in non-human animals has come from experiments with captive chimpanzees, and especially chimpanzees with a long history of training. Here I focus on Sarah Boysen's (summarized in Boysen 1993, 1996) work because of its explicit attempt to assess constraints on numerical competence. Initial research started with three young chimpanzees, living in a laboratory setting that provided them with experience on a suite of quite simple cognitive tasks, involving categorization, pointing and working with experimenters. The first experiment involving numerical assessment explored the chimpanzee's capacity to understand one-to-one correspondence by requiring them to place single pieces of monkey chow into each of the six compartments of a tray. The chimpanzees solved this problem quickly, helping to demonstrate that they had some understanding of one-to-one correspondence. In the subsequent experiment, subjects had to match the number of candies on a tray with a corresponding card marked with dots. Thus, for example, the subject would be presented with a tray of two candies (sample) and cards with either one, two or three marks. The task: select the two-mark card as the appropriate match for the two candies. Eventually, cards with Arabic numerals were substituted for the cards with marks, and the task repeated. Over several sessions, subjects acquired an ability to use numerical labels productively, although there were significant differences among the chimpanzees in both the number of trials required to reach criterion and the range of numerical labels acquired (e.g. Sheba learned 0–8, whereas Kermit learned 0–5). When this process was flipped around, such that subjects had to show comprehension of number (e.g. when presented with the Arabic number 2 on a video monitor, subjects were required to point to a tray of two candies), performance was remarkably poor, taking several hundred trials to reach criterion.

The next training phase focused exclusively on Sheba, since the other two chimpanzees were now too difficult to work with. In this experiment, Sheba was required to do some simple maths. Specifically, oranges were baited in one of three locations. Sheba explored each of the three locations and depending upon the total number of oranges encountered, provided her answer by pointing to the appropriate Arabic numeral on a platform; the Arabic numerals 1–4 were positioned in ordinal sequence. Not only was Sheba immediately successful at this task, but she generalized her knowledge in a transfer condition where the Arabic numerals were substituted for the oranges. Thus, if she found a number 2 in one location and a number 1 in another location, she would select the Arabic number 3 on the platform. Although Sheba's performance is impressive, it is not yet possible to discern the nature or content of her numerical representations (see Hauser & Carey 1997 for a more thorough discussion). That is, because the number of objects assessed are small, simple perceptual mechanisms may account for her performance, as well as the performance of other chimpanzees. In the test just described, for example, Sheba might be adding the number of oranges (or Arabic numerals) found in each location or she may be counting and using the last 'unnamed number' (Koehler 1950) to tag the total number of oranges encountered.

All of the tasks described thus far have been restricted to numerosities of four or less. Other studies with chimpanzees, however, have gone above four and as high as nine (Boysen 1993, Matsuzawa 1985, Rumbaugh & Washburn 1993); it is unclear whether the capacity to go beyond nine is a numerical problem, or a perceptual one involving, for example, a confusion between the integers 0 and 1, and values such as 10, 11 and so on which are perceptually similar. Although the chimpanzee's capacity to exceed what is likely to be the subitizing range (in humans at least) is impressive, such efforts have proved exceedingly difficult, requiring hundreds to thousands of trials, over periods of years. As Boysen (1993) summarizes: '. . . the establishment of such skills in a species even as intelligent as the chimpanzee requires a heroic effort . . . when compared to the seeming ease with which counting skills are acquired by pre-schoolers.' Nonetheless, the end-product of this research suggests a level of numerical competence that is close to the pre-schoolers, and in terms of underlying mechanism, appears to exceed the limits of subitizing; the latter can be debated if one wishes to argue that the subitizing range for humans is lower than it is for non-humans.

Due to their close phylogenetic proximity, chimpanzees provide an ideal system for understanding how numerical competence evolved in humans. It is possible, however, for comparable abilities to evolve in distantly related species due to processes such as convergence. One such candidate example is the African grey parrot, and in particular Pepperberg's star subject 'Alex'. In fact, Pepperberg's (1994) work is highly relevant to the conceptual issues raised in that it shows how an individual from a species with a relatively small brain (i.e. relative to humans and chimpanzees) can process numerosities that exceed the subitizing range. Such performance forces one to entertain the possibility that numerical competence in non-human animals involves more than low level perceptual processing mechanisms. In a nutshell, here is the key experiment.

Alex has a vocal gift: he is capable of using verbal labels to illustrate his grasp of categories and concepts. Thus, Pepperberg (1987a,b, 1988) has demonstrated that Alex can acquire concepts relevant to the physical attributes of objects (e.g. colour, texture, shape, labels), make use of these attributes to assess whether objects are the same/different and present/absent, and importantly, verbally label the number of objects in an array. The new experiments address, more precisely than before, what type of mechanism underlies Alex's performance on numerical tasks. As previously mentioned, studies of numerosity must exceed at least four objects (i.e. the constraint on humans) if subitizing is to be ruled out as a candidate mechanism. Moreover, and as Trick & Pylyshyn (1994) have argued, subitizing is an inefficient mechanism if the objects to be labelled represent a subset of the total number of objects and if the objects to be excluded from the final count have overlapping features. For example, consider a tray with 10 objects: two purple squares, four purple triangles, three green squares and one green triangle. In order to answer the question 'How many purple squares are there on the tray?', one must be able use the conjunction of two features (colour and shape) to give the correct answer. This is not possible with subitizing — at least for humans — and presumably involves focused attention, deliberate assessment of each object and an accounting of objects by feature. Alex was given such a task and was able to provide the correct numerical response with exceptional accuracy (83% overall). Although Pepperberg is appropriately cautious in her interpretation, proposing for example that non-human animals may use different mechanisms than humans even when the testing conditions are identical, these results strongly suggest that at least one individual from one non-human species exhibits numerical competence beyond subitizing.

Numerical thoughts without language

Baby mathematicians revealed by surprise

One problem with the current comparative literature on numerical abilities is that different methodological tools have been employed across species. Consequently, what may appear to be species differences in numerical competence (i.e. ability) may in fact represent differences in performance. The performance–ability distinction plagued work on animal learning and intelligence for 20 or so years (Macphail 1987a,b). To avoid re-entry into this intellectual quagmire, current research must derive a set of experimental procedures that can be used across species. To this end, we have embarked on a research programme that adopts a technique developed by cognitive psychologists interested in what prelinguistic human infants know about the physical and psychological world. This technique, known as the expectancy violation or preferential looking time (PLT) procedure, is powerful for it can be readily used in a variety of species with no modification. Since the logic underlying this approach has been detailed elsewhere (Hauser & Carey 1997, Spelke 1985), I provide only a brief sketch here.

In the recent remake of Jerry Lewis' *The Nutty Professor*, Eddie Murphy plays an extremely fat professor. Now Murphy's initial size is quite spectacular, but not off the chart of possible weights for humans. However, at one point, Murphy ingests a home-made concoction and is transformed overnight into a thin person. The concoction works for a while, but then Murphy experiences grave difficulties, his weight and body contorting, in real time, as if a grand puppeteer were playing mischief with the stretchability of his body parts.

When adults see Murphy's contortions on the big screen, they immediately know that it is not really happening to him. Rather, they appreciate the genius of the special effects team working the magic of the movie, playing with reality, stretching our imagination. And in the adult mind, this magic works because we understand what is physically and psychologically possible, while simultaneously entertaining plays on reality that move us into the world of fantasy. The adult's understanding of movie magic can be tapped by asking questions or by simply observing how intensely they look at events that violate expectations. Any magic show will capture the intense focus and attention of a willing adult because they enjoy being surprised.

In contrast with adults, we should not necessarily expect human infants to appreciate the appearance–reality distinction that is so often violated in the movies and during a magic show. Or, to be more specific, it is not clear whether human infants would understand the humour of Murphy's bodily transformations because they may perceive them as physically possible — consistent with the way the physical world works. It is precisely this distinction between possible vs. impossible, or expected vs. unexpected, that underlies the logic of the PLT procedure. By showing prelinguistic infants displays that are either consistent or inconsistent with properties of the physical and psychological world, developmentalists have begun to map the ontogeny of domain-specific knowledge systems, in the absence of significant environmental input and in the absence of significant linguistic talents.

Using the PLT procedure, there is now evidence that before the age of approximately 10 months, and as early as two to three months, human infants know that: (i) objects that have been moved out of sight continue to exist (i.e. an early form of object permanence); (ii) two solid objects cannot occupy the same space at the same time; (iii) moving objects follow a spatiotemporally continuous path unless obstructed; (iv) to avoid falling, an object must be supported by another object; (v) large objects can cause small objects to move further than the reverse; and (vi) stationary objects move if and only if they are contacted by another moving object or have an internal mechanism that permits self-propelled motion (Baillargeon 1994, Baillargeon & DeVos 1991, Gelman et al 1994, Leslie 1982, 1984, Leslie & Keeble 1987, Premack & Dasser 1991, Spelke 1994, Spelke et al 1995a,b,c). In addition to these findings, the PLT procedure has also been used to assess what human infants know about number, including the detection of cross-modal associations and simple arithmetical operations (Spelke 1994, Spelke et al 1995c, Starkey & Cooper 1980, Starkey et al 1990, Wynn 1992, 1996). I start here with a quick summary of human infant research that bears directly on the comparative work developed more fully below.

By the age of two to three months, human infants know implicitly (i.e. by means of PLT procedures, though not traditional Piagetian search tasks) that when one object is moved out of sight behind an opaque occluder it continues to exist, physically intact, behind the occluder. But do they know that when one object disappears behind a screen, followed by a second, that precisely two objects are present though out of sight? According to Karen Wynn (1992), the answer is 'yes'. Using the PLT procedure, Wynn showed that when a Mickey Mouse doll is occluded and then a second Mickey is placed behind the screen along with the first, five-month-old infants look longer when one or three Mickey Mouse dolls appear following removal of the occluder than when two appear. Based on these results, Wynn argued that human infants have an innate capacity to carry out simple arithmetic; similarly designed experiments also revealed that five-month olds can carry out simple subtraction (i.e. $2 - 1 = 1$).

Needless to say, Wynn's work has attracted considerable attention, criticism and an invigorated research programme into the ontogeny of human concepts (Carey 1995, Carey & Spelke 1994, Hauser & Carey 1997, Simon et al 1995). Of the many important issues raised, let me mention just a few. First, the original Wynn paradigm is known as an object-first design, meaning that an object is placed in view, occluded and then a second object is placed behind the occluder. Thus, the child directly sees the first object and must then deduce how many have been added (or subtracted) while the occluder is in place. This design appears to be simpler, computationally, then the screen-first design (Uller et al 1997a,b). Here, subjects first see an empty stage, followed by the introduction of the occluder, and then the placement of objects behind the occluder. Empirical support for the greater computational difficulty of this task comes from the fact that infants do not pass the screen-first version until eight to 10 months. Second, based on current results, it is not possible to discern whether the child's representation is of X distinct objects (e.g. where X = two Mickey Mouse dolls) or some Y stuff (e.g. where Y = some Mickey Mouse doll stuff). Third, since the number of objects presented is small, the infant's expectations may be violated on the basis of relatively low level perceptual mechanisms such as subitizing (Gallistel & Gelman 1991, 1992, Trick & Pylyshyn 1994) or, as Wynn and others (Carey 1995, Carey & Spelke 1994, Wynn 1992, 1996, Xu & Carey 1996) would argue, relatively higher level numerical-processing mechanisms.

Several other studies have been carried out since Wynn's experiments (Simon et al 1995, Uller et al 1997b, Wynn 1996, Xu & Carey 1996), replicating the original findings and extending them in ways that provide a more comprehensive understanding of the infant's conceptual representation. Specifically, studies of infants under the age of one year have shown that they: (i) find screen-first designs consistently harder than object-first designs (based on the age at which infants pass the test); (ii) consider non-solid objects (e.g. piles of sand) to consist of functionally different properties from solid objects, thereby making them subject to different kinds of arithmetical operations (e.g. one pile of sand plus one pile of sand need not result in two piles of sand — it could just as easily result in one bigger pile); (iii) have greater difficulty with tests of subtraction than tests of addition; (iv) ignore the specific properties of the objects during some arithmetical operations (Simon et al 1995), though not all (Xu & Carey 1996); and (v) fail on arithmetical tasks requiring property/kind level object individuation until they have acquired words for tagging the objects presented (Xu & Carey 1996). If babies come equipped with such numerical abilities then when during the course of evolution did the capacity emerge? Studies of non-human primates may hold some of the answers to this question.

Can other primates do maths?

In 1994 we initiated a research programme designed to make use of the PLT procedure to investigate what non-human primates know about the physical and psychological world. Our first studies focused on the problem of numerical representation and, in particular, the claims made by Wynn in her infant studies. Our focus was guided by three factors. First, and as reviewed above, a considerable amount of work had already been conducted on non-human primate numerical abilities. Thus, we were well equipped to make informed predictions and to design our experiments carefully. Second, the Wynn procedure was relatively simple, and could be imported to studies in the wild and in captivity. Third, the problem of numerical representation is of fundamental importance in figuring out how language contributes to the content and propositional structure of concepts.

The first experiment was conducted with semi-free-ranging rhesus monkeys living on the island of Cayo Santiago, Puerto Rico (Hauser et al 1996). This population was selected because the individuals are extremely well habituated to human observers, the potential sample size is large (900 individuals on the island) and consists of individuals from all age–sex classes; the subjects' habituation was critical for we required individuals who would sit and watch our displays in the absence of reinforcement. Although we wished to replicate Wynn's specific methods, we started out with a modified version: rather than run each individual on a set of familiarization trials (designed to remove the effects of object/action novelty) followed by a test trial (either possible $[1 + 1 = 2]$ or impossible $[1 + 1 = 1$ or $3]$), each subject received only one condition, and thus, one trial. We abbreviated our sessions because we were not convinced at the outset that subjects would sit and watch several trials, especially when no reinforcement was provided. In addition to this change, we also used a screen-first design, and instead of Mickey Mouse dolls, presented bright purple aubergines. Aubergines were selected as test items because they were colourful, bright and a novel food type; earlier attempts with coconuts and apples, familiar food items, caused almost all subjects to rush toward the display rather than sit and watch!

Results from the first experiments, using 61 adults for the three different test trial conditions, showed that subjects seeing the impossible outcome $(1 + 1 = 1)$ looked longer than subjects seeing the possible outcomes $(1 + 1 = 2; 1 + 0 = 1)$. Given these findings, and the ability of rhesus monkeys to sit and watch our displays, we tested an additional group of adult subjects using the more traditional approach for PLT studies of human infants. Specifically, each individual received two familiarization trials followed by one of the test trial conditions. Here again we found that subjects looked longer in the impossible than in the possible conditions. Finally, we ran a $2-1$ subtraction test as in the Wynn studies and, as with five-month olds, found that adult rhesus monkeys looked longer when the outcome was 2 (impossible) than when it was 1 (possible). In summary, our results were consistent with those presented by Wynn, but we were missing at least one important condition: $1 + 1 = 3$. In the absence of this condition, it is impossible to determine whether the rhesus monkeys expected precisely two aubergines or whether any quantity of two or more would be sufficient to match their expectations.

We recently returned to Cayo Santiago and ran an additional group of adult rhesus monkeys on the $1+1=2$ versus 3 test, in addition to a new test designed to address the 'more stuff' argument presented above (C. Uller, M. D. Hauser & S. Carey, unpublished data 1996). Specifically, subjects saw first one aubergine and then a second placed behind the screen. When the screen was removed, they saw one large aubergine, equal in volume to the two smaller ones; subjects were familiarized to the large aubergine prior to being tested, so this object's novelty was removed as a confound. In the $1 + 1 = 2$ vs. 3 test, subjects looked longer at three aubergines than two, thus completing the replication of Wynn on a screen-first design, and showing that changes in looking time were not merely guided by the presentation of more aubergines. In the $1 + 1 = 2$ vs. Big 1 test, subjects looked longer at the single large aubergine. If subjects had been attending to amount of aubergine stuff (e.g. the amount of purple in the display box) then one would expect no change in looking time. The fact that subjects looked significantly longer at the one large aubergine suggests that they did not anticipate this change — that two small aubergines are meaningfully different from a large one in this test procedure.

Based on the rhesus monkey results, the following conclusions are warranted:

(i) Given the similarity in procedural details, adult rhesus monkeys and eight- to 10-month-old human infants appear to have comparable abilities with regard to simple arithmetical computations.

(ii) Although it is still unclear what such studies tell us about the structure and content of their numerical representations, some low level perceptual accounts can be ruled out (e.g. spatial arrangement of objects, and coarse-grained changes in colour, size and volume).

At approximately the same time that we were completing the rhesus monkey studies, we embarked on an analogous set of experiments, using a New World monkey, the cotton-top tamarin (*Saguinus oedipus oedipus*), and a captive test environment (Uller 1997, C. Uller, M. D. Hauser & S. Carey, unpublished data 1996). A New World monkey was selected because we wished to assess when, evolutionarily, such numerical capacities emerged. A captive test environment was selected because it provided a more controlled environment for exploring which features of the experimental paradigm most significantly contributed to the results.

We started with a split-screen test, providing the tamarins with spatiotemporal information for individuation. Specifically, and as with the paradigm used by Spelke et al (1995c), subjects saw one object (a column of brightly coloured cereal known as 'Fruit Loops') emerge from behind one screen and then saw it return behind the screen. They then saw a second object move out from behind the adjacent screen and then return. In no case did they see either object in the open gap between the two screens. When humans see this display they assume that there are two objects, one behind each screen. So did the tamarins. When the screens were removed following the presentation, subjects looked longer when only one object was visible than when two objects were visible.

Having passed this simple test — simple because it provided spatiotemporal information for individuation — we then ran the tamarins on Wynn's addition test, using both the object-first and screen-first designs. The tamarins passed each test, looking longer after impossible than after possible outcomes. Tests of subtraction are in progress.

At this point, we must conclude that adult cotton-top tamarins, adult rhesus monkeys and human infants under the age of approximately 10 months have comparable numerical skills. Needless to say, each of these species may have more sophisticated skills as well. What was therefore needed was a set of tests that might allow us to distinguish between the abilities of these species, especially since we can be quite confident that no non-human animal will ever derive equations for solving problems in non-Euclidean geometry.

A crucial test presented itself. Xu & Carey (1996) had recently completed a study suggesting that the development of arithmetical skills requires not only the sortal concept of 'object', but also a property/kind distinction. Thus, if we pass by a window and see an apple pie, and then moments later pass by the same window and see a strawberry pie, we assume that there are two pies in the house, not an apple pie that has been transformed into a strawberry pie. (Note: I will here assume that most human adults do not, in such situations, engage in the philosophical contemplation of whether an apple pie that has its contents gutted and replaced by strawberry is, nonetheless, the same pie!) Importantly, when children pass arithmetical tasks involving property/kind distinctions, they do so at an age when first words have appeared in both the production and comprehension vocabulary. More succinctly, children are not capable of understanding that $1p + 1q = 2$ distinct objects until they acquire the verbal labels for p and q. If this hypothesis is correct, then two corollaries present themselves:

(1)   Non-human animals, lacking verbal/linguistic labels for objects, should fail the test (i.e. should be incapable of acquiring the particular numerical ability tapped).
(2)   If non-human animals pass the test, then either we must credit them with some form of symbolic labelling or tagging system, or passing the test does not require language; note that language may nonetheless facilitate passing the test in humans.

In the Xu and Carey test subjects are presented with a single screen. In succession, they first see one object (e.g. a duck) emerge and then return behind the left side of the screen and then see a second object (e.g. a truck) emerge and then return behind the right side of the screen. In this condition (property/kind test), subjects never see both objects at the same time; in a spatiotemporal control, both objects sit on the side of the screen at the same time, providing stronger evidence that at least two objects were behind the screen. When the screen is removed, they either see a duck and a truck (possible condition) or they see only one object (i.e. either a duck or a truck; impossible condition). Infants under the age of 10 months look longer at one object in spatiotemporal control but not in the property/kind test. It is not until children have acquired unambiguous word comprehension and production skills for the objects tested (i.e. approximately 12–13 months), that they look longer at the impossible condition.

We (Uller 1997, Uller et al 1997a) ran the Xu and Carey design using a group of adult rhesus monkeys on Cayo Santiago. Instead of ducks and trucks, we used a large orange carrot and a large slice of yellow squash. As with year-old human infants, rhesus monkeys also looked longer when only one object (either the carrot or squash) was revealed then when both objects were revealed, and did so in the spatiotemporal control and the property/kind tests. These results suggest that language, *sensu stricto*, is not necessary for property/kind individuation, at least as revealed by the particular PLT procedure used. It is, of course, possible that rhesus monkeys have evolved a system for silently tagging objects in the environment, but thus far, there is no evidence for such a system.

Levels of mathematical sophistication: the next generation

PLT tests of numerical abilities in primates, human and non-human, are limited for the following reasons: (i) the number of objects presented in a display are small, falling well within the range of subitization; (ii) factors that are likely to influence the nature of the numerical representation have yet to be fully fleshed out, thereby leaving considerable uncertainty with respect to which models (i.e. numeron, accumulator, object files, some combination of these or a new alternative) best account for current results; and (iii) new methods are needed, both modifying existing PLT tests and developing procedures other than PLT to complement current findings and more carefully explore the structure and content of the representation, its development and evolution. To address some of these issues, I turn next to some preliminary findings and new experimental approaches.

Going beyond subitization.   As several authors have suggested (Gallistel & Gelman 1991, 1992, Trick & Pylyshyn 1994), when the number of objects assessed is less than some critical value (e.g. somewhere around four, but perhaps as high as six), a rapid, sub-attentive and effortless processing mechanism known as subitizing is most likely responsible for how subjects represent the visual array. Studies of numerical ability in non-human animals, involving procedures other than PLT, have gone well beyond the potential magic number four and even six, suggesting that at least this form of numerical representation is unlikely to be accounted for by subitizing processes (see in particular Pepperberg's [1994] work with an African grey parrot). But in all of these tasks, the objects counted remain in view. In contrast, PLT procedures require subjects to represent the objects in working memory, and thus subitization as articulated by Trick & Pylyshyn (1994) cannot account for the results. We have been developing a number of approaches designed to circumvent the subitization problem, and to explore the difference between experiments that tax working memory and those that do not.

Rumbaugh & Washburn (1993) conducted a study with chimpanzees that required them to choose one out of two possible food treat arrays. Each array consisted of small clusters of treats. Thus, one side might have a cluster of five and two, the other a cluster of four and two. Subjects consistently picked the array yielding the greatest number of treats, even though they were not punished or prevented from taking the smaller amount. Because the spatial configuration of each array consistently varied, one cannot account for these results by invoking some simple perceptual mechanism (e.g. if the treats were only presented in a vertical or horizontal line, subjects could use a simple length assessment). Nor can one account for these results by a subitizing process since the number of objects presented far exceeds the human adult limit, and there is no a priori reason to think that chimpanzees should surpass this limit.

In a similar, but importantly modified, experimental procedure Boysen (1993, 1996) tested two well-trained chimpanzees (Sheba and Sarah) in a task that made one individual the selector of food and the other the receiver. Whichever food quantity the selector picked first, the receiver would obtain; the selector would obtain the remaining food quantity. For example, if Sheba (selector) was presented with a choice between one and six food treats, selection of six would provide Sarah (receiver) with six treats and Sheba with one. Thus, if Sheba is thinking selfishly, she should always pick one treat. However, neither Sheba nor Sarah picked one treat, or any of the smaller quantities. They consistently selected six (or the larger of the two quantities), and thereby only obtained one food treat for themselves. These are extraordinary findings on at least three levels. First, the failure to pick the smaller quantity of food continued to emerge even though this strategy failed to provide the desired goal — the larger food quantity. Second, both of these chimpanzees had an exceptional amount of experimental training over the course of their lives and thus they were by no means garden variety chimpanzees. Their unusual training histories makes the failure all the more puzzling. Finally, and perhaps most interestingly of all, Sheba was able to pass this test if Arabic numerals replaced the food (i.e. instead of two food treats, the Arabic number 2 was presented). This was possible because Sheba had been taught the association between Arabic numbers and food quantities in an earlier experiment. To account for the failure in the food task, Boysen invoked an argument from current thinking in theory of mind research. In particular, she suggested that the chimpanzees failed to solve the problem because they could not rule out competing hypotheses: 'If I pick what I want, I don't get it' vs. 'If I pick what I don't want, I get what I want'. A much simpler explanation, however, is that chimpanzees have difficulty inhibiting an intuitively correct response: select the quantity of food you want. In this sense, chimpanzees are like human infants under the age of nine months who, when presented with a transparent box with a toy inside, will repeatedly reach straight ahead for it even though this action is blocked by a perspex front, and even though they have been shown an open entrance on the side (Diamond 1988). Further support for this explanation comes from the fact that at least Sheba was able to solve the problem when, by pairing food with Arabic numerals, the presumed motivational pull of the food was partially detached from the task.

We (Hauser et al 1997) have recently completed experiments with cotton-top tamarins that are procedurally and conceptually similar to those of Rumbaugh & Washburn (1993) and Boysen (1993, 1996). In the first series, we presented subjects with a choice between two pieces of cloth, each piece separated by a partition. A food pellet was located 'ON' one cloth but 'OFF' the other piece of cloth. The subject's task was to pull the piece of cloth to gain access to the pellet (i.e. a means–end task). Over the course of several sessions, we systematically modified a number of factors that might influence the tamarins success at this task: (i) cloth colour, shape, size; (ii) food colour, shape, size; and (iii) relative position of the food and cloth (e.g. some pieces of food were off the cloth and separated by a gap whereas other pieces were off but touching the cloth; from a functional perspective, food pellets that were not completely on the cloth, could not be attained). The tamarins performed exceptionally well at this task, consistently selecting an ON pellet over an OFF pellet. Only one condition slowed all subjects ($n = 9$) down, even though they had been performing on previous conditions with perfection over the course of several sessions: food size. Here, we presented a choice between a small pellet on the cloth and a large pellet off the cloth. Subjects appeared highly motivated to obtain the large pellet, as evidenced by the fact that their scores dropped below chance for several sessions in a row. Ultimately, however, they were able to override this motivation, picking the accessible small piece of food over the inaccessible large piece.

Taking advantage of the fact that the tamarins understand the means–end cloth task, and have some understanding of the concept 'ON', we modified the task to look at their capacity for quantity assessments. Specifically, half of our subjects were put on an experiment to look at three vs. two whereas the other subjects were put on a six vs. four version (Fig. 2). In brief, each experimental session consisted of 20 trials, 18 controls involving a choice between one pellet ON the cloth and one or more pellets off, and two probe trials involving either three vs. two pellets ON or six vs. four pellets ON the cloth; as in the experiments with chimpanzees, the tamarins were not punished or prevented from picking the smaller of the two quantities. The 18 control trials were run for two reasons. First, if subjects were consistently presented with trials where pellets were placed ON both cloths, they might stop making choices since both sides are rewarded. Second, many of the control trials required the tamarins to pick the accessible pellet (but avoid the inaccessible) and presumably the more desirable alternative of several pellets; in some cases, they had to reject six pellets touching (but not on) the side of one cloth and select the one pellet on the opposite cloth. Again, subjects readily passed the control conditions, and thus far, have passed many of the probe trials comparing three vs. two and six vs. four. That is, the tamarins are typically selecting the larger of the two quantities, and some simple perceptual strategies have already been ruled out (e.g. length or spatial density assessments). Moreover, in several control trials they are selecting the one accessible pellet while inhibiting what we suggest is a strong motivation to select the larger inaccessible quantity. This procedure is simple, involves no explicit training and allows us to explore numerosities that clearly exceed the limits of subitization. Moreover, given

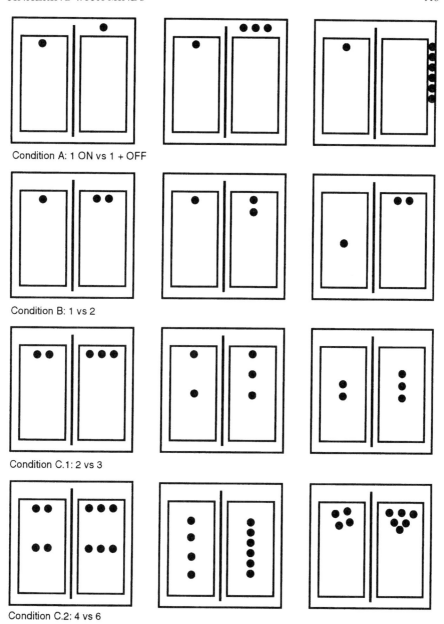

Condition A: 1 ON vs 1 + OFF

Condition B: 1 vs 2

Condition C.1: 2 vs 3

Condition C.2: 4 vs 6

FIG. 2. A numerical discrimination task involving a means–end relationship. Two pieces of cloth (white rectangles) sit on a tray separated by a partition. Either on or off each piece of cloth is one or more food pellets. The subject's task is to pull a piece of cloth to access the food pellet.

that the tamarins are able to pick the smaller of two quantities if the larger one is inaccessible, it will be possible to convert this task into one that is analogous to Boysen's selector/receiver task.

A more recent experiment, which is only in the earliest testing stages, has the potential to push the limits of subitization further, and has the advantage that it taps into working memory in some of the same ways as PLT procedures. It is also a spin-off from the human infant studies, and thus contributes to some of the comparative issues that have been raised in this review and elsewhere (Hauser & Carey 1997). The tamarins are shown two opaque boxes with closing lids; each box is placed on opposite sides of an opaque partition (Fig. 3). Subjects first see that the boxes are empty. In phase 1, subjects see a piece of food placed in one box and the lid closed, followed by the placement of a non-food item into the second box, and the lid closed; trials are counterbalanced with respect to order of food vs. non-food placement. Subjects are then allowed to approach only one box, and remove its contents by lifting the lid. Subjects attained accuracies of 90% or better on the first session, picking the food over the non-food item. In phase 2, 15 of 20 trials are identical to phase 1, whereas five trials involve placing one food item in one box and two food items in the second box. The two-food item box is loaded by placing the first item out of sight, followed by the placement of a second item. In this sense, subjects see an addition operation (i.e. 1 + 1) that is analogous to the screen-first design described previously. As in phase 1, subjects achieved accuracies of approximately 90% for the 15 old trials, and generally picked two food items over one food item on at least four out of five of the new trials; the preference for two food items occurred even on trials where they were the first food items placed, thus making the box with one food item the most recently loaded location for food. Having passed phase 2, we are now setting up a series of tests that will enable us to go beyond the subitizing range, explore the differences between subtraction and addition, and determine how specific properties of the object together with memory constraints affect success at numerical tasks. For example, in phases 1 and 2, subjects are allowed to retrieve the food immediately after the last item has been placed, and the lid on the boxes closed. If the numeron or accumulator models best account for the type of numerical representation recruited to solve the task, then changes in the period of time between concealment of the food and selection should have little effect. In contrast, the delay between concealment and selection should have a more significant effect on performance if it is guided by an object file type of representation. Along similar lines, we will manipulate the order in which food items are placed in each box, thereby varying how often the representation associated with each box is updated. For example, we will contrast two conditions. In condition A, one item is placed in box 1, then one and a second item placed in box 2, then one and another item in box 1 (i.e. three in box 1 with two updates, two in box 2 with one update). In condition B, one item is placed in box 1, one in box 2, one in box 1, one in box 2 and then one in box 1 (i.e. three updates in box 1, two updates in box 2). For each of these conditions we will explore how performance is influenced by the quantity of objects concealed. In this way, we can parametrically vary and assess how

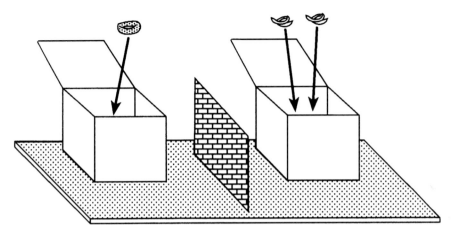

FIG. 3.   A search task for numerical assessment and object individuation. One or more pieces of food are placed in each box, and the subject is allowed to search and retrieve the food.

different factors contribute to performance, and thereby shed insight into the nature of the numerical representation recruited.

A final experiment in this section is based on some work that Premack & Premack (1994b) conducted several years ago with Sarah, the language-trained chimpanzee, and three other non-language-trained chimpanzees. The aim of the experiment was to assess the chimpanzee's capacity for causal inference, but it is also an experiment that has potential as a study of numerical abilities. I present here, in brief, the design because it is relevant to the procedure we adopted with our tamarins.

A chimpanzee was stationed with an experimenter (A) at a distance from a second experimenter (B). To the right and left of experimenter B were two opaque, and initially empty, buckets. The chimpanzee watched the following sequence of events. Experimenter B removed an apple and a banana from his pockets; these are both highly preferred food items. The apple was placed in one bucket, the banana in the other. A screen was then raised, occluding experimenter B from the chimpanzee. Experimenter A distracted the chimpanzee for a few minutes and then the screen was removed, revealing experimenter B who was now eating either the apple or banana. Upon completion, experimenter B walked away. The chimpanzee was now allowed to approach and choose a bucket. The task, obviously, is to pick the bucket with the remaining piece of food. Only one chimpanzee passed this test with consistency, the other three falling at or below chance. The task appears to tap some form of causal inference because the chimpanzee must infer that experimenter B is responsible for depleting one of the two food items. However, it is also a subtraction test, requiring object individuation, with experimenter B serving as the agent for taking one object away from two.

We were initially interested in Premack and Premack's test because we were surprised at the chimpanzee's general failure. It seems like such an easy task. In thinking about the design, we came up with some potential problems that may have caused difficulties for the chimpanzees (M. D. Hauser & K. Schecter, unpublished data 1996). Specifically, if the chimpanzee assumes that experimenter B has more than one apple or banana in his pockets, then eating one apple or banana does not mean that they have been removed from the buckets or, if they have, that the experimenter cannot replace them; and replacement seems like a strong possibility given the fact that the chimpanzee was being distracted while the occluder was put into place. What was needed, therefore, was a design that would allow subjects to provide unambiguous feedback to the experimenters that they understood how many objects, precisely, have been placed in the buckets, and that once placed, no more are added until the next trial. Our design comes close to meeting these criteria.

Tamarins sat in a transparent box with a sliding front door. Behind them, we placed an opaque food bowl and in front of them, an apparatus with two doors (Fig. 4); connected behind each door was a tube where food could be placed, concealed by the apparatus door. Opening the door provided access to the food. In phase 1 the tamarins watched an experimenter open both doors, reveal empty tubes, and then close the doors. The door to the tamarin's box was then opened and the subject allowed to open doors to the test apparatus. This step in phase 1 was designed to teach the tamarin that unless food was removed from behind its box and placed in one of the tubes of the apparatus, opening doors was fruitless. If the subject failed to open doors, or once door opening stopped and the subject returned to its test box, we moved on to the second step of phase 1. Here, the doors of the apparatus were opened and the experimenter indicated that both tubes were empty. The experimenter then reached behind the tamarin, removed a single piece of food from the bowl, and then placed it in one of the tubes; the food was not placed in the tube unless the subject visually tracked the removal of food and placement into the tube. Both doors were then closed simultaneously and the tamarin allowed to select one door. Once subjects reached criterion on this phase, they were moved to phase 2. Phase 2 was identical to phase 1 with one exception: once the food was placed inside the apparatus, and the doors closed, subjects were forced to wait for eight seconds before being allowed to select a door. This time delay was imposed because during the key test phase they would have to wait for no more than eight seconds before being allowed to pick a door. Subjects were not allowed to move on to the test phase until they performed at criterion and failed to open doors in step 1. Refraining from opening doors was critical because it provided us with some confidence in the tamarin's understanding of the relationship between taking food out of the food bowl and the presence of food behind the doors.

The test phase also involved step 1, where the tamarins were allowed to open doors even though no food had been placed inside the apparatus. If no doors were opened, they then watched the following sequence of events. (1) The experimenter removed a piece of raisin and a piece of differently coloured Froot Loop from the food bowl (both highly preferred food items), placing one item into one tube and the other in the opposite tube. As in phase 2, subjects were required to track food removal and

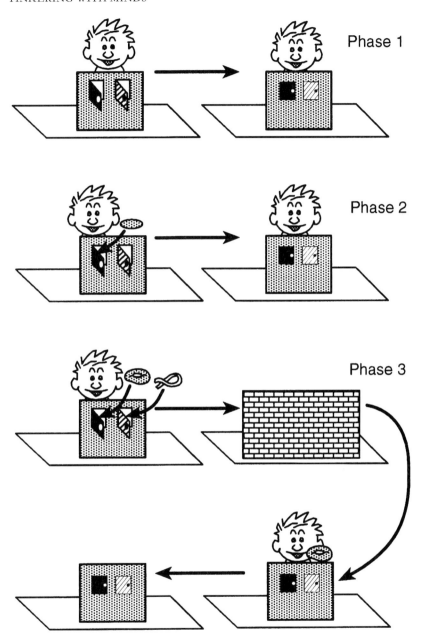

FIG. 4.   A numerical task involving object individuation and search. The actor removes one item, leaving the second item behind. The subject's task is to bind object properties with specific spatial locations and use this information to search for the available piece of food.

placement; they were also given the opportunity to look inside each tube (from their test box) before the doors were closed. (2) The doors were closed and a screen put in place, occluding the experimenter and test apparatus from the subject. (3) Out of view, the experimenter removed one of the food items. (4) The screen was lowered and the tamarin watched the experimenter eating the food; the experimenter made sure that the tamarin saw the food prior to eating it. (5) The subject was allowed to leave its test box and select one door of the apparatus; if the wrong door was selected, the subject was gently escorted back to its test box. Thus, a cost was imposed for the wrong choice. We tested a total of six subjects on this phase of the experiment, and only two subjects consistently passed the test. That is, out of 10 trials, two tamarins selected the door with the remaining piece of food on eight or more trials, and did so on two consecutive sessions. The other tamarins either performed at or above criterion on one session and then below criterion on two or more additional sessions, or failed to reach criterion on two consecutive sessions; three subjects never made it to the test phase.

Although the tamarins' performance on this task was by no means stellar, it appears that at least some individuals in the species have the capacity to solve the problem. Importantly, the procedure provides another technique for looking at numerical assessment, and perhaps in a context that is more ecologically valid than the others: when a human, or other tamarin, removes food from a location, this is analogous to the types of situations that arise in natural foraging contexts. One might therefore consider a slight modification to our design and Premack and Premack's that would involve a large, patchily distributed resource area, and two subjects — a forager and an observer. The observer watches as an experimenter hides X items of food in the environment. The forager is then introduced and allowed (by the experimenter) to remove n out of X food items. The observer is then allowed to enter the foraging environment to obtain food. In one sense, this test is similar to the kinds of procedures used by Balda, Kamil and others with caching birds (Balda & Kamil 1988). It has the potential, however, to be modified in interesting ways to look at numerical quantification. For example, some food locations could be loaded with multiple food items, and the forager allowed to remove only a subset. Consequently, each food location represents a mini-mathematical problem, coupled of course with a memory problem — i.e. the subject must tag each location with a particular value and store this value and spatial location. For some species, or under some conditions within a given species or individual, the value stored may be something simple like 'present/ absent' whereas under other conditions, a more precise value may be stored such as 'five items of food'.

Tweaking the preferential looking time procedure. The logic of the PLT procedure is that events violating physical or psychological properties of the world recruit our attention and cause us to look longer at the display — we are surprised. If this logic is correct then it follows that subjects should be surprised at a violation on the first trial, should show surprise in the absence of familiarization and should maintain interest in looking at this violation for a longer period of time than presentation of an event that is not in violation. We have recently run an experiment to explore the logic of these claims. Specifically, we took 11 cotton-top tamarins and divided them into two groups. One group (possible) was presented with a $1+1=2$ display on 10 consecutive trials. The second group (impossible) was presented with a $1+1=1$ display on 10 consecutive trials. This design generated two predictions: (i) on the first trial, subjects should look longer at the impossible condition than the possible condition; and (ii) they should maintain a higher level of attention (i.e. longer looking) throughout the 10-trial session. This is precisely what we observed (Fig. 5). Specifically, five out of six subjects looked longer on the first trial of the impossible test than did subjects presented with the possible test and, on average, subjects on the impossible condition looked longer over the course of the 10 trials. Thus, although it is possible that novelty alone may cause a ceiling effect in looking time studies, the habituation procedure described is easy to run, and it provides additional support for the logic of the PLT technique.

Current PLT procedures are problematic in at least two ways, both of which we have been addressing with new experimental procedures. The first concerns the potential for Clever Hans effects. Consider the typical PLT design. Subjects are initially familiarized to the set of objects and events to be used in the test trials. Following familiarization and, therefore, general decrease in interest to the display, subjects are either tested with a display that is consistent with properties of the physical or psychological world, or

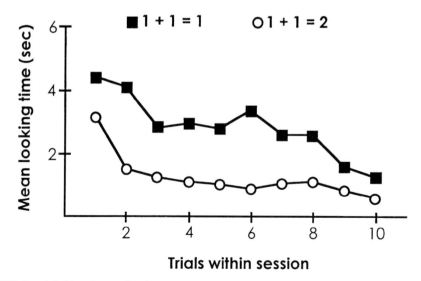

FIG. 5. A habituation study of $1+1=2$ versus 1. One group of subjects received 10 trials of $1+1=1$. A second group received 10 trials of $1+1=2$. The y-axis shows the mean amount of time spent looking (seconds) at the display.

they are presented with a display that is inconsistent. Each subject's looking time is either scored on-line by two observers (using a stop watch), who are blind to the conditions presented, or each video record is digitized onto a computer and then scored blind by independent observers. Because they are blind to the conditions, there is little room for bias. But the potential for bias is not completely absent. Specifically, observers know in advance that each subject will receive familiarization trials first, and test trials later. This is particularly the case when looking time is scored on-line with stop watches; the advantage of digitized video is that trials can be stored as files and scrambled with regard to order within a session. Thus, some aspect of the experimental design is transparent, and thus biases may enter.

A more significant bias lies in the fact that the experimenter responsible for presenting the test stimuli knows precisely what each condition is designed to achieve. Consequently, it is possible that objects are presented differently in possible versus impossible trials and this, in and of itself, might alter looking times. For example, in the $1 + 1 = 2$ vs. 1 or 3 test, objects may be jiggled more in the 1- or 3-outcome test. One solution to this problem is to convert all presentations to video, either real or animated. This shift would not only eliminate the possibility of Clever Hans effects, but would also provide significantly greater control over the relevant features of the presentation. We have recently embarked on an experiment using video animations (Macromind Director, version 4.0), attempting to replicate our earlier findings on $1 + 1 = 2$ vs. 1 or 3. As with studies of human infants (E. S. Spelke & S. Carey, personal communication 1996), it has proven extremely difficult to get the tamarins to attend to a video monitor, especially when there are no rewards forthcoming. However, given the fact that our tamarins watch a video monitor during operant procedures, and that chimpanzees, rhesus monkeys, chickens and jumping spiders all attend to video presentations, we have begun to pursue a set of techniques that will make the monitor more salient. For example, we filmed one tamarin running down a Y-maze to get food from a red box but failing to obtain food from the adjacent arm with the green box. We show this video tape loop to other tamarins, and then place them in the Y-maze without any prior experience. If subjects consistently go to the red box we can be reasonably confident that they extracted some relevant information from the video footage. At this point, the monitor has gained salience and we will attempt to replicate the addition experiments. If we replicate this finding, then we will be in a strong position to conduct many more PLT experiments, but in the absence of Clever Hans problems.

Conclusion

The human psychological adaptation is characterized by many extraordinary capacities. Such capacities have a history. It is the goal of work in cognitive evolution to determine when such capacities emerged and how particular pressures shaped their ultimate architecture. This goal can only be accomplished by implementing a comparative approach that derives its power from an interdisciplinary marriage between evolutionary biology and cognitive science. Importantly, however, this marriage must be guided by a set of methodological tools that allows for direct inter-specific comparisons.

In this chapter I have provided one approach to fulfilling the goals of this research agenda, focusing quite narrowly on the evolution of numerical competence. Others have already contributed vast amounts of information to this problem. My own contribution has been to show how one method — the PLT procedure — can help in making the comparative approach to this problem even more powerful than it has been because differences in performance are unlikely to cloud the fundamental interest in differences or similarities in ability.

Down the line, research on numerical competence in human infants and non-human animals will have to focus greater efforts on the nature of the underlying representation, and how the numerical capacity is enriched by the acquisition of language. In so doing, we are likely to uncover what features of the human brain are really critical in supporting the language organ, and its spectacular communicative output.

Acknowledgements

I thank Susan Carey for comments on the manuscript. Research was supported by a Carey National Science Foundation Young Investigator Award and by equipment from Psion Inc.

References

Baillargeon R 1994 A model of physical reasoning in infancy. In: Rovee-Collier C, Lipsitt L (eds) Advances in infancy research, vol 9. Ablex, Norwood, NJ

Baillargeon R, DeVos J 1991 Object permanence in young infants: further evidence. Child Dev 62:1227–1246

Balda RP, Kamil AC 1988 The spatial memory of Clark's nutcrackers (*Nucifraga columbiana*) in an analogue of the radial arm maze. Anim Learn Behav 16:116–122

Baron-Cohen S 1995 Mindblindness. MIT Press, Cambridge, MA,

Bertram BCR 1975 Social factors influencing reproduction in wild lions. J Zool (Lond) 177:463–482

Bickerton D 1990 Species and language. Chicago University Press, Chicago, IL,

Bickerton D 1995 Language and human behavior. University of Washington Press, Seattle, WA

Bloom P 1994 Generativity within language and other cognitive domains. Cognition 51:177–189

Boysen ST 1993 Counting in chimpanzees: nonhuman principles and emergent properties of number. In: Boysen ST, Capaldi EJ (eds) The development of numerical competence: animal and human models. Lawrence Erlbaum Associates Inc., Hillsdale, NJ, p 39–59

Boysen ST 1996 'More is less': the distribution of rule-governed resource distribution in chimpanzees. In: Russon AE, Bard KA, Parker ST (eds) Reaching into thought: the minds of the great apes. Cambridge University Press, Cambridge, p 177–189

Boysen ST, Capaldi EJ 1993 The development of numerical competence: animal and human models. Lawrence Erlbaum Associates Inc., Hillsdale, NJ

Campbell JID 1994 Architectures for numerical cognition. Cognition 53:1–44

Carey S 1995 On the origin of causal understanding. In: Sperber D, Premack D, Premack A (eds) Causal cognition. Clarendon, Oxford, p 113–146

Carey S, Spelke E 1994 Domain specific knowledge and conceptual change. In: Hirschfeld LA, Gelman SA (eds) Mapping the mind: domain specificity in cognition and culture. Cambridge University Press, New York, p 162–201

Cheney DL, Seyfarth RM 1990 How monkeys see the world: inside the mind of another species. Chicago University Press, Chicago, IL

Chomsky N 1957 Syntactic structures. Mouton, The Hague

Chomsky N 1966 Cartesian linguistics. Harper & Row, New York

Chomsky N 1986 Knowledge of language: its nature, origin, and use. Praeger, New York

Chomsky N 1988 Language and problems of knowledge: the Managua Lectures. MIT Press, Cambridge, MA

Corballis M 1992 On the evolution of language and generativity. Cognition 44:197–226

Corballis M 1994 The generation of generativity: a response to Bloom. Cognition 51:191–198

Cosmides L, Tooby J 1994a Beyond intuition and instinct blindness: toward an evolutionarily rigorous cognitive science. Cognition 50:41–77

Cosmides L, Tooby J 1994b Origins of domain specificity: the evolution of functional organization. In: Hirschfeld LA, Gelman SA (eds) Mapping the mind: domain specificity in cognition and culture. Cambridge University Press, Cambridge, p 85–116

Cummins D, Allen C 1997 The evolution of mind. Oxford University Press, New York, in press

Davis H, Perusse R 1988 Numerical competence in animals: definitional issues, current evidence, and new research agenda. Behav Brain Sci 11:561–615

Diamond A 1988 Differences between adult and infant cognition: is the crucial variable presence or absence of language? In: Weiskrantz L (ed) Thought without language. Clarendon, Oxford, p 337–370

Elgar MA 1986 House sparrows establish foraging flocks by giving chirrup calls if the resources are divisible. Anim Behav 34:169–174

Gallistel CR 1992 The organization of learning. MIT Press, Cambridge, MA

Gallistel CR, Gelman R 1991 Subitizing: the preverbal counting process. In: Kessen WE, Ortony A, Craik FIM (eds) Thoughts, memories, and emotions: essays in honor of George Mandler. Lawrence Erlbaum Associates Inc., Hillsdale, NJ, p 65–91

Gallistel CR, Gelman R 1992 Preverbal and verbal counting and computation. Cognition 44:43–74

Gelman R, Gallistel CR 1986 The child's understanding of number (2nd edn). Harvard University Press, Cambridge, MA

Gelman SA, Coley JD, Gottfried GM 1994 Essentialist beliefs in children: the acquisition of concepts and theories. In: Hirschfield LA, Gelman SA (eds) Mapping the mind: domain specificity in cognition and culture. Cambridge University Press, Cambridge, p 341–366

Goodall J 1986 The chimpanzees of Gombe: patterns of behavior. Harvard University Press, Cambridge, MA

Gould J, Gould C 1995 The animal mind. Scientific American, New York

Griffin DR 1992 Animal minds. Chicago University Press, Chicago, IL

Grinnell J, McComb K 1996 Maternal grouping as a defense against infanticide by males: evidence from field playback experiments on African lions. Behav Ecol 7:55–59

Hauser MD, Carey S 1997 Building a cognitive creature from a set of primitives: evolutionary and developmental insights. In: Cummins D, Allen C (eds) The evolution of mind. Oxford University Press, New York, in press

Hauser MD, Wrangham RW 1987 Manipulation of food calls in captive chimpanzees: a preliminary report. Folia Primatol 48:24–35

Hauser MD, MacNeilage P, Ware M 1996 Numerical representations in primates. Proc Natl Acad Sci USA 93:1514–1517

Hauser MD, Kralik J, Botto-Mattan C 1997 Using means–end problem solving to explore abstract relational concepts in nonhuman animals: experiments with a nonhuman primate. J Comp Psych, submitted

Heinsohn R, Packer C, Pusey AE 1996 Development of cooperative territoriality in juvenile lions. Proc R Soc Lond Ser B 263:475–479

Kahneman D, Treisman A, Gibbs B 1992 The reviewing of object files: object specific integration of information. Cognit Psychol 24:175–219

Koehler O 1950 The ability of birds to 'count'. Bull Anim Behav 9:41–45

Leslie AM 1982 The perception of causality in infants. Perception 11:173–186

Leslie AM 1984 ToMM, ToBy, and Agency: core architecture and domain specificity. In: Hirschfeld LA, Gelman SA (eds) Mapping the mind: domain specificity in cognition and culture. Cambridge University Press, New York, p 119–148

Leslie AM, Keeble S 1987 Do six-month old infants perceive causality? Cognition 25:265–288

Macphail E 1982 Brain and intelligence in vertebrates. Clarendon, Oxford

Macphail E 1987a The comparative psychology of intelligence. Behav Brain Sci 10:645–695

Macphail E 1987b Intelligence: a comparative approach. In: Blakemore C, Greenfield S (eds) Mindwaves: thoughts on intelligence, identity and consciousness. Basil Blackwell Ltd., Oxford, p 177–194

Macphail E 1994 The neuroscience of animal intelligence. Columbia University Press, New York

Matsuzawa T 1985 Use of numbers by a chimpanzee. Nature 315:57–59

McComb K, Pusey A, Packer C, Grinnell J 1993 Female lions can identify potentially infanticidal males from their roars. Proc R Soc Lond Ser B 252:59–64

McComb K, Packer C, Pusey A 1994 Roaring and numerical assessment in contests between groups of female lions, *Panthera leo*. Anim Behav 47:379–387

Meck WH, Church RM 1983 A mode control model of counting and timing processes. J Exp Psychol Anim Behav Processes 9:320–334

Nagel T 1974 What is it like to be a bat? Philos Rev 83:2–14

Packer C, Pusey AE 1983 Adaptations of female lions to infanticide by incoming males. Am Nat 121:716–728

Pepperberg IM 1987a Acquisition of the same/different concept by an African gray parrot (*Psittacus erithacus*): learning with respect to color, shape, and material. Anim Learn Behav 11:179–185

Pepperberg IM 1987b Evidence for conceptual quantitative abilities in the African gray parrot: labeling of cardinal sets. Ethology 75:37–61

Pepperberg IM 1988 'Comprehension of absence' by an African gray parrot: learning with respect to questions of same/different. J Exp Anal Behav 50:553–564

Pepperberg IM 1994 Numerical competence in an African gray parrot (*Psittacus erithacus*). J Comp Psych 108:36–44

Povinelli DJ 1993 Reconstructing the evolution of mind. Am Psych 48:493–509

Premack D 1986 Gavagai! or the future history of the animal language controversy. MIT Press, Cambridge, MA

Premack D, Dasser V 1991 Perceptual origins and conceptual evidence for theory of mind in apes and children. In: Whiten A (ed) Natural theories of mind. Blackwell, Oxford, p 253–266

Premack D, Premack AJ 1994a Origins of human social competence. In: Gazzaniga M (ed) The cognitive neurosciences. MIT Press, Cambridge, MA, p 205–218

Premack D, Premack A 1994b Levels of causal understanding in chimpanzees and children. Cognition 50:347–362

Quine WV 1973 On the reasons for the indeterminacy of translation. J Phil 12:178–183

Rumbaugh DM, Washburn DA 1993 Counting by chimpanzees and ordinality judgements by macaques in video-formatted tasks. In: Boysen ST, Capaldi EJ (eds) The development of numerical competence. Animal and human models. Lawrence Erlbaum Associates Inc., Hillsdale, NJ, p 87–108

Simon T, Hespos S, Rochat P 1995 Do infants understand simple arithmetic? A replication of Wynn (1992). Cognit Dev 10:253–269

Spelke ES 1985 Preferential looking methods as tools for the study of cognition in infancy. In: Gottlieb G, Krasnegor N (eds) Measurement of audition and vision in the first year of postnatal life. Ablex, Norwood, NJ, p 85–168

Spelke ES 1994 Initial knowledge: six suggestions. Cognition 50:431–445

Spelke ES, Phillips A, Woodward AL 1995a Infants' knowledge of object motion and human action. In: Sperber D, Premack D, Premack A (eds) Causal cognition. Clarendon, Oxford, p 44–76

Spelke ES, Vishton P, von Hofsten C 1995b Object perception, object-directed action, and physical knowledge in infancy. In: Gazzaniga M (ed) The cognitive neurosciences. MIT Press, Cambridge, MA, p 165–179

Spelke ES, Kestenbaum R, Simons DJ, Wein D 1995c Spatio-temporal continuity, smoothness of motion and object identity in infancy. Br J Dev Psychol 13:113–142

Starkey P, Cooper R 1980 Perception of numbers by human infants. Science 210:1033–1035

Starkey P, Spelke ES, Gelman R 1990 Numerical abstraction by human infants. Cognition 36:97–127

Stephens DW, Krebs JR 1986 Foraging theory. Princeton University Press, Princeton, NJ

Studdert-Kennedy M 1997 The origins of generativity. In: Hurford J, Knight C, Studdert-Kennedy (eds) Evolution of human language. Edinburgh Press, Edinburgh, in press

Trick L, Pylyshyn Z 1994 Why are small and large numbers enumerated differently? A limited capacity preattentive stage in vision. Psychol Rev 101:80–102

Uller C 1997 Origins of numerical concepts: a comparative study of human infants and nonhuman primates. PhD thesis, Massachusetts Institute of Technology, Cambridge, MA

Uller C, Carey S, Hauser MD 1997a Is language needed for constructing sortal concepts? A study with nonhuman primates. Cognit Dev, in press

Uller C, Hauser MD, Carey S 1997b Evolutionary roots of the human numerical capacity: experiments with two nonhuman primate species. Cognition, in press

Waser PM 1977 Sound localization by monkeys: a field experiment. Behav Ecol Sociobiol 2:427–431

Wittgenstein L 1953 Philosophical investigations. Blackwell Scientific, Oxford

Wrangham RW 1977 Feeding behaviour of chimpanzees in Gombe National Park, Tanzania. In: Clutton-Brock TH (ed) Primate ecology: studies of feeding and ranging behaviour in lemurs, monkeys and apes. Academic Press, London, p 503–538

Wynn K 1992 Addition and subtraction by human infants. Nature 358:749–750

Wynn K 1996 Infants' individuation and enumeration of actions. Psych Sci 7:164–169

Xu F, Carey S 1996 Infants' metaphysics: the case of numerical identity. Cognit Psychol 30:111–153

DISCUSSION

Shepard: Please can you clarify how you effected the substitution of the vegetables.

Hauser: In the property/kind condition, the subject sees a piece of squash coming out and then going back behind the screen. Next, they see a carrot coming out and then going back behind the opposite side of the screen. They are therefore shown only the carrot or the squash at any one time, and thus are never given evidence that two objects exist at the same time. In the spatiotemporal condition, they see the same sequence as in property/kind, but then also see the squash and carrot placed on opposite sides of the screen at the same time, and then see them placed simultaneously behind the screen. This final sequence provides evidence that at least two objects exist behind the screen.

Gangestad: I also have a question of clarification. As I understand, in the human studies you did with Susan Carey, the infants who failed the tasks did so because they required language to label and solve the tasks. Monkeys don't use language to solve the tasks successfully, so is it your view that humans have lost an adaptation that their ancestors had?

Hauser: It is debatable, even without the rhesus monkey data, whether language is actually necessary for what the child is doing. Xu & Carey (1996) interviewed parents about whether their children had comprehension and production skills for those particular words, and they found that those of them that did would pass this test. Therefore, it's merely a correlation. The rhesus monkey results suggest that the problem can be solved without language.

Møller: I have a question about phylogeny. It's no coincidence that you're studying the New World monkeys and humans, but is this really a homologous phenomenon?

Hauser: This is a difficult problem to address given our lack of understanding of basic neurocognitive mechanisms across species. We would like, however, to run comparable experiments with chimpanzees, especially given the recent fossil evidence that the most ancient hominid had both human- and chimp-like features. I would like to claim, based on that evidence, that if the numerical processing of chimpanzees is like that of humans, then it is most likely a homology. It's much more difficult in the case of the rhesus monkeys because we could be observing evolutionary convergence.

Møller: Would you make a stronger case if you had any back-up from studies on the brains of these species?

Hauser: If, for example, prefrontal cortex activity showed similar kinds of processing in rhesus monkeys and humans, I'm not sure that this would provide more compelling evidence. If I've got comparable methods and comparable behavioural results what more can studying the neurobiology tell me about the behaviour? It seems to be a reductionist view and I'm not sure I want to buy into it.

Daly: If the results showed that we're using different structures to do the same job, then this would suggest some degree of convergence.

Hauser: Yes. A similar example is in the area of face recognition. People have done a lot of work at the single-unit level in the superior temporal sulcus of the rhesus monkey brain. For example, Perrett & Mistlin (1990) have shown that certain cells are responsive to faces but not non-face visual stimuli. All the new imaging data on human face recognition, however, suggest that different areas of the brain are involved in face processing.

Rogers: I was interested that this ability in humans appears at the same time that verbal representations appear. Steve Pinker's book *The language instinct* convinced me that people don't think with verbal representations (Pinker 1994). If that's true then mentalese must be developing at the same time as verbalese.

Pinker: I would take the evidence to suggest that language isn't involved. In fact I would turn the story on its head by asking, how do children learn a word to begin with if they don't have the concept of a particular kind of object? As with any correlation, one can flip it around and sometimes get a more plausible explanation. Your results show that adult monkeys, who don't have words, pass the test, so language is obviously not necessary. It's more plausible that the object concepts are driving the word learning, rather than the other way around. That is, as soon as children have the concept of an object, they have something that they can attach the verbal label to.

Hauser: Agreed. But our results also don't rule out the possibility that language facilitates the development of these concepts.

Shepard: It seems plausible to imagine that the monkey has a mental image of the objects coming out and going back in, and that they can preserve this representation even when the objects are hidden behind the screen. Therefore, what would Susan Carey's response be to this, and how does this possibility that mental images are being compared bear on Karen Wynn's claim that addition and subtraction are being carried out as arithmetic operations (Wynn 1996)?

Hauser: As far as I know, Susan Carey has only recently shown an interest in animal data. The current model that we would like to propose, which accounts for all the results so far, is a slight modification of Anne Treisman's object file model, i.e. that what the child and the monkey are doing when they see an object is opening up an object file with certain types of properties and are binding it to that object. The problem is that at some point the object file model fails because it is not possible to open up too many object files. All the experiments we've done so far are with small numbers of objects, and therefore the object file model can account for all of these. The competing models are numerical symbol models; for example, the numeron model of Gelman & Gallistel (1991, 1992), which is not linguistic but involves non-numerical symbols in the brain. Another example is the accumulator model of Meck & Church (1983), which is an analogue representation such that if the animal or infant is in a counting mode it opens up a certain amount of energy for the first object, some more energy for the second etc. resulting in the accumulation of information.

Gigerenzer: David Hume (1975) once wrote that the human mind is so fine-tuned to numerical frequencies that it could discriminate an event that occurred 10 000 times from one that occurred 10 001 times, which may be a little too optimistic. How fine-tuned is the ability of animals to discriminate between six and seven, for example? And how do they achieve this without any verbal labels?

Hauser: Meck & Church (1983) have observed that rats can be trained to press a lever precisely 49 times, such that presses of 48 or 50 fail to result in reinforcement. Moreover, the response seems to be some sort of counting since factors, such as time and energy, required to depress the lever were controlled for.

Pinker: Is there a normal distribution of responses centred around 49?

Hauser: Yes, but the distribution is narrow. Pigeons can also discriminate between 24 and 25. The interesting thing about discrimination in cases such as the chimpanzees, where they were taught Arabic numerals and the differences between the numbers 1, 2 and 3 etc., is that at each stage, for example learning the number 3, they seem to learn that this tag represents 3 or more. Similarly, the Arabic numeral 4 represents 4 or more, and so on. Unlike human infants, therefore, who eventually figure out that the Arabic numeral is equal to precisely one value and no more or less, the chimpanzees learn an approximation.

Gigerenzer: Does this means that rats can discriminate between 48 and 49 but not between 49 and 50?

Hauser: They are better at recognizing lower numbers, and 49 seems to be their maximum. They're good at discriminating between, for example, 24 and 25. There may also be differences across sensory domains. There have also been some experiments with human infants (a few months old) in which subjects are repeatedly presented with two objects, for example two pears or two carrots, and then presented with three objects. Having habituated to two, they dishabituate to three objects. But when this task exceeds four objects, the effect disappears. We've now done the same experiments with cotton-top tamarins and they show comparable patterns. The unfortunate assumption which some people are making is that all these numerical tasks involve low level perceptual phenomena such as subitizing, i.e. an effortless, preattentive processing system that provides rudimentary numerical assessments. Pepperberg (1994) has performed an elegant experiment with Alex the talking parrot in which Alex can count how many green trucks there are amongst a variety of green objects and different coloured trucks. This most likely rules out subitization because Alex has to do a conjunction of two features.

Maynard Smith: This may be totally irrelevant but I would like to bring up something that I was obsessed by some 20 years ago, which I call the counting problem. It has nothing to do with human counting, rather it addresses whether an animal in development can regularly produce 14 or 49 segments, or five fingers etc. There is a sharp distinction between cases where they obtain the correct number almost every time and cases where the number is centred around a particular number and has a distribution either side. The data suggest that animals can produce the correct number up to about six, seven or eight, but they can only produce the correct number sometimes if the number is larger than that, that is unless they cheat. *Drosophila* embryos produce 14 segments every time, but they achieve this by producing seven and then doubling. I had a rather nice theory as to why roughly eight was the limit, based on the accuracy with which organisms can in practice fix continuous variables — such as times, sizes or lengths. The idea was that they produced a number by taking the nearest integer to the ratio between two continuous variables. However, this wouldn't go anywhere near 49. I remember at the time being rather excited by the fact that the number I was coming up with from developmental biology was the same as the number that animals could learn to count. I was therefore astonished by your statement that rats can recognize the number 49. The only conceivable relevance in this is that the brain, when counting up to 49, must be using a different mechanism to the mechanisms that are being used in the development of segments.

Pinker: An interesting comparison is that in human short-term memory the magic number is seven plus or minus two. If you make people remember longer lists they also cheat. For example, they can't remember a list of 49 items, but they try to remember them as seven groups of seven.

Maynard Smith: I would like to explain why seven is an important number. If you look at the coefficient of variation of either linear measurements (such as the length of your limb or how tall you are) or of the times which animals take to do things, and if counting is performed by taking the nearest integer to the ratio of two numbers, the largest number you can generate accurately is somewhere around six, seven or eight. The coefficient of variation is extremely constant.

Thornhill: Marc Hauser, do you have any evidence that individuation in monkeys is a psychological adaptation of some sort?

Hauser: Yes, I would say that the ability to individuate is a crucial adaptation and that the fact that they deploy it in a numerical system is a by-product. Object individuation and object recognition represent crucial primitives, which number then has to link up with because number implies the ability to individuate objects. We are doing experiments with wild chimpanzees to see whether the ability to numerate by ear the number of individuals calling has any consequences for how they negotiate inter-community aggression.

Thornhill: But wouldn't they obtain as much to eat without individuating?

Hauser: Absolutely. It is possible that when a chimpanzee sees fruit on a tree and assesses whether there is enough for itself that it may not individuate at all. It probably does not count the fruit, but rather it makes some low level assessment of the amount of food available, possibly using colour as a guideline.

Thornhill: Then what is the ability to individuate for? Because we need to know its function before we can decide whether it is an adaptation.

Cosmides: Any process that requires the estimation of probabilities or rates requires counting, and counting requires the ability to individuate. Classical conditioning, for example, requires the ability to count events per unit time — i.e. to compute rates. The ability to compute rates or to make judgements under uncertainty (by estimating relative frequencies) is phylogenetically widespread and fundamental to many behavioural processes (Gallistel 1990); this would require adaptations that can both individuate and count objects and events.

Dawkins: Whenever ornithologists want to watch a bird, two of them go into a hide and only one comes out. This fools birds who, therefore, presumably can't count.

Hauser: If that's correct then the monkeys would not be fooled. Given the sociality of primates, I would guess that the pressure comes from social relationships — they clearly recognize individuals by face and by voice — so that they make a distinction which can be deployed in other domains.

References

Gallistel CR 1990 The organization of learning. MIT Press, Cambridge, MA

Gallistel CR, Gelman R 1991 Subitizing: the preverbal counting process. In: Kessen WE, Ortony A, Craik FIM (eds) Thoughts, memories and emotions: essays in honour of George Mandler. Lawrence Erlbaum Associates Inc., Hillsdale, NJ, p 65–91

Gallistel CR, Gellman R 1992 Preverbal and verbal counting and computation. Cognition 44:43–74

Hume D 1975 A treatise of human nature. Clarendon Press, Oxford

Meck WH, Church RM 1983 A mode control model of counting and timing processes. J Exp Psychol Anim Behav Processes 9:320–334

Pinker S 1994 The language instinct. Harper Collins, New York

Pepperberg IM 1994 Numerical competence in an African gray parrot (*Psittacus erithacus*). J Comp Psychol 108:36–44

Perrett DI, Mistlin AJ 1990 Perception of facial characteristics by monkeys. In: Stebbins WC, Berkley MA (eds) Comparative perception: complex signals. John Wiley & Sons, New York, p 187–216

Wynn K 1996 Infants' individuation and enumeration of actions. Psych Sci 7:164–169

Xu F, Carey S 1996 Infants' metaphysics: the case of numerical identity. Cognit Psychol 30:111–153

Dissecting the computational architecture of social inference mechanisms

Leda Cosmides and John Tooby

Center for Evolutionary Psychology, Department of Psychology, University of California, Santa Barbara, CA 93106, USA

Abstract. Scientists have been dissecting the neural architecture of the human mind for several centuries. Dissecting its computational architecture has proven more difficult, however. Within the cognitive sciences, for example, there is a debate about the extent to which human reasoning is generated by computational machinery that is domain specific and functionally specialized. While some claim that the same set of cognitive processes accounts for reasoning across all domains (e.g. Rips 1994, Johnson-Laird & Byrne 1991), others argue that reasoning is generated by several different mechanisms, each designed to operate over a different class of content (e.g. Baron-Cohen 1995, Cheng & Holyoak 1985, Cosmides & Tooby 1992, Leslie 1987). Indeed, it has recently been proposed that the human cognitive architecture contains a *faculty of social cognition*: a suite of integrated mechanisms, each of which is specialized for reasoning and making decisions about a different aspect of the social world. Candidate devices include a theory of mind mechanism, an eye direction detector, social contract algorithms, permission schemas, obligation schemas, precaution rules, threat detection procedures and others (e.g. Baron-Cohen 1995, Cheng & Holyoak 1985, Cosmides 1985, 1989, Cosmides & Tooby 1989, 1992, 1994, Fiddick et al 1995, Fiske 1991, Jackendoff 1992, Leslie 1987, K. Manktelow & D. Over, unpublished paper, 1st Int Conf on Thinking, Plymouth, UK 1988, Manktelow & Over 1990, M. Rutherford, J. Tooby, L. Cosmides, unpublished paper, 8th Annual Meeting Human Behav Evol Society, Northwestern Univ, IL 1996, J. Tooby & L. Cosmides, unpublished paper, 2nd Annual Meeting Human Behav Evol Society, Evanston, IL 1989). To decide among these sometimes competing proposals, psychologists need empirical methods and theoretical standards that let us carve social inference mechanisms at the joints. We will argue that the theoretical standards needed are those of the 'adaptationist programme' developed in evolutionary biology. To show how these standards can be applied in dissecting the computational architecture of the human mind, we will discuss some recent empirical methods and results.

1997 Characterizing human psychological adaptations. Wiley, Chichester (Ciba Foundation Symposium 208) p 132–161

The adaptationist stance

Focus on architecture

At a certain level of abstraction, every species has a universal, species-typical evolved architecture. For example, one can open any page of the medical textbook, *Gray's anatomy,* and find the design of this evolved architecture described down to the minutest detail — not only do we all have a heart, two lungs, a stomach, intestines and so on, but the book will also describe human anatomy down to the particulars of nerve connections. This is not to say there is no biochemical individuality: no two stomachs are exactly alike — they vary a bit in quantitative properties, such as size, shape and how much HCl they produce. But all humans have stomachs and they all have the same basic *functional* design — each is attached at one end to an oesophagus and at the other to the small intestine, each secretes the same chemicals necessary for digestion and so on. Presumably, the same is true of the brain and, hence, of the evolved architecture of our cognitive programmes — of the information-processing mechanisms that generate behaviour.

The cognitive architecture, like all aspects of the phenotype from molars to memory circuits, is the joint product of genes and environment. But the development of architecture is buffered against both genetic and environmental insults, such that it reliably develops across the (ancestrally) normal range of human environments. Characterizing the universal, species-typical architecture of human cognitive mechanisms is the central goal of psychology.

As psychologists, we are studying a system of fantastic complexity. Isolating a cog within an intricate machine of manifold interacting parts would be tremendously difficult, even if we had a large number of duplicates to experiment with. But we don't, because every time this fantastically complex system reproduces itself, sexual recombination injects variation into its design. Extracting what is invariant under these circumstances is a daunting task. It is not impossible, however. The structure of an evolved system reflects its function. Knowing a system's function can, therefore, illuminate its design. The 'adaptationist programme' is a research strategy in which theories of adaptive function are key inferential tools, used to identify and investigate the design of evolved systems.

Why does structure reflect function?

The evolutionary process has two components: chance and natural selection. Natural selection is the only component of the evolutionary process that can introduce complex *functional* organization into a species' phenotype (Dawkins 1986, Williams 1966).

The function of the brain is to generate behaviour that is sensitively contingent upon information from an organism's environment. It is, therefore, an information-processing device. Neuroscientists study the physical structure of such devices, and cognitive psychologists study the information-processing programmes realized by that structure. There is, however, another level of explanation — a functional level.

In evolved systems form follows function. The physical structure is there because it embodies a set of programmes; the programmes are there because they solved a particular problem in the past. This functional level of explanation is essential for understanding how natural selection designs organisms.

An organism's phenotypic structure can be thought of as a collection of 'design features' — micro-machines, such as the functional components of the eye or liver. Over evolutionary time, new design features are added or discarded from the species' design because of their consequences. A design feature will cause its own spread over generations if it has the consequence of solving adaptive problems: cross-generationally recurrent problems whose solution promotes reproduction, such as detecting predators or detoxifying poisons. If a more sensitive retina, which appeared in one or a few individuals by chance mutation, allows predators to be detected more quickly, individuals who have the more sensitive retina will produce offspring at a higher rate than those who lack it. By promoting the reproduction of its bearers, the more sensitive retina thereby *promotes its own spread over the generations* until it eventually replaces the earlier model retina and becomes a universal feature of that species' design.

Hence natural selection is a feedback process that 'chooses' among alternative designs on the basis of *how well they function*. It is a hill-climbing process, in which a design feature that solves an adaptive problem well can be outcompeted by a new design feature that solves it better. This process has produced exquisitely engineered biological machines — the vertebrate eye, photosynthetic pigments, efficient foraging algorithms, colour constancy systems — whose performance is unrivalled by any machine yet designed by humans.

By selecting designs on the basis of how well they solve adaptive problems, this process engineers a tight fit between the function of a device and its structure. To understand this causal relationship, biologists developed a theoretical vocabulary that distinguishes between structure and function.

Engineering standards

Those who study species from an adaptationist perspective adopt the stance of an engineer. In discussing sonar in bats, for example, Dawkins proceeds as follows: '...I shall begin by posing a problem that the living machine faces, then I shall consider possible solutions to the problem that a sensible engineer might consider; I shall finally come to the solution that nature has actually adopted' (Dawkins 1986, p 21–22).

Engineers figure out what problems they want to solve, and then design machines that are capable of solving these problems in an efficient manner. Evolutionary biologists figure out what adaptive problems a given species encountered during its evolutionary history, and then ask themselves, 'what would a machine capable of solving these problems well under ancestral conditions look like?' Against this background, they empirically explore the design features of the evolved machines that, taken together, comprise an organism. Definitions of adaptive problems do not, of course, uniquely specify the design of the mechanisms that solve them. Because there are often multiple ways of achieving any solution, empirical studies are needed to decide 'which nature has actually adopted'. But the more precisely one can define an adaptive information-processing problem — the 'goal' of processing — the more clearly one can see what a mechanism capable of producing that solution would have to look like. This research strategy has dominated the study of vision, for example, so that it is now commonplace to think of the visual system as a collection of functionally integrated computational devices, each specialized for solving a different problem in scene analysis — judging depth, detecting motion, analysing shape from shading and so on.

Design evidence

Because adaptations are problem-solving machines, they can be identified using the same standards of evidence that one would use to recognize a human-made machine: design evidence. One can identify a machine as a television rather than a stove by finding evidence of complex functional design: showing, for example, that it has many co-ordinated design features (antennas, cathode ray tubes, etc.) that transduce television waves and transform them into a colour bit map (a configuration that is unlikely to have arisen by chance alone), whereas it has virtually no design features that would make it good at cooking food. Complex functional design is the hallmark of adaptive machines as well. One can identify an aspect of the phenotype as an adaptation by showing that: (1) it has many design features that are complexly specialized for solving an adaptive problem; (2) these phenotypic properties are unlikely to have arisen by chance alone; and (3) they are not better explained as the by-product of mechanisms designed to solve some alternative adaptive problem. Finding that an architectural element solves an adaptive problem with 'reliability, efficiency and economy' is prima facie evidence that one has located an adaptation (Williams 1966).

Design evidence is important not only for explaining why a known mechanism exists, but also for discovering new mechanisms, ones that no one had thought to look for. Theories of adaptive function define what would count as a 'good design', and that allows one to generate testable hypotheses about the organization of a phenotypic structure. Thus, they can also be used heuristically, to guide investigations of phenotypic design.

Knowledge of adaptive function is necessary for carving nature at the joints

An organism's phenotype can be partitioned into adaptations, which are present because they were selected for; by-products, which are present because they are causally coupled to traits that were selected for (e.g. the whiteness of bone); and noise, which was injected by the stochastic components of evolution.

Lewontin (1979) defines adaptationism as 'that approach to evolutionary studies which assumes without further proof that all aspects of the morphology, physiology and behaviour of organisms are adaptive optimal solutions to problems'. By this definition, there are no adaptationists in the ranks of evolutionary biology. Not all aspects of an organism are functional, and every evolutionary biologist knows this (Williams 1966, Dawkins 1982).

Every machine, whether it was engineered by humans or by the evolutionary process, has non-functional aspects by virtue of being an ordinary causal system. The colour of the base of an overhead projector is unrelated to its function (projecting images on a screen) and so is the fact that the number of mirrors it has is a prime number. The colour is a by-product of a functional aspect (metals strong enough to support the projector happen to have a colour) and it has two mirrors because (given its function) the laws of reflectance require two — not because two is a prime number. The property, 'falls to earth when dropped', is a by-product of its having mass, and is most parsimoniously explained by appeal to the laws of gravitation.

Appeal to the laws of chemistry and physics are not sufficient, however, to explain why an overhead projector has mirrors, a light, a transparent surface and so on. These parts and properties are simultaneously present and arranged as they are *because* this configuration solves a problem. They are design features. To explain their presence and configuration, one needs to refer to the projector's function. Knowing its function is also necessary if one is to figure out which aspects of an overhead projector are *without* function. The same is true for organisms.

Like other machines, only narrowly defined aspects of organisms fit together into functional systems: most ways of describing the system will not capture its functional properties. Indeed, every organism has an infinite number of non-functional 'traits' because there are an infinite number of ways of carving a phenotype into 'parts' and 'properties' ('knee plus ear'; 'colour of mucus'; 'third epithelial layer of skin on the right arm plus salt receptors on tongue'; 'being less than 10 [or 20 or 2000...] feet tall'; 'can be burned by acid'). For this reason, the assertion that organisms have non-functional aspects is true, but trivial.

Theories are developed to explain phenomena. The phenomenon that Darwin was trying to explain is the presence of *functional* organization in the phenotypes of organisms — the kind of organization that one finds in artefacts that were designed by an intelligent engineer to solve a problem of some kind. Functional organization is the *explanandum*, the phenomenon that the theory was developed to explain. Figuring out how to 'dissect' the architecture of a species in a way that illuminates this organization and explains its presence is, therefore, the one task that no Darwinian can escape or evade. To arrive at the appropriate construal, one must conceptualize this architecture as composed of non-random parts that interact in such a way that they solve adaptive problems. And this, of course, requires theories of adaptive function. They are engineering specifications, which provide the criteria necessary to decide whether a property of an organism is a design feature, a functionless by-product, a kluge in the system or noise.

Reverse engineering an inference system

The adaptationist programme can be used to reverse engineer inference systems. Its application suggests that the computational architecture of the human mind might be considerably different than is usually assumed (Cosmides & Tooby 1987, 1992, 1994, Tooby & Cosmides 1992).

Psychologists have long known that the human mind contains circuits that are specialized for different modes of perception, such as vision and hearing. But until recently, it was thought that perception and, perhaps, language were the only activities caused by cognitive processes that are functionally specialized. Other cognitive functions — learning, reasoning, decision making — were thought to be accomplished by circuits designed to operate uniformly over every class of content. These circuits were thought to be few in number, content independent and general purpose, part of a hypothetical faculty that generates solutions to all problems: 'general intelligence' (e.g. Fodor 1983, Johnson-Laird & Byrne 1991, Piaget 1950, Rips 1994). Experiments were designed to reveal what computational procedures these circuits embodied; prime candidates were all-purpose heuristics and 'rational' algorithms — ones that implement formal methods for inductive and deductive reasoning, such as Bayes's rule or the propositional calculus. These algorithms are jacks of all trades: because they are content free, they can operate on information from any domain (their strength). They are also masters of none: to be content independent means that they lack any domain-specialized information that would lead to correct inferences in one domain but would not apply to others (their weakness).

This research programme has produced a formidable paradox. When given artificial, laboratory-administered reasoning problems, people perform in ways that seem inept, especially when compared to artificial intelligence systems. Such findings led many psychologists to conclude that the faculty of human reasoning is riddled with crippling defects: heuristics, biases and fallacious principles that violate canons of rationality derived from logic, mathematics and philosophy (e.g. Kahneman et al 1982). Yet natural reasoning systems — human and non-human minds alike — negotiate the complex natural tasks of their world with a level of operational success far surpassing that of the most sophisticated existing artificial intelligence systems. Although artificial systems are usually composed of programmes that embody exactly those 'rational' principles that human minds are thought to lack, none has yet been able to match the performance even of a normal four-year-old child on everyday inferential tasks: inducing a grammar, analysing scenes, detecting predators, inferring the meaning of a smile, the wishes of a potential friend or the intentions of a potentially hostile stranger.

The paradox evaporates when one considers two things: (1) the limitations of rational algorithms; and (2) the nature of the problems human inference mechanisms were designed to solve.

Natural competences

An adaptationist would expect information-processing mechanisms — including inference systems — to be *ecologically rational*: to embody principles that allow adaptive problems to be solved with reliability, economy and precision (Tooby & Cosmides 1997). As a result, one expects them to work well under conditions that resemble the ancestral ones that shaped their design. They are calibrated to these environments, and they embody information about the stably recurring properties of these ancestral worlds.

One can think of the human computational architecture as a collection of evolved problem solvers. Many of these are expert systems, equipped with 'crib sheets': inference procedures and assumptions that embody knowledge specific to a given problem domain. These generate correct (or, at least, adaptive) inferences that would not be warranted on the basis of perceptual data alone. For example, there is now at least some evidence for the existence of inference systems that are specialized for reasoning about objects, physical causality, number, the biological world, the beliefs and motivations of other individuals, and social interactions (e.g. Atran 1990, Baron-Cohen 1995, Brown 1990, Cheng & Holyoak 1985, Cosmides 1989, Cosmides & Tooby 1989, 1992, Fiske 1991, Frith 1989, Hatano & Inagaki 1994, Jackendoff 1992, Keil 1994, Leslie 1987, 1988, Leslie & Thaiss 1992, Spelke 1990, Springer 1992, Wynn 1992).

Different problems require different crib sheets. For example, an assumption that is useful for predicting the behaviour of people — that their movements are caused by internal states, such as intentions, beliefs and desires — would be misleading if applied to inanimate objects. Two inference machines are better than one when the crib sheet that helps solve problems in one domain is misleading in another. This suggests that many evolved computational mechanisms will be domain specific: they will be activated in some domains but not others. Some of these may embody rational methods, but others will have special-purpose inference procedures that respond not to logical form but to content types — procedures that work well within the stable ecological structure of a particular domain, even though they might lead to false or contradictory inferences if they were activated outside of that domain.

An algorithm that is free of content is ignorant of the world. As a result, machines limited to executing Bayes's rule, *modus ponens,* and other procedures derived from mathematics or logic cannot go beyond the data of the senses. Having no crib sheets, there is little they can deduce about a domain; having no privileged hypotheses, there is little they can induce before their operation is hijacked by combinatorial explosion. The difference between domain-specific methods and domain-independent ones is akin to the difference between experts and novices: experts can solve problems faster and more efficiently than novices because they already know a lot about the problem domain.

Identifying domain-specific mechanisms

A major criterion for establishing the existence of a domain-specific inference system is whether, at an information-processing level, it appears to constitute a functionally isolable computational unit. Is it activated independent of other units? Does it produce inferential steps unavailable to other units? Does it contain systems of procedures that are complexly specialized for processing information about a particular domain? Sometimes the neurological basis of a specialization can be identified and dissociated from other competences (see, for example, Leslie & Thaiss 1992, Baron-Cohen 1995, on dissociations between the theory of mind mechanism and other specializations), adding to the credibility of the cognitive-level characterization. But the primary criterion for distinguishing specializations is functional or cognitive, not neurological.

In reverse engineering a computational system composed of domain-specific inference engines, there are a number of questions that one must address.

(1) *Existence.* Does a hypothesized reasoning specialization actually exist, or are reasoning patterns better explained as the product of a domain-general mechanism in interaction with individual experiences and knowledge databases?

(2) *Scope.* What is the correct definition of the boundaries of the domain that the hypothesized reasoning unit operates over?

(3) *Proper cognitive description.* What is the correct specification of the specialization's procedures and representational formats?

(4) *Adaptive function.* Do these procedures and representational formats show the fit between form and function that one expects of a cognitive adaptation designed to solve the adaptive problem under consideration?

(5) *Universality.* Does it develop in all normal humans, regardless of culture, or is it sensitive to cultural variation and dependent on the details of individual experience?

(6) *Ontogenetic timing.* When does it develop — does it have a regular ontogenetic schedule?

(7) *Activation.* What conditions regulate its activation and deployment (e.g. can it be turned off and on by the presence or absence of a particular type of social situation)?

(8) *Regulation and function.* What activities are regulated or supported by the specialization? What other functions or behaviours are dependent on its output?

(9) *Inter-relationships.* What role does it play in a larger network of computational units? Do other specializations share a common database, or use its outputs as inputs or otherwise depend on its operation?

(10) *Neural basis.* Is the operation of the specialization (or its impairment) associated with particular regions of the brain? Is the specialization differentially activated or impaired by various hormones, drugs, or physiological and emotional states?

(11) *Role in real-world events.* What role does reasoning about a specific domain play in social interactions or other events? Does it explain aspects of real world phenomena (e.g. food sharing, gang-related violence)?

(12) *Health implications.* Does malfunctioning of the specialization (e.g. overactivation, underactivation, inappropriate activation) play a role in identifiable clinical disorders (e.g. autism, paranoia)?

As a result of research addressing these questions, there is growing evidence that the human cognitive architecture contains expert systems specialized for reasoning about the social world (e.g. Baron-Cohen 1995, Bugental & Goodnow 1997, Cheng & Holyoak 1985, Cosmides 1985, 1989, Cosmides & Tooby 1989, 1992, 1994, Etcoff et al 1991, Ekman 1992, Fernald 1992, Fiddick et al 1995, Fiske 1991, Jackendoff 1992, Leslie 1987, Mann 1992, M. Rutherford, J. Tooby, L. Cosmides, unpublished paper, 8th Annual Meeting Human Behav Evol Society, Northwestern Univ, IL 1996, J. Tooby & L. Cosmides, unpublished paper, 2nd Annual Meeting Human Behav Evol Society, Northwestern Univ, IL 1989). An adaptationist would expect their inference procedures, representational primitives and default assumptions to reflect the structure of adaptive problems that arose when our hominid ancestors interacted with one another. This expectation has guided our own research on human inference. We have proposed that reasoning about social exchange, precautions and threats is generated by three, functionally distinct, mechanisms; that each has a computational design that is specialized for solving the adaptive problems that typified its respective domain; and that each of these mechanisms is a component of the evolved architecture of the human mind — a reliably developing, species-typical set of cognitive procedures (e.g. Cosmides 1989, Cosmides & Tooby 1989, 1992, J. Tooby & L. Cosmides, unpublished paper, 2nd Annual Meeting Human Behav Evol Society, Evanston, IL 1989). We will use the literature on conditional reasoning to illustrate the role that the adaptationist programme can play in evaluating competing claims about the architecture of domain-specific reasoning mechanisms, and focus on social exchange as an example.

Characterizing the computational architecture that generates social inferences

Social computation and conditional reasoning

In categorizing social interactions, there are two basic consequences humans can have on each other: helping or hurting, bestowing benefits or inflicting costs. Some social behaviour is unconditional: for example, a mother nurses her infant without exacting a favour in return. However, most social acts are conditionally delivered. Indeed, much of the substance of human life is shaped by 'conditionals': statements or behaviours that express an intention to make one's behaviour contingent upon that of another. People conditionally help each other in reciprocal, dyadic, co-operative interactions; they conditionally threaten each other; and they form coalitions, defined by mutually understood contingencies of within-group co-operation and between-group competition. The inferential processes and decision rules that operate on conditionals make these activities possible and regulate their outcomes. This creates a selection pressure for cognitive designs that can detect and understand social conditionals reliably, precisely and economically (Cosmides 1985, 1989, Cosmides & Tooby 1989, 1992).

One important category of social conditional is social exchange — conditional helping — carried out by individuals or groups on individuals or groups. A social exchange involves a conditional of the approximate form: if person A provides the requested benefit to or meets the requirement of person or group B, then B will provide the rationed benefit to A (herein, a rule expressing this kind of agreement to co-operate will be referred to as a *social contract*).

Applying the adaptationist programme

Step 1: characterizing an adaptive problem. Our first step was to analyse the nature of such conditional interactions, including the structure of inferences that both: (1) make them possible; and (2) are necessary to guide an individual through these situations to successful outcomes or pay-offs. Using such analyses and the existing literature in economics, game theory and evolutionary biology, Cosmides (1985) and Cosmides & Tooby (1989) developed a task analysis or computational theory (in David Marr's sense) of the information-processing problems that arise in situations of social exchange. For example, economists and evolutionary biologists had already explored constraints on the emergence or evolution of social exchange using game theory, modelling it as a repeated Prisoners' Dilemma. One important conclusion was that social exchange cannot evolve in a species or be stably sustained in a social group unless the cognitive machinery of the participants allows a potential co-operator to detect cheaters (i.e. individuals who accept a benefit without satisfying the requirements that provision of that benefit was made contingent upon), so that they can be excluded from future interactions in which they would exploit co-operators (e.g. Axelrod 1984, Axelrod & Hamilton 1981, Boyd 1988, Trivers 1971, Williams 1966). Such analyses provided a principled basis for generating detailed hypotheses (called *social contract theory*) about reasoning procedures that would be capable, because of their domain-specialized nature, of detecting the presence of these social conditionals, interpreting their meaning and successfully solving the inference problems they pose. These hypotheses can be tested using standard methods from cognitive psychology.

Step 2: searching for design evidence. Using the foregoing as a starting point, it was possible to engage the rich literature that already existed on how people reason about conditional rules — a literature that, in the early 1980s, lacked a single theory or set of theories that persuasively accounted for the known body of experimental findings. One of the principal tools reasoning researchers have used to explore conditional reasoning is the Wason selection task, a paper-and-pencil test in which subjects are asked to identify possible violations of a conditional rule of the form 'if P then Q'. Complex patterns in reasoning performance are elicited by differences in the content and context of conditional rules in the Wason selection task (e.g. Wason 1983, Wason & Johnson-Laird 1972). Such content effects are what one would predict if some reasoning was generated by domain-specific reasoning specializations, such that different procedures are activated by different contents.

Content-dependent performance on the Wason selection task.   The Wason selection task was originally designed to see whether people are intuitive Popperians: whether they spontaneously attempt to falsify conditional rules by applying content-independent rules of logic. It is a word problem in which subjects are asked what additional information they would need to see to determine whether a conditional rule of the form 'if P then Q' has been violated by any one of four instances. Each instance is represented by a card. One side of a card tells whether the antecedent is true or false (i.e. whether P or not-P is the case), and the other side of that card tells whether the consequent is true or false (i.e. whether Q or not-Q is the case). The subject, who is only allowed to see one side of each card, is asked which card(s) must be turned over to see if any of them violate the rule. The four cards the subject must choose from show terms representing the logical categories P, not-P, Q and not-Q (Fig. 1). The rules of logical inference are content free so, no matter what P and Q stand for, the logically correct response is to choose the P card (to see if it has a not-Q on the other side) and the not-Q card (to see if it has a P on the other side).

There is a large body of literature showing that people are not good at detecting potential violations of conditional rules, even when these rules deal with *familiar content drawn from everyday life*. For example, descriptive rules — conditionals describing some state of the world — typically elicit a fully correct response (P and not-Q) from only 5–25% of subjects tested (Cosmides 1985, Wason 1983).

The Wason selection task is a convenient tool for testing hypotheses about reasoning specializations designed to operate on social conditionals because: (1) it tests reasoning about conditional rules; (2) the task structure remains constant while the content of the rule is changed; (3) content effects are easily elicited; and (4) there is already a body of existing experimental results against which performance on new content domains can be compared. For example, to show that people who ordinarily cannot detect violations of conditional rules can do so when that violation represents cheating on a social contract would constitute initial support for the view that people

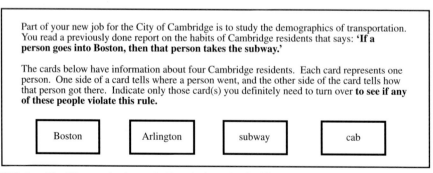

Part of your new job for the City of Cambridge is to study the demographics of transportation. You read a previously done report on the habits of Cambridge residents that says: **'If a person goes into Boston, then that person takes the subway.'**

The cards below have information about four Cambridge residents. Each card represents one person. One side of a card tells where a person went, and the other side of the card tells how that person got there. Indicate only those card(s) you definitely need to turn over **to see if any of these people violate this rule.**

| Boston | Arlington | subway | cab |

FIG. 1.   The Wason selection task (descriptive rule, familiar content).

have reasoning procedures specialized for detecting cheaters in situations of social exchange. To find that violations of conditional rules are spontaneously detected when they represent bluffing on a threat — a computationally different problem — would, for similar reasons, support the view that people have reasoning procedures specialized for analysing threats. Our general research plan has been to use subjects' inability to spontaneously detect violations of conditionals expressing a wide variety of contents as a comparative baseline against which to detect the presence of performance-boosting reasoning specializations. By seeing what content manipulations switch on or off high performance, the boundaries of the domains within which reasoning specializations successfully operate can be mapped. For example, there are now a number of experiments comparing performance on Wason selection tasks in which the conditional rule either did or did not express a social contract: a situation in which one is entitled to a benefit from a party only if one has satisfied the requirement that the offer of this benefit was made contingent upon (e.g. 'If a man eats cassava root [described as an aphrodisiac], then he must have a tattoo on his face'). Although very few subjects correctly identify potential violations of descriptive conditionals, 65–80% of subjects do so when the conditional rule expresses a social contract and a violation represents cheating. Subjects routinely check for cheating by choosing the cards that represent a person who has accepted the benefit ('ate cassava root') and a person who has not satisfied the requirement ('has no tattoo'). Furthermore, it is not just a question of how much 'facilitation' a conditional rule elicits: different hypothesized reasoning specializations predict different choices on the Wason selection task. The *pattern* of choices subjects make, given the content of the problem, can be used to test alternative hypotheses about the nature of the reasoning procedures activated. For example, the inference mechanisms that generate responses to social contracts *do not apply content-free logical rules*: they cause subjects to choose the benefit accepted card and the requirement not satisfied card regardless of their logical category (Fig. 2).

In fact, experiments that systematically manipulate problem content demonstrate a series of domain-specific effects predicted by our computational theory of social exchange, thereby providing a substantial body of design evidence. They show that the mechanisms activated by social contract content have many components that appear to be functionally specialized for reasoning about social exchange, including procedures that are well designed for detecting cheaters. For instance: (1) these procedures operate so as to detect cheaters, even when the social contract expressed is highly unfamiliar; (2) they do not operate unless the representation of the rule satisfies the cost–benefit constraints of a social contract; (3) the only violations they detect are ones that represent illicitly taken benefits (cheating) — they are not good at detecting innocent mistakes; (4) they are sensitive to whose perspective is being taken in an exchange; (5) they do not embody a content-independent formal logic — they identify cheaters even when this leads to answers that violate the strictures of, for example, the propositional calculus; (6) they cause people to 'read in' deontic operators such as 'may' and 'must', corresponding to obligation and entitlement,

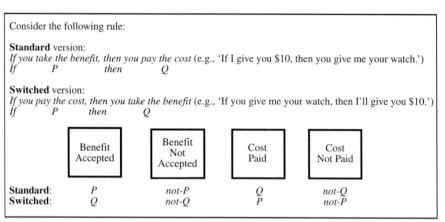

FIG. 2.   Generic structure of a social contract.

even when these are not explicitly stated in the rule; (7) they can be primed separately from other kinds of deontic rules; and (8) the ability to correctly identify potential cheaters can be intact in individuals suffering from schizophrenia, even though schizophrenia causes impairments in logical and other deliberative reasoning tasks — a result that indicates that social exchange procedures are neurologically dissociable from mechanisms that govern 'general intelligence'. (For review, see Cosmides & Tooby 1992, 1989, Cosmides 1985, 1989, Fiddick et al 1995, Gigerenzer & Hug 1992, Maljković 1987, Platt & Griggs 1993). Wason selection tasks involving social exchange elicit a pattern of results so distinctive that we have proposed that reasoning in this domain is governed by computational units that are domain specific and functionally distinct: *social contract algorithms* (Cosmides 1985, 1989, Cosmides & Tooby 1992).

Step 3: eliminating by-product hypotheses.   The human cognitive phenotype has many features that appear to be complexly specialized for solving the adaptive problems that arise in social exchange. However, demonstrating this is not sufficient for claiming that these features are cognitive adaptations *for* social exchange. One also needs to show that these features are not more parsimoniously explained as the by-product of mechanisms designed to solve some other adaptive problem or class of problems.

For example, Cheng & Holyoak (1985, 1989) also invoke content-dependent computational mechanisms to explain reasoning performance that varies across domains. But they attribute performance on social contract rules to the operation of a permission schema (and/or an obligation schema; these do not lead to different predictions on the kinds of rules usually tested; see Cosmides 1989), which operates over a larger class of problems. They propose that this schema consists of four production rules:

(1)   if the action is to be taken, then the precondition must be satisfied;
(2)   if the action is not to be taken, then the precondition need not be satisfied;
(3)   if the precondition is satisfied, then the action may be taken; and
(4)   if the precondition is not satisfied, then the action must not be taken;

and that their scope is any permission rule, that is, any conditional rule to which the subject assigns the following abstract representation: 'if action A is to be taken, then precondition P must be satisfied'. All social contracts are permission rules, but not all permission rules are social contracts. The conceptual primitives of a permission schema have a larger scope than those of social contract algorithms. For example, 'a benefit taken' is a kind of 'action taken', and a 'cost paid' (i.e. a benefit offered in exchange) is a kind of 'precondition satisfied'. They take evidence that people are good at detecting violations of precaution rules — rules of the form, 'if hazardous action H is taken, then precaution P must be met' — as evidence for their hypothesis (on precautions, see K. Manktelow & D. Over, unpublished paper, 1st Int Conf on Thinking, Plymouth, UK 1988, Manktelow & Over 1990). After all, a precaution rule is a kind of permission rule, but it is not a kind of social contract. We, however, have hypothesized that reasoning about precaution rules is governed by a functionally specialized inference system that differs from social contract algorithms and operates independently of them (Cosmides & Tooby 1992, Fiddick et al 1995).

In other words, there are two competing proposals for how the computational architecture that causes reasoning in these domains should be dissected. Application of the adaptationist programme suggests at least five kinds of evidence that can be used to decide between them.

(1) Over which transformations is the behaviour of the system invariant?

Transformations of input variables.   Native speakers of English recognize that: (a) 'furry brown bears sleep soundly'; and (b) 'colourless green ideas sleep furiously' are both grammatical sentences with identical syntactic structures. Transformations that substitute alternative words from the same part of speech are irrelevant to this judgement. Transformations of phrase structure are not: although the words of 'furry soundly bears brown sleep' are identical to those in (a), it is not a grammatical sentence. With examples like these, Chomsky (1957) showed that English syntax uses arguments such as *noun* and *verb*, which must bear a certain relationship to one another. Moreover, the grammatical system must be independent of meaning systems, because its operation is invariant over transformations that change word meaning, but it must preserve syntactic structure.

Transformations of input should be irrelevant to the operation of *any* syntactic system, as long as they fall within the range of input variables that its arguments accept: i.e. as long as they do not violate its *argument structure*. By systematically varying input, one should be able to discover the rules of a syntactic system and the arguments they take. This principle should allow one to discover which proposed syntax — that of the permission schema or the social contract algorithms — better describes the behaviour of the inference systems activated by the problems discussed above.

For example, according to the grammar of social exchange, a rule is not a social contract unless it contains a *benefit to be taken*. Transformations of input should not matter, as long as the subject continues to represent an action or state of affairs as beneficial to the potential violator, and the violator as illicitly obtaining this benefit. This is true: performance on social contract rules is just as good when the benefit to be taken is highly unfamiliar (e.g. eating cassava root, getting an ostrich eggshell) as when it is familiar (e.g. drinking beer, being assigned to a good high school).

The corresponding argument of the permission schema — *an action to be taken* — has a larger scope: not all 'actions taken' are 'benefits taken'. If this construal of the rule's argument structure is correct, then the behaviour of the reasoning system should be invariant over transformations of input that preserve it. But it is not. For example, consider two rules: (a) 'if one goes out at night, then one must tie a small piece of red volcanic rock around one's ankle'; and (b) 'if one takes out the garbage at night, then one must tie a small piece of red volcanic rock around one's ankle'. Most undergraduate subjects perceive the action to be taken in (a) — going out at night — as a benefit, and 80% of them answered correctly. But when one substitutes a different action — taking out the garbage — into the same place in the argument structure, then performance drops to 44%. This transformation of input preserves the *action to be taken* argument structure, but it does not preserve the *benefit to be taken* argument structure — most people think of taking out the garbage as a chore, not a benefit. If the syntax of the permission schema were correct, then performance should be invariant over this transformation. But a drop in performance is expected if the syntax of the social contract algorithms is correct.

We have been doing similar experiments with precaution rules (e.g. 'if you make poison darts, then you must wear rubber gloves'). All precaution rules are permission rules (but not all permission rules are precaution rules). We have been finding that the degree of hazard does not affect performance, but the nature of the precaution does — even though all of the *precautions taken* are instances of *preconditions satisfied*. Performance drops when the precaution is not perceived as a good safeguard given the hazard specified (M. Rutherford, J. Tooby & L. Cosmides, unpublished paper, 8th Annual Meeting Human Behav Evol Society, Northwestern Univ, IL 1996, J. Tooby & L. Cosmides, unpublished paper, 2nd Annual Meeting Human Behav Evol Society, Evanston, IL 1989). This is what one would expect if the syntax of the rules governing reasoning in this domain take arguments such as *facing a hazard* and *precaution taken*; it is not what one would expect if the arguments were *action taken* and *precondition satisfied*.

Transformations of context. A syntactic system has certain operators and conceptual primitives. For example, in both the permission schema and social contract algorithms *must* is a deontic operator indicating obligation (not a modal indicating necessity). But social contract algorithms contain certain conceptual primitives that the permission schema lacks. For example, *cheating* is taking a benefit that one is not entitled to; we have proposed that social contract algorithms have procedures that are specialized for detecting *cheaters*. This conceptual primitive plays no role in the operation of the permission schema. For this schema, whenever the action has been taken but the precondition has not been satisfied, a *violation* has occurred. People should be good at detecting violations, whether that violation counts as cheating (the benefit has been illicitly taken by the violator) or a mistake (the violator does not get the benefit stipulated in the rule).

Given the same social contract rule, one can manipulate contextual factors to change the nature of the violation from cheating to a mistake. When we did this, performance changed radically, from 68% correct in the cheating condition to 27% correct in the mistake condition. Gigerenzer & Hug (1992) found the same drop in response to a similar context manipulation. (Their interesting 'perspective change' experiments provide another example of how transformations of context can be used to decide between competing proposals.)

(2) Neurological dissociations

Cognitive neuroscientists have been using data about double dissociations to dissect the cognitive architecture. Strokes, head traumas, diseases and developmental disorders (such as autism) sometimes damage one mechanism without affecting another. For example, as a result of brain damage, some people lose the ability to recognize individual faces, but they can still recognize emotional expressions on faces. Others lose the ability to recognize emotional expressions, but they can still discriminate individual faces (Etcoff 1984). This is called a *double dissociation*. It is prima facie evidence that performance on these two tasks is caused by two different mechanisms, rather than one. If it had been caused by one mechanism, then damaging that mechanism should depress performance on both tasks: it should not be possible to find people in which performance on each is *selectively* impaired.

The same logic can be applied to the study of reasoning. If reasoning about social contracts and precautions is caused by one and the same mechanism — a permission schema — then neurological damage to this schema should lower performance on both rules equally. But if reasoning about these two domains is caused by two, functionally distinct mechanisms, then one could imagine neurological damage to the social contract algorithms that leaves the precaution mechanisms unimpaired and vice versa. Selective damage of social contract algorithms should depress performance on social contracts, but not on precaution rules. Selective damage of precaution mechanisms should depress performance on precaution rules, but not on social contracts.

Strokes, head traumas, diseases and developmental disorders often produce extensive brain damage. So even if a two-mechanism hypothesis is correct, there is no guarantee that one will find a patient in which the damage is localized enough to impair one mechanism but not the other. But if such patients *can* be found, they provide prima facie evidence that reasoning in these two domains is caused by two separate mechanisms. In collaboration with Valerie Stone, we have been giving Wason selection tasks to individuals with focal brain damage. Although this effort has just begun, it would appear that we have found a patient who performs at high levels on precaution rules, but not on social contract rules (V. Stone, L. Cosmides & J. Tooby, unpublished paper, 8th Annual Meeting Human Behav Evol Society, Northwestern Univ, IL 1996). We will continue to screen people, to see if people with the opposite dissociation — performing at high levels on social contracts, but not on precaution rules — can also be found.

Complex machines can be 'broken' in a number of different ways, and the logic of double dissociations can be applied to impairments caused by genetic variation as well. Natural selection tends to eliminate genetic variation, so the heritability of adaptations is usually low, not high (for discussion of the relationship between heritability and adaptationism see Tooby & Cosmides 1990). Nevertheless, sexual recombination injects 'noise' into phenotypic development. As a result, 'normal' genetic variation has many random effects — sometimes ones that impair the functioning of one mechanism while leaving another intact (e.g. most people who are genetically predisposed to myopia speak normally). Such effects can assist in the dissection of computational architecture. For example, a mutation involving a single dominant autosomal gene impairs the acquisition of certain grammatical rules, yet has no effect on spatial reasoning and other forms of non-verbal intelligence (Gopnik 1990). Individuals with a different genetic disorder, Williams syndrome, speak fluently and grammatically, yet are severely retarded (Pinker 1994). This double dissociation indicates that the mechanisms responsible for non-verbal 'intelligence' are not sufficient to explain the acquisition of grammar, and vice versa: more than one mechanism must be invoked to explain performance in these two domains.

Genetic variation could, in principle, selectively impair one reasoning mechanism while sparing another, making behaviour genetic data relevant. It is informative to know, for example, whether the concordance of a particular reasoning dissociation is higher in identical than fraternal twins. An advantage of this method is that it vastly expands the pool of potential subjects, to encompass many people who would not ordinarily be classified as having neurological damage.

(3) Functional dissociations: evidence from priming

The logic of double dissociations can be extended to encompass *functional* (rather than neurological) dissociations. If one can create experimental conditions that temporarily activate (or deactivate) mechanisms in a selective way, one can see whether levels of performance on two different 'target' tasks can be dissociated. One way of doing this is through *priming*.

Priming experiments have been used extensively in memory research, sometimes to decide whether performance on two different tasks was caused by a single memory system, or two different ones. The subject engages in an activity that temporarily activates ('primes') a mechanism or representation; one then sees whether this influences performance on a target task that immediately follows.

In a typical priming experiment, the subject is given two tasks in sequence. The first is called the 'prime', and the second is called the 'target'. To see whether the initial task influences performance on the target task, one compares performance on the target when it follows the prime to performance on the target when it follows a control task. When performing the initial task enhances performance on the target task (by some measure, such as reaction time or percent correct), this is called 'priming': the initial task primed performance on the target task. Using the Wason selection task, we have adapted this method to the study of reasoning, and found that we can: (1) selectively activate functionally distinct inference mechanisms; and (2) thereby elicit reasoning by analogy.

If domain-specific inference engines structure how cognitive architectures construe similarity across situations, then one social situation will be categorized as the same as another if both can be mapped onto the same set of domain-specific representations. For example, an ambiguous rule such as 'if a person wears a grey shirt, then that person is 19 years old' will be categorized as a social contract if the context makes it clear that the people under discussion: (i) think of wearing grey shirts as a rationed benefit (e.g. a military honour); and (ii) construe the statement about age as a requirement. Otherwise, the subject will categorize the rule as one which simply describes the habits of people over 19, and it will elicit the low levels of performance generally found for descriptive rules. If context causes an otherwise ambiguous rule to be categorized as a social contract, then cheater-detection procedures can operate on it, generating high levels of performance. Previous experiments on social contract reasoning have confirmed this prediction (see, for example, Cosmides 1989, Cox & Griggs 1982).

The same theoretical considerations provide a principled basis for predicting instances of transfer, or 'priming'. Although reasoning by analogy appears to be common in everyday life, it has been difficult to produce in the laboratory. The most common laboratory result is to find no transfer from a successfully solved problem to a target problem. But if one's model of a domain-specific inference mechanism is correct, then one should be able to reliably produce transfer to an ambiguous problem by first activating the appropriate mechanism with a clear instance of a problem drawn from that domain. Because this transfer is caused by the activation of a specific inference engine, this phenomenon can be called *inference priming*. Using the Wason selection task, we have been able to produce transfer to an ambiguous target problem in exactly this way (Fiddick et al 1995). Moreover, we have been able to prime social contract and precaution reasoning separately: i.e. we have been able to produce a double dissociation.

In these experiments, subjects were asked to solve two Wason selection tasks. The first—the prime—was either a clear social contract, a clear precaution rule or an ambiguous rule (as a control condition against which performance following the other two primes could be compared). In the second task — the target — the rule was always an ambiguous one that, when presented either alone or after another ambiguous rule, elicits low levels of performance. In some conditions, the target rule had the following two properties: (1) it could, in principle, be interpreted as a precaution rule; but (2) it would be difficult to interpret as a social contract. Let's call these 'ambiguous precaution rules'. An example would be 'if one empties the garbage cans, then one first eats red clay'. Emptying garbage cans and eating clay are both negative things — neither can be readily construed as a benefit. This should block a social contract interpretation. On the other hand, one could imagine situations in which emptying garbage could be hazardous (e.g. it could contain glass or disease-ridden materials), and sometimes people ingest substances to inoculate themselves from harm (e.g. penicillin). In other conditions, the target rule had opposite properties: (1) it could, in principle, be interpreted as a social contract; but (2) it would be difficult to interpret as a precaution rule. Let's call these 'ambiguous social contracts'. An example would be 'if one goes to the festival, then one lives in the village'. One can easily imagine situations in which going to the festival (or living in the village) could be rationed benefits, but neither of these activities sounds particularly hazardous.

We found the following. (1) A clear social contract strongly primed (i.e. elevated) performance on an ambiguous social contract target. Moreover, this was due to the activation of social contract categories, not logical ones: when the prime was a switched social contract, in which the correct cheater detection answer is not the logically correct answer (see Fig. 2), subjects' matched their answers on the target to the prime's benefit/requirement categories, not its logical categories. (2) A clear precaution rule strongly primed an ambiguous precaution target. Most importantly, (3) these effects were caused by the operation of two mechanisms, rather than one: a clear precaution rule produced little or no priming of an ambiguous social contract target; similarly, a clear social contract produced little or no priming of an ambiguous precaution target.

This should not happen if permission schema theory were correct. In that view, it shouldn't matter which rule is used as a prime, because the only way in which social contracts and precautions can have an effect on ambiguous rules is through activating the more general permission schema. Because both types of rules strongly activate this schema, an ambiguous target should be primed equally by either one.

(4) Cross-cultural evidence

If an inference mechanism is part of the human cognitive architecture, then it should reliably develop in individuals (of a particular age/sex morph) across the ancestrally normal range of human environments. (Facultative adaptations would, of course, be an exception to this generalization; Williams 1966.)

According to Cheng & Holyoak (1985), the permission schema is not a component of the evolved architecture of the human mind. It is induced by content-independent mechanisms of an unspecified kind, which are (presumably) evolved components of our computational architecture. The schema is induced when these mechanisms operate on information gleaned from the environment while the individual is attempting to achieve various goals. They do not specify what kind of information or goals are relevant; the implication is, however, that different cultural environments could lead to the induction of different schemas. The theory provides no a priori reason to expect the permission schema to be present in every human culture. Moreover, it suggests that the design of a schema should reflect the exigencies of life in the modern world, even if these bear no correspondence to the exigencies of life for ancestral hunter–gatherers.

In contrast, we have argued that reasoning about social contracts, precautions and threats is generated by three functionally distinct mechanisms; that each has a computational design that is specialized for solving the adaptive problems that typified their respective domains; and that each of these mechanisms is a component of the evolved architecture of the human mind — a reliably developing, species-typical set of cognitive procedures.

Statistical distribution of mechanisms.   If our proposal is correct, then one should find manifestations of the same inference mechanisms across cultures. Moreover, their design features should reflect the statistical distribution of adaptive problems they evolved to solve. In contrast, structures built by content-independent mechanisms should reflect the statistical distribution of modern problems faced by the population under investigation. This is because the world experienced by an individual organism is the only source of content for a mechanism that starts out content free. For example, content-free learning mechanisms cannot account for the distribution of phobias: people rarely (if ever) develop phobias to electric sockets, cars and other dangers of the modern world. But they readily develop phobias of snakes and spiders — dangers faced by ancestral hunter–gatherers — even though these pose no significant danger in their personal lives (Marks 1987).

Analogously, the design and statistical distribution of reasoning mechanisms can provide evidence for or against alternative hypotheses about their genesis and structure. Even though the nature of the process that sculpts the permission schema is underspecified, one wonders why this process has not built reasoning mechanisms that are good at detecting violations of causal rules, for example. We live in a technological society. When an appliance stops working, we hypothesis test: the toaster heats up only if it is plugged in; is it plugged in? No, so it won't heat up ... and so on (Cosmides 1989).

Universality of mechanisms.   Finding a culture in which the permission schema is absent would not count against that hypothesis. Finding a culture in which social contract algorithms were lacking would count against our hypothesis, however. Cognitive experiments supporting the hypothesis that there is a reasoning specialization for cheater detection have been conducted in different parts of the world (e.g. USA, UK,

Germany, Hong Kong), but these sites were all in industrialized nations. Although each instance is informative, the evidence for species-typicality gains strength in proportion to the *diversity* of subject populations tested.

It is, of course, impossible to test for social contract algorithms in every human culture. An alternative is to test individuals from a culture that is different from our own along as many dimensions as possible. With Larry Sugiyama, we have been testing social contract reasoning among the Shiwiar, a population of hunter–horticulturalists in the rain forests of the Ecuadorian Amazon (Sugiyama et al 1995). The Shiwiar live in a culture that is about as different from industrialized society as currently exists, and which, in many ways, mirrors the kind of social environment in which humans evolved.

Shiwiar in our study area have no everyday direct contact with outsiders. They depend on hunting, fishing, gardening and foraging for their livelihood. Men continue to use traditional blowguns in hunting, although some use muzzle-loading shotguns as well when shot and powder is available. Relatively few Shiwiar speak Spanish. In day-to-day life Shiwiar is the dominant language, ties of kinship and affinity dominate social relationships mediated by gossip, witchcraft and the threat or use of violence, and distribution of goods is largely controlled through traditional systems of trade and kinship obligations. In short, although it is impossible to find a group of people who are not subject to some influence from the industrialized world, the Shiwiar in our study villages are at the far end of the spectrum in this regard. To the extent that they have been influenced by outsiders, it has been largely a material and not a psychological influence. Given that Shiwiar speak a non-western language, live in small isolated villages where they hunt, gather and practice swidden horticulture for their livelihood, and continue to interpret life with a Shiwiar world view, it would be difficult to argue that any convergence of experimental results between them and subjects in the industrialized world is due to western acculturation.

Sugiyama administered oral versions of the Wason selection task, testing social contract rules and descriptive rules. The Shiwiar tested showed the same pattern of responses as one finds in American college students.

(5) Developmental timetable

Ontogenetic evidence can also be used in dissecting cognitive architecture. In general, one expects cognitive adaptations to manifest a reasonably uniform timetable. Language acquisition unfolds in a uniform manner between 18 months and four years of age, for example (Pinker 1994); indeed, the acquisition of phonemes during the first year follows a uniform time-course whether they are being spoken by a hearing child or signed by a deaf child (Petitto & Marentette 1991). Uniform emergence is not, of course, a hard and fast rule: some adaptations mature in response to cues encountered by different individuals at diverse points in the life cycle and/or ones that some individuals never experience at all. Certain fish change sex in response to a social cue, for example. If a female blue-headed wrasse happens to be the largest in her group when the resident male dies, she turns into a male. Otherwise, she stays female (Warner et al 1975). But even in these cases, when one understands the design of the adaptation, one can predict its developmental timetable.

The same cannot be said for knowledge acquired through content-free inference procedures or as a by-product of adaptations designed to process information from other domains. Writing — a by-product of cognitive adaptations for language — is learned at many different ages and sometimes not at all. So are cooking, calculus and agricultural techniques. If they are induced via content-general procedures operating in goal-defined contexts, then there is no particular reason to expect the development of permission schemas (or social contract algorithms) to follow the same timetable across individuals or cultures.

So far, not much is known about the development of social contract algorithms and precaution rules. Preliminary evidence suggests, however, that they emerge fairly early. When given age-appropriate versions of the selection task, seven-year olds correctly detect violations of both social contracts and precaution rules (Girotto et al 1988); moreover, Cummins (1997) has found that three- and four-year olds correctly detect violations of precaution rules. (Children of this age have not yet been tested on rules that can be interpreted *only* as a social contract.) What is interesting about these data is the uniformity of emergence, not the early age at which this occurs — an adaptation can emerge at any point in the life cycle (e.g. beards, teeth, breasts).

Precocious performance is neither necessary nor sufficient for sustaining an adaptationist hypothesis. It is, however, relevant for evaluating claims of content-free learning (e.g. Markman 1989). The early age at which children solve these Wason tasks undermines the hypothesis that the domain-specific reasoning mechanisms responsible were constructed by content-independent procedures operating on individual experience. Pre-schoolers, who have a limited experience base, are not noted for the accuracy and consistency of their reasoning in many other domains, even ones with which they have considerable experience. For example, many children this age will say that a racoon can change into a skunk; that the word 'needle' is sharp; that there are more daisies than flowers; that the amount of liquid changes when poured from a short fat beaker into a tall thin one; and that they have a sister but their sister does not (Boden 1980, Carey 1984, Keil 1989, Piaget 1950). When a child has had experience in a number of domains, it is difficult to explain why a domain-general mechanism would cause the early and uniform acquisition of a reasoning schema for one domain yet fail to do so for others.

Conclusion

Dissecting computational architecture is, in essence, discovering its functional organization. This requires theories of function. Whether one is discussing human-made artefacts or biological systems, to characterize something as a *mechanism* is to commit oneself to the proposition that it has a given design because that design solves some problem. Cognitive scientists do not always acknowledge this, and so the assumptions about function that motivate their methods are sometimes left implicit. This does not always distort the study of information-processing adaptations: research on the eye has progressed, for example, because its function is so obvious and its design so closely parallels that of a machine designed by human engineers to solve a similar problem (the camera). But the eye is an exception. The function (if any) of most components of the human computational architecture is either unknown or so vaguely defined that hypotheses about their design cannot be motivated by reference to human-made machines. Absent a theory of function, there is no basis for deciding which machines are functionally analogous.

There are undoubtedly further methods that could be turned to the task of dissecting the architecture of our minds. But the human mind is a biological system, and the only process that creates functional organization in biological systems is natural selection. Methods rooted in the logic of adaptationism are the most efficient way to find that organization, because knowing the problem is halfway to knowing the solution.

Acknowledgements

For many clarifying discussions, we thank Clark Barrett, Don Brown, Larry Fiddick, Gerd Gigerenzer, Rob Kurzban, Steven Pinker, Melissa Rutherford, Roger Shepard, Valerie Stone, Larry Sugiyama and Don Symons. We also thank the James S. McDonnell Foundation and the National Science Foundation (NSF grant BNS9157-449 to J. T.) for their financial support.

References

Atran S 1990 The cognitive foundations of natural history. Cambridge University Press, New York

Axelrod R 1984 The evolution of cooperation. Basic Books, New York

Axelrod R, Hamilton WD 1981 The evolution of cooperation. Science 211:1390–1396

Baron-Cohen S 1995 Mindblindness: an essay on autism and theory of mind. MIT Press, Cambridge, MA

Boden M 1980 Jean Piaget. Viking, New York

Boyd R 1988 Is the repeated prisoner's dilemma a good model of reciprocal altruism? Ethol Sociobiol 9:211–222

Brown A 1990 Domain-specific principles affect learning and transfer in children. Cognit Sci 14:107–133

Bugental D, Goodnow J 1997 Socialization processes. In: Eisenberg N (ed) Handbook of child psychology: social, emotional and personality development. John Wiley & Sons Inc., New York

Carey S 1984 Cognitive development: the descriptive problem. In: Gazzaniga MS (ed) Handbook of cognitive neuroscience. Plenum, New York, p 37–66

Cheng P, Holyoak K 1985 Pragmatic reasoning schemas. Cognit Psychol 17:391–416

Cheng P, Holyoak K 1989 On the natural selection of reasoning theories. Cognition 33:285–313

Chomsky N 1957 Syntactic structures. Mouton, The Hague

Cosmides L 1985 Deduction or Darwinian algorithms? An explanation of the 'elusive' content effect on the Wason selection task. Doctoral dissertation, Harvard University, Cambridge, MA (University Microfilms #86-02206)

Cosmides L 1989 The logic of social exchange: has natural selection shaped how humans reason? Studies with the Wason selection task. Cognition 31:187–276

Cosmides L, Tooby J 1987 From evolution to behavior: evolutionary psychology as the missing link. In: Dupre J (ed) The latest on the best: essays on evolution and optimality. MIT Press, Cambridge, MA, p 277–306

Cosmides L, Tooby J 1989 Evolutionary psychology and the generation of culture. II. Case study: a computational theory of social exchange. Ethol Sociobiol 10:51–97

Cosmides L, Tooby J 1992 Cognitive adaptations for social exchange. In: Barkow J, Cosmides L, Tooby J (eds) The adapted mind. Oxford University Press, New York, p 163–228

Cosmides L, Tooby J 1994 Beyond intuition and instinct blindness: toward an evolutionarily rigorous cognitive science. Cognition 50:41–77

Cox J, Griggs R 1982 The effects of experience on performance in Wason's selection task. Mem Cognit 10:496–502

Cummins D 1997 Evidence of deontic reasoning in 3- and 4-year old children. Mem Cognit, in press

Dawkins R 1982 The extended phenotype. Freeman, San Francisco, CA

Dawkins R 1986 The blind watchmaker. Norton, New York

Ekman P 1992 Facial expressions of emotion: new findings, new questions. Psychol Sci 1:34N–38N

Etcoff N 1984 Selective attention to facial identity and facial emotion. Neuropsychologia 22:281–295

Etcoff N, Freeman R, Cave K 1991 Can we lose memories of faces? Content specificity and awareness in a prosopagnosic. J Cognit Neurosci 3:25–41

Fernald A 1992 Human maternal vocalizations to infants as biologically relevant signals: an evolutionary perspective. In: Barkow J, Cosmides L, Tooby J (eds) The adapted mind. Oxford University Press, New York, p 391–428

Fiddick L, Cosmides L, Tooby J 1995 Priming Darwinian algorithms: converging lines of evidence for domain-specific inference modules. Seventh annual meeting of the Human Behavior and Evolution Society. University of California, Santa Barbara, CA

Fiske A 1991 Structures of social life: the four elementary forms of human relations. Free Press, New York

Fodor J 1983 The modularity of mind: an essay on faculty psychology. MIT Press, Cambridge, MA

Frith U 1989 Autism: explaining the enigma. Blackwell, Oxford

Gigerenzer G, Hug K 1992 Domain-specific reasoning: social contracts, cheating and perspective change. Cognition 43:127–171

Girotto V, Light P, Colbourn C 1988 Pragmatic schemas and conditional reasoning in children. Q J Exp Psychol Sect A Hum Exp Psychol 40:469–482

Gopnik M 1990 Dysphasia in an extended family. Nature 344:715

Hatano G, Inagaki K 1994 Young children's naive theory of biology. Cognition 50:171–188

Jackendoff R 1992 Languages of the mind. MIT Press, Cambridge, MA

Johnson-Laird P, Byrne R 1991 Deduction. Lawrence Erlbaum Associates Inc., Hillsdale, NJ

Kahneman D, Slovic P, Tversky A 1982 Judgment under uncertainty: heuristics and biases. Cambridge University Press, New York

Keil F 1989 Concepts kinds and cognitive development. MIT Press, Cambridge, MA

Keil F 1994 The birth and nurturance of concepts by domain: the origins of concepts of living things. In: Hirschfeld L, Gelman S (eds) Mapping the mind: domain specificity in cognition and culture. Cambridge University Press, New York, p 234–254

Leslie A 1987 Pretense and representation: the origins of 'theory of mind'. Psychol Rev 94:412–426

Leslie A 1988 Some implications of pretense for the development of theories of mind. In: Astington JW, Harris PL, Olson DR (eds) Developing theories of mind. Cambridge University Press, New York, p 19–46

Leslie A, Thaiss L 1992 Domain specificity in conceptual development: neuropsychological evidence from autism. Cognition 43:225–251

Lewontin R 1979 Sociobiology as an adaptationist program. Behav Sci 24:5–14

Maljković V 1987 Reasoning in evolutionarily important domains and schizophrenia: dissociation between content-dependent and content independent reasoning. Undergraduate honors thesis, Harvard University, Cambridge, MA, USA

Manktelow K, Over D 1990 Deontic thought and the selection task. In: Gilhooly KJ, Keane MTG, Logie RH, Erdos G (eds) Lines of thinking, vol 1. Wiley, Chichester, p 153–164

Mann J 1992 Nurturance or negligence: maternal psychology and behavior preference among preterm twins. In: Barkow J, Cosmides L, Tooby J (eds) The adapted mind. Oxford University Press, New York, p 367–390

Markman E 1989 Categorization and naming in children. MIT Press, Cambridge, MA

Marks I 1987 Fears phobias and rituals. Oxford University Press, New York

Petitto L, Marentette P 1991 Babbling in the manual mode: evidence for the ontogeny of language. Science 251:1493–1496

Piaget J 1950 The psychology of intelligence. Harcourt, New York

Pinker S 1994 The language instinct. Harper Collins, New York

Platt R, Griggs R 1993 Darwinian algorithms and the Wason selection task: a factorial analysis of social contract selection task problems. Cognition 48:163–192

Rips L 1994 The psychology of proof. MIT Press, Cambridge, MA

Spelke E 1990 Principles of object perception. Cognit Sci 14:29–56

Springer K 1992 Children's awareness of the implications of biological kinship. Child Dev 63:950–959

Sugiyama L, Tooby J, Cosmides L 1995 Cross-cultural evidence of cognitive adaptations for social exchange among the Shiwiar of Ecuadorian Amazonia. Seventh annual meeting of the Human Behavior and Evolution Society. University of California, Santa Barbara, CA

Tooby J, Cosmides L 1990 On the universality of human nature and the uniqueness of the individual: the role of genetics and adaptation. J Pers 58:17–67

Tooby J, Cosmides L 1992 The psychological foundations of culture. In: Barkow J, Cosmides L, Tooby J (eds) The adapted mind. Oxford University Press, New York, p 19–136

Tooby J, Cosmides L 1997 Ecological rationality in a multimodular mind. In: Cummins D, Allen C (eds) The evolution of mind. Oxford University Press, New York, in press

Trivers R 1971 The evolution of reciprocal altruism. Q Rev Biol 46:35–57

Warner R, Robertson D, Leigh E 1975 Sex change and sexual selection. Science 190:633–638

Wason P 1983 Realism and rationality in the selection task. In: Evans J St BT (ed) Thinking and reasoning: psychological approaches. Routledge, London, p 44–75

Wason P, Johnson-Laird P 1972 The psychology of reasoning: structure and content. Harvard University Press, Cambridge, MA

Williams G 1966 Adaptation and natural selection. Princeton University Press, Princeton, NJ

Wynn K 1992 Addition and subtraction by human infants. Nature 358:749–750

DISCUSSION

Buss: You have extracted the social contract from the natural relationships in which they evolved. For example, mateships and friendships are two different types of relationship in which social contracts come into play, and what constitutes cheating is different in those two relationships: having sex with someone outside a relationship would be constituted as cheating or a violation of a contract but this wouldn't be the case in the context of a friendship, whereas failure to reciprocate immediately within a friendship may be a violation but this may not be the case in a mateship. How can you deal with the issue of even more domain specificity than you're been arguing for?

Cosmides: First one needs to develop a theory of the adaptive problems involved in mating and friendship. At that point one can ask, like an engineer, 'What properties would we expect a mechanism well designed for detecting cheating in mateships to have? Are these the *same* properties we would expect of a mechanism well designed for detecting cheating in friendships? Could one, more general, social contract mechanism solve both adaptive problems? Or should we expect to find one mechanism that deals with social contracts outside the domain of mateships, and another specialized for mating relationships?' If we decided that there might be a mechanism specialized for detecting cheating in mateships, then we could conduct reasoning experiments transforming content and context, and see, empirically, whether there is a syntax specialized for that adaptive domain. Developing the theory is the crucial step, and it can reveal dimensions of an adaptive problem that common sense notions of 'reciprocity' or 'cheating' would not. For example, John Tooby and I have been working on a model of selection pressures based on what we call the 'Banker's Paradox'. The model suggests that friendship is not based on reciprocation, but rather on a form of deep engagement that has more in common with economic models of insurance. This leads to different psychological predictions. For example, if mechanisms that govern friendship were shaped by selection pressures for reciprocity, then people should pick friends who have *different* tastes than themselves, because this provides more opportunities for gains in trade. In contrast, the Banker's Paradox model predicts you will pick friends with tastes similar to your own; moreover, it predicts that reciprocating immediately will be construed as a sign that you are not a person's friend.

Buss: But in some social contracts, such as mateships, the nature of the contract is to tag to specific types of commodities, such as sexual commodities, and the way you develop the theory is content independent. A complete theory of social contracts would have to specify the nature of the content of the exchange, which may differ from species to species and from relationship to relationship.

Cosmides: I doubt that the social exchange mechanisms that I have been describing apply to the mating domain at all. (Indeed, excessive concern with reciprocation is probably the death knell of a mateship.) In my view, there is probably a separate set of mechanisms that govern reasoning, decision making, partner evaluation, preferences and memory for information involving mating than for other kinds of relationships — i.e. there are probably mechanisms that are specialized for mateships.

Nisbett: I have a comment on the opposite topic. David Buss is talking about highly specific domains, whereas I believe there are also general inferential rules, some of which I would argue we share with animals. For example, we seem to share temporal considerations for assessing causality: the ability to learn an association between two arbitrary stimuli doesn't seem to last beyond a few seconds. Holyoak et al (1989) have argued for an 'unusualness heuristic'. For decades it was assumed that many trials were necessary before animals could learn associations between arbitrary events. But Leon Kamin showed that one-trial learning was possible. Holyoak et al (1989) argued that a highly general heuristic can account for such results. That is, organisms will form an association when one unexpected event is followed by another unexpected event. This unusualness heuristic is absolutely content free.

Cosmides: I'm not sure that it is necessarily content free. There is domain specificity in what gets defined as unexpected versus not unexpected. Therefore, in the Garcia & Koelling (1966) experiments, if rats are nauseated after eating food they infer that it is the food that is nauseating them, but if they are nauseated after seeing a red light they don't infer that the red light is causing this.

Nisbett: This is the opposite of what we've been talking about: this is clearly domain specific.

Cosmides: I'm saying that a red light followed by nausea is an unexpected event, but they do not become conditioned by it.

Nisbett: It's true that a red light followed by nausea would not be expected, but it would never be noticed either because you can't produce instant nausea. Even if you could, you would probably not have one-trial learning, as you do with novel tastes. Organisms are 'counter prepared' to learn associations such as those between light and illness, whereas they are so well prepared to learn associations between novel foods and nausea that you can have a gap between the food and the nausea of 24 h or more and still get learning. In contrast, the maximum gap for arbitrary associations — that is, those for which the organism is neither prepared nor counter-prepared — is a matter of seconds. What Holyoak et al (1989) are referring to is these non-prepared associations for which it is possible to build up expectations within the context of the experiment. For example, consider the situation in which a rat learns over many trials that a buzzer signals imminent shock, and then has a trial in which a bright light immediately precedes the buzzer but the buzzer is not followed by shock. On the very next occasion on which the light precedes the buzzer, the rat acts as if it 'holds the hypothesis' that the buzzer does not signal shock as usual. Thus, the unusual event of the bright light preceding the buzzer is somehow linked in the rat's mind to the unusual event of the buzzer not predicting shock. Moreover, within just a few trials, the rat is at asymptote for learning. This is not like standard Skinnerian learning, in which many pairings are necessary to reach asymptote. Hence, Holyoak et al (1989) maintain that the rat has a heuristic it is applying rather than relying on mere brute associative learning.

Other general inferential rules are probably limited to humans. One example of this is the sunk-cost rule, which describes the tendency for people to consume something just because they have paid for it. You can teach people that they're following this rule by making them realize that consuming something with negative value is not a good idea, and you can change their behaviour across a wide range of domains. It's difficult to imagine that this is an inferential rule that we share with animals.

Tooby: We're not arguing that there are no general rules. We are just suggesting that psychologists should consider the hypothesis that a given performance is generated by domain-specific mechanisms on an equal basis with the hypothesis that it is generated by domain-general mechanisms, rather than either ruling it out a priori or accepting a lower standard of evidence for the domain-general hypothesis.

Cosmides: Adaptationism can help one understand when a cognitive system will be relatively content independent as well. For example, there are certain kinds of problems where you have to track the frequency of events ontogenetically, so that the phylogenetic specification of those frequencies won't really help: one can't specify phylogenetically that there's going to be more elk in one canyon versus another canyon. So different, more content-general mechanisms exist to track that kind of information. So far, the only limitation on the frequency computation system that we have found is the ability to individuate an object. For example, if you give somebody something that they don't token individuate, such as the sides of cards, they do badly on probabilistic reasoning tasks.

Hauser: I would like to raise a comment about our inability to detect deception. Have you tried running these kinds of experiments with logicians, who presumably have no problem with conditionals? If they showed a faster response to a social contract than a standard modus tolens problem wouldn't that be even stronger evidence for your claim about specialization?

Cosmides: That's an interesting idea. I haven't yet tried that, although logicians had difficulty with the original abstract selection test. There are some differences in performance between populations, which I suspect are due to differences in the ability to solve pencil-and-paper problems. If you give German undergraduates the same battery of problems, the difference in performance between social contracts and descriptive problems is identical to when they are given to American undergraduates, but their scores are shifted up slightly. These studies were performed by Gerd Gigerenzer and it's interesting that only 6% of subjects gave the logically correct answer on every single problem, even when the logically correct answer would fail to detect cheaters (Gigerenzer & Hug 1992). When Gerd asked them about this in the debriefing they all said that they solved the problems by applying the rules of inference of the propositional calculus. Furthermore, they said it was difficult to do this for some of the problems: it turned out that those were the switched social contracts, where the correct answer for detecting cheaters is not the logically correct answer.

Hauser: Has anyone timed how long it takes them to give the correct answer?

Cosmides: No.

Tooby: In many of these experiments we are tracking the proper adaptive responses. Your prediction that logicians would do this faster may not necessarily hold true. Logicians would, because of their training, have the tendency to give the answer that is correct from the point of view of formal logic, which in many cases would be an inappropriate response, rather than the answer that is correct from an adaptive point of view.

Gigerenzer: Logicians usually don't believe that everything can be reduced to one form of logic, such as first-order logic. Some try to develop domain-specific forms of logic. These logicians wouldn't even think of treating conditionals in a content-independent way.

I would like to raise a more general point. If psychological adaptations are, to some important degree, modular then we need to ask, how should we think of a module? The examples we have heard suggest two ways. The first is to assume that modules are psychological adaptations designed to solve important adaptive problems, such as a module for mate choice and one for social contracts. For instance, Leda Cosmides and John Tooby's social contract module is of that type — a module that integrates a specific mix of tools, including mechanisms for face or voice recognition, cheater detection, and characteristic emotions and behaviours. A second way to think about a module is not in terms of a characteristic mix of tools but in terms of a single tool. Proposals such as that of a number module, or a language module, seem to fall under this second view.

Cosmides: There are two different ways of thinking about this. The first addresses the question of whether the visual system is just one adaptation or a collection of adaptations, the outputs of which are functionally integrated to produce scene analysis. Scene analysis, i.e. knowing what things are present in the world and where they are, is just as much an adaptive problem as social exchange, but solving that problem supports many other activities. The same is true of language. For example, if you have language then you can make contracts for the future. One example that illustrates the interrelationship between adaptations is the theory that autism is a selective impairment of a theory of mind mechanism (Baron-Cohen et al 1985). There are different components of social exchange, and some clearly need a theory of mind. For example, if I am going to offer you something that you may want I have to be able to model what your desires are. But to detect cheating, it's not clear whether you can't just look for certain behavioural events. Similarly, it is not clear how much 'general intelligence' one needs for reasoning about social exchange. Maljković (1987) showed that people with schizophrenia have impaired general reasoning, but their ability to detect cheaters is intact. Finding double dissociations can help clarify these questions. For example, people with autism may have a selectively impaired theory of mind but they can have normal IQs, whereas people with Williams' syndrome seem to have an intact theory of mind but can be profoundly retarded and do badly in spatial tests. One can ask, will a person with autism have trouble with social exchange but be good at detecting violations of precaution rules, which are not social rules? Would we find the opposite in people with Williams' syndrome, i.e. that they are good at detecting cheaters but not at detecting rotations of precaution rules? Ultimately, it will be interesting to find out how these are related and whether the output of one mechanism provides input to another mechanism.

References

Baron-Cohen S, Leslie A, Frith U 1985 Does the autistic child have a 'theory of mind'? Cognition 21:37–46

Garcia J, Koelling R 1966 Relations of cue to consequence in avoidance learning. Psychon Sci 4:123–124

Gigerenzer G, Hug K 1992 Domain-specific reasoning: social contracts, cheating and perspective change. Cognition 43:127–171

Holyoak K J, Koh K, Nisbett RE 1989 A theory of conditioning: inductive learning within rule-based default hierarchies. Psychol Rev 96:315–340

Maljković V 1987 Reasoning in evolutionarily important domains and schizophrenia: dissociation between content-dependent and content-independent reasoning. Undergraduate honours thesis, Harvard University, Cambridge, MA, USA

Language as a psychological adaptation

Steven Pinker

Department of Brain and Cognitive Sciences, Massachusetts Institute of Technology, Cambridge, MA 02139, USA

Abstract. Language is the remarkable faculty by which humans convey thoughts to one another by means of a highly structured signal. Language works by two principles: a dictionary of memorized symbols, that is, words, and a set of generative rules organized into several subsystems, that is, grammar. The machinery of language appears to have been designed to encode and decode propositional information for the purpose of sharing it with others. Language is universally complex and develops reliably throughout the species, partly independently of general intelligence. I suggest that language is an adaptation for sharing information. It fits with many other features of our zoologically distinctive 'informavore' niche, in which people acquire, share and apply knowledge of how the world works to outsmart plants, animals and each other.

1997 Characterizing human psychological adaptations. Wiley, Chichester (Ciba Foundation Symposium 208) p 162–180

If an alien biologist studying the human species were to observe us in this room today, it would be struck by the fact that you are sitting quietly doing nothing but listening to a man make noises as he exhales. This remarkable habit of our species is called language. You are listening to my exhaling noises because I have coded information into some of the acoustic properties of that noise which your brains have the ability to decode. By this means I am able to transfer ideas from inside my head to inside your head. In this chapter I will say a few words about the design of the human language faculty — the tricks behind the ability to transfer information by making noise — and about whether language is an adaptation, and if so, what it is for.

The design of human language

What is the secret behind my ability to cause you to think specific thoughts by means of the vocal channel? There is not one secret, but two, and they were identified in the 19th century by continental linguists.

Words

The first principle was articulated by Ferdinand de Saussure and lies behind the mental dictionary, our finite memorized list of words. A word is an arbitrary symbol, a connection between a signal and an idea shared by all members of the community. The word *duck* doesn't look like a duck, walk like a duck or quack like a duck. However, I can use it to convey the idea of a duck because we all have, in our developmental history, formed the same connection between the sound and the meaning. Therefore, I can bring the idea to mind in you virtually instantaneously simply by making that noise. If instead I had to shape the signal to evoke the thought in your mind using some sensible connection between its form and its content, every word would require the amusing but inefficient contortions of a game of Charades.

The symbols underlying words are bi-directional. Generally, if I can say something, I can understand it, and vice versa. When children learn words, their tongues are not moulded into the right shape by parents, and they do not need to go through a process of being rewarded for successive approximations to the target sound for every word they learn. Instead, children have an ability upon hearing somebody else use a word to know that they in turn can use it to that person or to a third party and expect to be understood.

Another noteworthy feature of the mental dictionary is its size. One can use dictionary sampling techniques to estimate how many independent memorizations must have taken place to install a typical adult vocabulary. For example, if you take the largest dictionary you can find, pick, say, the third word down on the right hand side of every 10th page, put the word in a multiple-choice test, correct for guessing, and then multiply the performance on the test by the size of the dictionary, you obtain estimates for a typical high school student of about 60 000 independent words, probably twice that number for a highly literate high school student. The learning rate for the smaller estimate works out to be about one word every 90 minutes, starting at the age of one. Considering that every one of these words is as arbitrary as a telephone number or a date in history, it is remarkable that children pick one up every hour and a half, at the same time as they are struggling over multiplication tables, the date of the Treaty of Versailles and so on.

Grammar

Of course we don't just learn individual words. We combine them into strings when we talk, and that leads to the second trick behind language, grammar. The principle behind grammar was articulated by Wilhelm von Humboldt as 'the infinite use of finite media'. Inside everyone's head there is a finite algorithm — it has to be finite because the head is finite — with the ability to generate an infinite number of potential sentences, each of which corresponds to a distinct thought. For example, our knowledge of English incorporates rules that say 'a sentence may be composed of a noun phrase (subject) and a verb phrase (object)' and 'a verb phrase may be composed of a verb, a noun phrase (object) and a sentence (complement)'. That pair of rules is *recursive*: an element is introduced on the right hand side of one rule which is also on the left hand side of the other rule, creating the possibility of a loop that could generate sentences of any size, such as 'I think that she thinks that he said that I wonder whether . . . .' We can generate an infinite number of sentences, and each sentence expresses a distinct thought. Therefore, this system gives us the ability to put an unlimited number of distinct thoughts into words, and other people to interpret the string of words to recover the thoughts.

Grammar can be thought of as a discrete combinatorial system, which can be opposed to a blending system. In a blending system all the possibilities that you get when you combine the ingredients lie on a continuum between the two endpoints. For example, if you mix red paint and white paint you obtain various degrees of pink paint, which differ in an analogue fashion. In a discrete combinatorial system, such as DNA, atoms, molecules and sentences, each one of the combinations can be qualitatively different from the other combinations and from the ingredients. Therefore, there is an infinite number of thoughts, but the set is infinite in a discrete way as opposed to a continuous way.

Grammar can express a remarkable range of thoughts because our knowledge of languages resides in an algorithm that combines abstract symbols, such as noun and verb, as opposed to concrete concepts, such as man and dog or eater and eaten. This gives us an ability to talk about all kinds of wild and wonderful ideas. We can talk about a dog biting a man, or, as in the journalist's definition of 'news', a man biting a dog. We can talk about aliens landing at Harvard. We can talk about the universe beginning with a big bang, or the ancestors of native Americans immigrating to the continent over a land bridge from Asia during an Ice Age or Michael Jackson marrying Elvis's daughter. All kinds of novel ideas can be communicated because our knowledge of language is couched in abstract symbols like 'noun' and 'verb' which can embrace a vast set of concepts, and then can be combined freely into an even vaster set of propositions. How vast? In principle it is infinite; in practice it can be crudely estimated by assessing the number of word choices possible at each point in a sentence (roughly, 10) and raising it to a power corresponding to the maximum length of a sentence a person is likely to produce and understand, say, 20. The number is 10^{20} or about a hundred million trillion sentences.

Let me say a bit more about the design of the human grammatical system. Grammar is not just a single pair of rules, as I listed above, but hundreds or thousands of rules, which fall into a number of subsystems. The most prominent is syntax, the component of language that combines words into phrases and sentences. One of the tools of syntax is word order. That is the tool that allows us to distinguish *Man bites dog* from *Dog bites man*. Order is the first thing people think of when they think about syntax, but in fact it is a relatively superficial manifestation; there are complex principles that underlie its ability to convey ideas.

Far more important than linear word order is constituency. A sentence has a hierarchical structure. It can be represented as a bracketed string of phrases embedded within phrases, which allow us to convey complex propositions of ideas embedded inside ideas. A simple demonstration of the brain's ability to parse sentences into hierarchical phrase structures is an (unintentionally) ambiguous sentence published in *TV Guide*: *On tonight's programme Dr Ruth will discuss sex with*

Dick Cavett. We have a single string of words in a particular order, but it has two very different meanings. The intended meaning was that Dick Cavett is who you discuss sex with, and the alternative meaning is that sex with Dick Cavett is what you discuss. In this particular sentence they are, unfortunately, expressed by the same string of words; the different interpretations correspond to the different phrase structure bracketings our brain can impose on that string. Thankfully, not all sentences are as blatantly ambiguous as that one, and the brain's ability to compute phrase structure is put to use in recovering the single meaning that the speaker intended. Much as in other symbolic systems that encode logical information, such as arithmetic, propositional calculus and computer programming, it is important in linguistic communication to get the parentheses right, and that's what phrase structure is for.

Syntax also involves predicate–argument structure, the component of language that encodes the information that a logician would express as a predicate, that is, a relationship among a set of participants, and its arguments, that is, the particular participants in a given instance of that relationship. In particular, what your grammar school teacher called the predicate of the sentence, namely the verb, is similar in many ways to what a logician would call a predicate. To understand a sentence you cannot merely pay attention to the order of words, or even just group them. You have to look up information associated with the predicate. A simple demonstration is the pair of sentences *Man fears dog* and *Man frightens dog*. The word *man* is the subject of both of these sentences, but the semantic role that *man* is playing in the first sentence is different from the role it is playing in the second sentence: in one case the man is causing the fear; in the other the man is being affected by fear. That shows that in understanding a sentence you have to look up information stored with the mental dictionary entry of the verb and see whether it says 'my subject is the one doing the fearing' or 'my subject is the one causing the fear'.

A fourth trick of syntax is the operation called the transformation, which is associated most strongly with Noam Chomsky's theory of transformational grammar. After you have generated a hierarchical tree structure into which the words of a sentence are plugged, a further set of operations can mangle the order of words in precise ways. For example, the sentence *Dog is bitten by man* contains the verb *bite*, which ordinarily requires a direct object. But in this sentence the object is missing from its customary location; it has been moved to the front of the sentence. This gives us a means of shifting the emphasis and the quantification of a given set of participants in a relationship. The sentence *Man bites dog* and *Dog is bitten by man* both express the same information about who did what to whom — a man does the biting, a dog gets bitten — but one of them is a comment about the man and the other is a comment about the dog. Similarly, sentences in which a phrase is replaced by a *wh*-word and moved to the front of a sentence, such as *Who did the dog bite?*, allow the speaker to seek the identity of one of the participants in a given interaction. Transformations, therefore, give us a layer of meaning above and beyond who did what to whom. That layer emphasizes or seeks information about one participant or another, while keeping the actual event that one is talking about constant.

Syntax, for all that complexity, is only one component of grammar. All languages have a second combinatorial system, morphology, in which bits of words are assembled to produce whole words. In English we don't have much morphology, compared with other languages, but we have some. The noun *duck* comes in two forms — *duck* and *ducks* — and the verb *quack* in four — *quack, quacks, quacked* and *quacking*. In other languages morphology plays a much greater role. In Latin, for example, there is a rich inflectional system which plays an important role in expression. By placing suffixes onto the ends of nouns, one can convey information about who did what to whom, allowing one to scramble the left-to-right order of the words for emphasis or style. For example, *Canis hominem mordet* and *Hominem canis mordet* have the same non-newsworthy meaning, and *Homo canem mordet* and *Canem homo mordet* have the same newsworthy meaning.

In addition to syntax and morphology, language comprises a third combinatorial system — a third layer of assembly of elements into larger elements by rules, called phonology. The rules of phonology govern the sound pattern of a language. In no language do people form words by associating them directly with articulatory gestures like a movement of the tongue or lips. Instead, a fixed set of articulatory gestures is combined into sequences, each sequence defining a word. The combinations are governed by phonological rules that people have to acquire as they acquire a language. In English, for example, we know that *bluck* is not a word but could be a word, whereas *nguck* is not a word and could not be a word because the English rules of word formation don't allow the consonant *ng* at the beginning. In other languages, *ng* can be placed there. Interestingly, whereas syntax and morphology are semantically compositional, that is, you can predict the meaning of the whole by the meanings of the elements and the way they are combined, this is not true of phonology. You cannot predict the meaning of *duck* from the meaning of *d*, the meaning of *u* and the meaning of *k*. The combinatorial system called phonology simply allows us to have large vocabularies, for example, 100 000 words, without having to pair each one of them with a different simple noise coming out of the mouth.

Phonology also consists of a set of adjustment rules which, after the words are defined and combined into phrases, smooth out the sequence of articulatory gestures to make them easier to pronounce and comprehend. One of those rules in English causes us to pronounce the same morpheme *-ed* for the past tense in three different ways depending on what it is attached to. In *jogged* it is pronounced as *d*. In *walked* it is pronounced as a *t*, thanks to a rule that keeps the consonants at the end of a word either all voiced (larynx buzzing) or all unvoiced. And in *patted* the suffix is pronounced with the neutral schwa vowel before it, thanks to a rule that inserts a vowel to separate two *d*-like sounds. Therefore, even though the actual morpheme is the same in all cases, that is, *d*, there are rules that fiddle with the pronunciation pattern before it is articulated. These adjustments are not just peoples' effort to be clear, or for that matter their tendency to be lazy, as they put words together. There is a set of regulations for each language that dictate when you are allowed to be lazy and when you are not, and they are partly arbitrary in that you acquire them as you acquire the sound pattern of a language. (An accent is what happens when someone applies the phonological adjustment rules of one language to the content of another language.) Phonological rules have the function of helping people achieve a mixture of clarity and ease of pronunciation, but they are a distinct part of one's knowledge of language.

Interfaces of language with other parts of the mind

Grammar is only one component of language. It has to look towards the rest of the mind in three different directions. Grammar has to be connected to the ear, so that we can understand; to the mouth, so that we can articulate; and to the rest of the mind, so that we can say sensible things in the context of a conversation. Each requires an interface.

The first interface is the speech articulation or articulatory phonetic system. One of the salient properties of this system is the actual anatomy of the vocal tract, which seems to have evolved in the human lineage in the service of the language. Darwin pointed to the fact that every mouthful of food we swallow has to pass over the trachea, with some chance of getting lodged in it and causing death by choking. The human vocal tract has a low larynx by mammalian standards; this placement compromises a number of physiological functions but allows us to articulate a large range of vowel sounds. Because our larynx is so low in the throat it gives room for the tongue to move both back and forth and up and down independently. This defines a two-dimensional space in which the tongue can move. Because there are two resonant cavities, defined by the position of the tongue with respect to the throat, on one hand, and the mouth, on the other, we can produce a two-dimensional space of vowel sounds, which multiplies out the number of distinct discriminable signals we can articulate. One can argue that given the physiological cost, that is, the risk of death by choking, there must have been a corresponding benefit in our evolutionary history, presumably the benefit of rapid, expressive communication.

The second interface is speech comprehension. Information being received by the ear has to be unpacked into a meaning. One of the remarkable features of speech comprehension is the way in which the brain can unpack a stream of sound into its component words, which are not physically demarcated in the sound stream by little silences analogous to the small spaces that separate words on a printed page. When we hear language as a string of words we are the victims of an illusion. We realize this only when we hear speech in another language: to our ears it sounds like a continuous ribbon of sound, which is exactly what it is, physically speaking. Another demonstration bringing this feat to our attention is a kind of wordplay seen in doggerel such as 'Mairzey doats and dozey doats' (Mares eat oats and does eat oats) and 'Fuzzy Wuzzy was a bear', which are designed to exploit the fact that speech is not a discrete chain of words separated by silence. Another remarkable feature of speech comprehension is the rate at which information can be conveyed. A rapid talker can convey about 40 phonemes per second, and even a more leisurely talker can reach 10 to 20 phonemes per second. Twenty cycles per second is the lower limit of pitch perception in humans. We hear 20 beats per second not as 20 rapid events but as a low tone or a buzz. Clearly, when we are listening to speech at 25 phonemes per second we are not registering 25 separate auditory events, because that is neurologically impossible. There must be some sort of multiplexing or compression of the information, in which the phonemes are superimposed in the process of speaking and the brain has to unpack them in the process of understanding.

Finally, language has an interface with more general inference systems. The decoding of the literal information conveyed by words and grammar is just the first step of a long chain of inference by which we try to guess what the speaker wants us to think he or she is trying to say. We engage in this process of inference even in understanding simple sentences. A nice example from the linguist Jim McCawley shows the knowledge we must apply to something as simple as assigning referents to pronouns such as *he* and *she*. Imagine a dialogue in which Marsha says 'I'm leaving' and John says 'Who is he?' We all know who 'he' is, not by any information that is explicitly encoded in the sentence, but by our knowledge about human behaviour. Those expectations are brought to bear in understanding a sentence, in conjunction with the particular rules of grammar.

Is language an adaptation?

With that summary of the design of the human language faculty in mind, we can now turn to the question: is language an adaptation? Darwin wrote, 'Man has an instinctive tendency to speak, as we see in the babble of our young children, while no child has an instinctive tendency to bake, brew or write.' This is probably the first statement that human language is an adaptation. What are the alternatives, and what is the evidence?

One alternative is that language is not an adaptation itself, but is a manifestation of more general cognitive abilities, some form of 'general intelligence', in which case general intelligence would be the adaptation, not language. There is a reasonable amount of evidence against this possibility.

First, language is universal across societies and across all neurological normal people within a society. There may be technologically primitive peoples, but there are no primitive languages. And the language of uneducated, working class and rural speakers has been found to be systematic and rule-governed, though the rules may belong to a dialect that did not have the good fortune of becoming the standard dialect of Britain and its former colonies.

Second, languages conform to a universal design. The languages of the highlands of New Guinea, for example, use computational machinery that is identical to that described earlier in this chapter, even though that machinery was motivated by an examination of English and other European languages.

A third kind of evidence was alluded to by Darwin: the ontogenetic development of language. There is a uniform sequence of stages that children pass through all over the world. That sequence culminates in mastery of the local language, despite the computational difficulty of programming a computational system to take in a finite sample of sentences from a couple of speakers (the child's parents) and induce a grammar for the infinite rest of the language. Moreover, children's speech patterns, including their errors, are highly systematic, and often conform to linguistic universals for which there was no direct evidence in parents' speech.

A fourth kind of evidence also comes from the study of language acquisition. If children are thrown together without a model language, such as in a multilingual plantation or, if the children are deaf, a school that does not have people using sign language, the children will develop a systematic, rule-governed language of their own, a phenomenon called creolization.

A fifth kind of evidence is that language and general intelligence, to the extent we can make sense of that term, seem to be doubly dissociable in neurological and genetic disorders. In aphasias and in a developmental syndrome (probably genetic in origin) called Specific Language Impairment, intelligent people can have extreme difficulties speaking and understanding. Conversely, in what clinicians informally call 'chatterbox syndromes', severely retarded children may talk a stream of fully grammatical English but with a content that is highly childlike or is confabulated, bearing no relation to the world.

A different alternative to the hypothesis that language is an adaptation is the possibility that language indeed is a separate system from general intelligence, but that it evolved by non-selectionist mechanisms. Perhaps, on this view, language evolved all at once as the product of a macromutation, or as a by-product of some other evolutionary development such as evolving a large head. The main reason to doubt this theory is the standard argument for the operation of natural selection, the argument from adaptive complexity. The information-processing circuitry necessary to produce, comprehend and learn language must involve a great deal of organization and detail. As with other complex biological systems that accomplish improbable feats, this circuitry is unlikely to have evolved by something as crude as a single mutation or some other evolutionary force that is insensitive to what the circuit accomplishes.

What did language evolve for?

If language is an adaptation, what is it an adaptation for? Note that asking this question is different from asking what language is typically *used* for, especially what it is used for at present. The question concerns the engineering design of language and the extent to which it informs us about the selective pressures that shaped it.

What is the machinery of language trying to accomplish? The system looks as if it was put together to encode and decode digital propositional information — who did what to whom, what is true of what, when, where and why — into a signal that can be conveyed from one person to another.

It is not difficult to think of why it would have been a good thing for a species with the rest of our characteristics to evolve the ability to do this. The structures of grammar are well suited to conveying information about technology, such as which two things may be put together to produce a third thing; about the local environment, such as where things are and which people did what to which others; and about one's own intentions, such as 'If you do this, I will do that', which convey relationships of exchange and of dominance, as in threats.

Gathering and exchanging information of this kind is, in turn, integral to the larger niche that modern *Homo sapiens* has filled, which George Miller has called the informavore niche and which John Tooby and Irven DeVore previously called the cognitive niche. Tooby and DeVore have assembled a theory that tries to explain the list of properties of the human species that a biologist would consider zoologically unusual, such as our extensive manufacture of and dependence on complex tools, our wide range of habitats and diets, our long childhoods and long lives, our hypersociality, and our division into groups or cultures each with a set of distinctive local variations in behaviour. Their explanation is that the human lifestyle is a consequence of a specialization for overcoming the evolutionary fixed defences of plants and animals by cause-and-effect reasoning, which is driven by intuitive theories about various domains of the world, such as objects, forces, paths, places, manners, states, substances, hidden biochemical essences, and other people's beliefs and desires.

The information captured in these intuitive theories is reminiscent of the information that the machinery of grammar is designed to convert into strings of noises. It is probably not a coincidence that what is special about humans is that we outsmart other animals and plants by cause-and-effect reasoning, and language seems to be a way of converting information about cause-and-effect and action into a signal.

An unusual feature of information is that it can be duplicated without loss. If I give you a fish, I don't have the fish, as we know from sayings like 'you can't eat your cake and have it'. Information, however, *can* be both eaten and had: if I tell you how to fish, it is not the case that I now lack the knowledge of how to fish because I've given it away to you; we can both have it. There is a brilliant, eccentric computer programmer, Richard Stallman at the Massachusetts Institute of Technology, who started a free software foundation based on the idea that no one should charge for software. It's perfectly reasonable, he argues, for a baker to charge for bread, since there's only a finite amount of flour and once it is given to someone then someone else cannot have it. But once software is developed it can be copied, Stallman argues, so there is no reason that it should not be free. If it is given to one person, that does *not* mean that someone else cannot have it (with the exception of the floppy disk itself).

Tooby and DeVore have pointed out that in a species like ours that lives on information, it is quite natural that in conjunction with evolving the ability to gather this information we evolved a means to exchange it. Having language multiplies the benefit of knowledge. Knowledge is not only useful to oneself as a way of figuring out, for example, how to build snares to catch rabbits, but it is also useful as a trade good: I can exchange it with somebody else at a low cost to myself and hope to get something in return. It can also lower the original acquisition cost. I can learn about how to catch a rabbit from someone else's trial and error; I don't have to go through it myself.

Language, therefore, fits with other features of the informavore niche. The zoologically unusual features of *Homo sapiens* can be united by the idea that humans have evolved an ability to encode information about the causal structure of the world and to share it among themselves. Our hypersociality makes sense because information is a particularly good commodity of exchange, one that makes it worth people's while to hang out together. Our long childhood is an apprenticeship — before we go out in the world, we spend a lot of time learning what everyone else around us has figured out already. The existence of culture can be seen as a kind of pool of local expertise. Many traditions develop locally because many of the requirements to deal with the various aspects of the world have been acquired by other people, resulting in a network of information sharing that is close to what sociologists and anthropologists have called 'culture'. Humans have long life spans because once you've had an expensive education you might as well make the most out of it by conserving your bodily resources and stretching out your life span so that your expertise can be put to use. Humans inhabit a wide range of habitats because we don't have knowledge that is highly specialized, such as how to catch a rabbit; our knowledge is more abstract, such as how living things work and how objects bump into each other. That machinery for construing the world can be applied to many kinds of environments; it is not specific to a particular ecosystem.

People have occasionally raised objections to the hypothesis that language is an adaptation for sharing information. One objection is that organisms are competitors, so sharing information is in fact costly by virtue of the advantages it gives to one's competitors. If I teach someone to fish, they may overfish the local lake, leaving no fish for me. The argument, however, just boils down to the standard problem of the difficulties facing the evolution of co-operation or altruism, and the solution in the case of language is the same. By sharing information with our kin, we help copies of our genes inside those kin, and when it comes to sharing information with non-relatives, if we ensure that we inform only those people who return the favour, we both gain the benefits of trade. Certainly we do use our faculties of social cognition to ration our conversation with those with whom we have established a non-exploitative relationship; hence the expression 'to be on speaking terms'.

A second objection is that language may be used to deceive, so perhaps language evolved as a means of manipulation rather than as means of communication. The answer to the objection is, once again, that language surely co-evolved with our faculties of social cognition. We apply those faculties as we listen to others; we are constantly vigilant for whether we are being lied to. And I find it hard to imagine any coherent account by which language evolved to allow us to manipulate others. Unlike signals with the physiological power to manipulate another organism directly, such as loud noises or chemicals, the signals of language are impotent unless the recipient actively applies complicated neural machinery to decode them. It is impossible to use language to manipulate someone who doesn't understand the language, so if language is an adaptation to manipulate others, how could it have gotten off the ground? What would have been the evolutionary incentive for the designated targets to evolve those exquisitely complex mental algorithms for unpacking the speech wave into words and assembling them into trees? Like a shop owner who makes sure he is not around when the gangster selling 'protection' comes by, or a negotiator who remains incommunicado until a deadline passes, hominids in the presence of the first linguistic manipulators would have done best by refusing to allow their nascent language systems to evolve further, and language evolution would have been over before it began.

Further reading

Pinker S 1994 The language instinct. Harper Collins, New York
Pinker S 1997 How the mind works. Norton, New York
Tooby J, DeVore I 1987 The reconstruction of hominid evolution through strategic modeling. In: Kinzey WG (ed) The evolution of human behavior: primate models. State University of New York Press, Albany, NY, p 183–237

DISCUSSION

Dawkins: Presumably, if somebody were to challenge you with reading and writing as, on the face of it, looking like adaptations but too recent to be so, you would say that we have dissected the world wrongly and that reading and writing just count as language. On the other hand, they use completely different sensory systems, which is slightly worrying. One may also find different pathologies such as dyslexia which, if it is real, suggests that a special brain mechanism is present.

Pinker: Many of the criteria that I listed for language being an innate mental system would not apply to reading. Reading is not universal either across societies or within societies, and the ontogeny of reading is different from the ontogeny of language: children have to be badgered to read and usually they don't succeed unless attentive adults are encouraging them to, and often not even then. Dyslexia, in most cases, is a symptom of a much larger deficit, such as difficulty with language, or with processing fine spatial detail. Even when it appears in relative isolation, it may be caused not by damage to a cohesive brain system but by damage to a set of connections between brain systems. Reading requires a transfer of information from the visual system to the rest of the language system, and dyslexia may involve a disconnection of that information pipeline. That is how Norman Geschwind, for example, characterized aphasia: a disconnection syndrome as opposed to damage to a computational system itself.

Dawkins: Do dyslexics have a visual impairment?

Pinker: That's unclear. The causes of dyslexia are heterogeneous, probably because reading is a complicated, multistage activity and if any link in the chain is damaged the entire ability disappears. In many cases, dyslexia is indeed associated with difficulty in processing fine visual detail, and some people who are called dyslexics do have problems in seeing small objects composed of multiple parts. But, in general, an answer to the question of whether dyslexia is as coherent a system as spoken language would strengthen or weaken the idea that language is a separate system.

Hauser: A partial answer to this is that language doesn't seem to be modality specific: i.e. one can do sign language.

I also have a different question on the issue of arbitrariness. Darwin (1872) has pointed out that large animals produce low frequency vocalizations and smaller animals produce high frequency vocalizations. John Ohala (1983, 1984) has also commented that in languages there are sound–meaning associations; for example, in French the word for small is *petit*, which has a higher pitch than the word *gros*, meaning large. The key question is do you find any counter examples to that, where the word for a large thing has a high pitch? Because if not then there are constraints at that level as well.

Pinker: Phonetic symbolism is a weak statistical phenomenon. You can give words for 'large' and 'small' in a set of languages to people who don't speak those language and ask them which one means 'large' and which one means 'small'. With that binary choice, people can answer with at best about 60% accuracy, usually less. Phonetic symbolism, then, is an asterisk to the statement that words are arbitrary. Indeed, even that footnote might be deletable, because the real experiment would be to give someone a set of randomly selected words in a language they don't speak and ask them to guess something about their meanings. I suspect that in this case the hit rate would be close to zero. For this reason I think Saussure's arbitrary symbol is the overriding design principle in the words of human languages.

Kacelnik: The evidence that language is an adaptation is overwhelming. What would be extremely useful is an example of something we've learnt about the way language works in terms of an adaptationist or any other evolutionary perspective. That is, can you tell us something that we know about language that we wouldn't know if we didn't think in an adaptationist way?

Pinker: Here is an example from my own work. I study regular and irregular inflectional morphology in children, adults and neurological patients. Languages often have two distinct mechanisms for expressing the same grammatical information. In one, such as *sing/sang* and *bring/brought*, you change a verb in unpredictable ways to express the past tense. In the other, such as *walk/walked* and *talk/talked*, you add a suffix at the end of every word. The second regular inflection seems to be a combinatorial process of a kind that characterizes grammar. The first is arbitrary, and the child has to memorize the words one by one. Why do languages have an irregular system at all? Regular and irregular systems are different in many ways, and that raises a puzzle. It seems bizarre that languages should have things that are just there because they violate a rule — it just doesn't fit an evolutionary perspective in which psychological systems are well-designed machines that serve some function.

That attitude forced me to reformulate the problem. I thought that perhaps there is no system of irregular inflection *per se*; perhaps it is a case in which the word system is being applied to conveying a kind of information that is ordinarily done by the grammar system. *Brought* is just a word, like *duck*, that you memorize. If you equip a species with the ability to learn *duck*, you get, as a spandrel or by-product, the ability to have irregular words. And that leads to a set of predictions: anything we know to characterize the psychology of words should characterize the psychology of irregular and inflectional forms in English and other languages. For example, neurological disease that damages the mental dictionary should selectively damage irregular verbs in comparison with regular verbs. We can test this by looking at a kind of aphasic patient who is constantly in a tip-of-the-tongue state (such a patient would say 'The guy did the thing and then he went out and he did another thing.'). In such cases the content words don't come to mind fast enough because of damage to the posterior parietal regions of the brain. And sure enough, these people have more trouble converting *bring* to *brought* than *walk* to *walked*. A second prediction along these lines is that the *form* of irregular verbs should follow the formation rules in a given language for a simple word. In English the basic word is a monosyllable. And as predicted, all the irregular words are monosyllabic or have monosyllabic roots. So we have two criteria for wordhood — there are others as well — that also characterize irregularity. This in turn supports the idea that there isn't a complex but useless brain system for irregularity, just a word system being over-applied.

This in turn raises another question. Why should we have a word system? Why don't we have rules all the way down? We can imagine a system that has a set of rules for combining little atoms of thought. In fact, in the 17th and 18th centuries during the Enlightenment there were a number of proposals for 'perfect languages' in which you would utter the equivalent of a sentence just to convey a concept like *duck*: 'a feathered animate being that is bilaterally symmetrical . . . ' and so on. We laugh when we read such proposals because they would force people into long-winded circumlocutions for common, simple thoughts; clearly real languages did better by dividing the labour of expression into a system of memorized words for common and simple concepts and a system of combinatorial grammar for novel and complex concepts. Thinking of language as a system with some kind of engineering logic allowed me to reconceptualize and seek new data that would otherwise have been attributed to sheer quirkiness.

Sperber: You asked the question: is language an adaptation? A more appropriate question, however, would be: is the language faculty an adaptation? The language faculty is a mental organ and a clear biological phenomenon, whereas language itself is, as Chomsky would argue, an abstraction, or, possibly, a social phenomenon. The language faculty is what allows the child to acquire the particular language of his or her community. Still, there is an apparent paradox: on the one hand, actual languages cannot develop unless the individual has a language faculty; on the other hand, a language faculty is plausible as an adaptation only in a social environment where languages are already present. I would like to know your present thoughts on this problem (which extends to other biologically evolved social competencies).

Pinker: I was actually referring in my presentation to the language faculty, not to language. Everywhere I said language I could, and perhaps should, have said the language faculty, i.e. the neurocircuitry that allows us to acquire and use language.

Your other comment — that there is no point having the ability to learn language in an environment in which there is no language to be learnt — can be restated by asking who the first language mutant talked to. One answer, which I don't really believe, is that s/he talked to the 50% of his/her relatives who shared that gene. A better answer, however, is that his/her listeners already had other means of communication and that language plugged into them. For instance, if an individual had 27% of the modern system and talked to someone who had only 22%, they could have closed the gap by using the non-language parts of the mind to reason out what the other person meant, much in the same way as when you were in a foreign city and you can decode newspaper headlines and signs on buses by knowing some cognates and general rules of thumb. Still better would have been an ability to understand all of a person's speech automatically and reliably, and that could have set the stage for the evolution of the next increment of an ability to process language effortlessly and automatically.

Sperber: This suggests a relationship between the language faculty and another evolved adaptation, namely the 'theory of mind', or the ability to attribute mental states to others. Members of a species that can recognize the mental states of one another can also attempt, in a variety of ways, to indicate their mental states to others and cause them thereby to adopt similar mental states. In such a species, some form of language could evolve as a particularly efficient — but not unique — way of making one's mind known to others.

Miller: If you posit that language is an adaptation for conveying propositional information from one mind to another, it will satisfy syntax theorists and inflectional morphologists because it describes and explains the features of language that they pay attention to. However, it won't satisfy sociolinguists, who want to know what people talk about, under what conditions, who they talk to, and how language interacts with gender, power and other aspects of social behaviour. Why stop at the somewhat non-specific level of saying it's about propositional information? Wouldn't a more specific approach enable sociolinguists, as well as syntax theorists, to be adaptationists?

Pinker: The idea I have been suggesting would encompass all the uses to which we put language, because the set of propositions the grammatical machinery is designed to express embraces messages such as gossip and the exchange of intentions and threats.

Miller: But that's too non-specific to make theoretical predictions that explain the observed distribution of content, language use and social behaviour.

Pinker: The grammatical machinery doesn't seem to be so concrete as to be specific to sexual interactions, for example. Languages don't have grammatical machinery specific to content domains such as sexuality or food. As far as I can tell, the kinds of information easily conveyed by grammar is one level more abstract: action in general, not sexual action in particular; and desire, not specifically sexual desire. My hunch is that what sociolinguists study may not be a separate domain; it may be the intersection of sociality and language. In other words, there may not be a separate psychological faculty for the social use of language; if you have a species with a social psychology and give them language, then what sociolinguists study will fall out as they interact.

Daly: If one's hypothesis were that the utility of a language in our evolutionary history had been primarily in order to convey social information, would this lead to any different propositions about the structure of language that one could test? Different, that is, from what you'd expect if your hypothesis were instead that we had mainly used language to convey information about the whereabouts or state of non-human resources? Has anyone tried to ascertain whether there is any sense in which language design features reflect its primary uses at that kind of level?

Pinker: Yes. If the only information that you were given was the semantics that is shared by all human languages, and you tried to figure out the human niche from that, you would guess that we are a species with a theory of mind because we have grammatical constructions that convey attitudes toward propositions. You would guess that we are a social species concerned with issues of altruism because there are elements dedicated to deontic information such as *should*, *ought* and *must*. And you would guess that we are obsessed with space and force because there are constructions for things being at a location and moving from one location to another. This rules out many hypotheses in which language evolved to express narrow and specific kinds of messages. Language could not have *just* been a means of conveying gossip; if it were, why would we have propositions for talking about where objects are located? As best as we can determine from the interface between language and thought, that is, semantics, that there seems to be a set of contents that the human species is obsessed with conveying to other people, of which space, action and social relations seem to be the main players.

Shepard: With regard to the origin of language, you contrasted macromutation versus microevolution via selection. I'm not sure why these two possibilities should be opposed. That is, if language is an adaptation doesn't that leave it open as to what extent it's through micro- or macromutation? Language is a complex system which has presumably emerged over a relatively short period of time (compared to spatial competency, which presumably evolved over a much longer period of time). Is it possible that some system in the brain that evolved in the service of something else, such as spatial cognition, becomes pre-empted for a new function?

Pinker: Macromutations do not generate jumps in complexity; they enable a parametric variation of complexity that's already present. In the evolution of a complex organ by natural selection, many, many micromutations must be involved, simply on the basis of probability considerations. Since mutations are random with respect to function, one large mutation is extremely unlikely to give you all the pieces you need to convey who did what to whom in a stream of noise. Specifically, if you took a computer program or a neural network that was designed to process spatial information, and changed a single aspect in a random way, it is unlikely to change into a system for encoding information into a complex sound wave which another human can decode. More generally, one random change to a pre-existing system is not going to result in a well-engineered system capable of performing a new function. Also, though the emergence of language is recent compared to the eye and the ear, it is still a matter of about six to eight million years, based on the assumption that chimpanzees have zero language, and that span of time encompasses 300 000 generations. Therefore, it is not necessary to invoke macromutations just to be consistent with the chronology.

Shepard: I wouldn't argue with that. It's just that there are so many things in language that are in one way or another isomorphic to spatial cognition.

Pinker: I think 'isomorphic' is too strong a word. I feel silly saying this to you, of all people, but surely a lot of spatial cognition is analogue, in a way that language emphatically is not. Language is wonderful for many things but it is spectacularly bad for other things, such as conveying motor skills and spatial arrangements. Giving directions is a chore and it is virtually impossible to tell someone how to assemble something over the telephone. Most of our spatial sense is far richer than what language is designed to express.

Dawkins: The point I want to make follows on from this. For people whose livelihoods depend on it, the ability to read animal footprints in the ground is extraordinarily well developed and is sometimes compared with the ability to read. This may also have been turned into other abilities. If you can read footprints, for example, you have the ability to translate a history of events into a two-dimensional map. Perhaps this could lead to the development of art. It wouldn't be a large step from reading footprints in the sand to picking up a stick and drawing a picture of the animal's path. There's also the ability to infer from clues that have been left behind something about what's happened in the past. The mental abilities required for reading footprints in the ground could eventually have been turned into something far grander, perhaps even the scientific method.

Pinker: It is indeed astonishing how foragers can look at what appears to us to be a random scratch and infer the size, age, species, physical condition and whereabouts of the animal that made it. That ability nicely answers the commonly asked question, isn't human intelligence hyperdeveloped for what our ancestors could have used it for?

Bouchard: It would be better if you could use a different word other than 'reading'. However, you could actually be talking about the process that led on to the development of reading in its literary sense.

Pinker: My hunch is that there isn't a mental module for reading animal tracks. Rather, the foragers use their faculties of intuitive geography, intuitive biology and possibly also intuitive psychology, since animals share our psychology and foragers predict the behaviour of animals partly by trying to think like them.

Dawkins: I wanted to suggest that it is specifically a module for reading footprints.

Hauser: It's much simpler than that because many animals build associations in the world; for instance, animals can pick up smells or other cues to find out the location of other animals. This may not necessarily be 'reading', rather it may be the understanding of associations between particular cues and the environment.

Kacelnik: It is possible to conceive many scenarios, such as in the reading of footprints, where the abilities required for language to evolve existed previously and had a different function. An analogous situation to this is the development of the eye from sensing light. This is fine as a generality but if you are more ambitious and propose footprint reading as a precise hypothesis, then you would have to think of ways of differentiating that hypothesis from multiple possible alternatives.

Maynard Smith: There is a specific way in which one can ask what the language organ was doing before it was doing language. Many organs were originally doing something else; for example, our arms are modified fins. Therefore, it is possible that the mental machinery for analysing visual input has divided up and some of it has gone on to service language. The way of answering the question, which we may not be able to use yet but will soon be able to, is that if there is a language organ there must be many genes concerned with specific components of that system. Those genes didn't come from nowhere. They're homologues of other genes which were involved with the visual system, the auditory system or whatever. Therefore, we have to find and identify these specific genes and their homologues.

Daly: It is also possible, to a degree, to identify homology in neurochemistry and neuroanatomy.

Pinker: Yes, quite right. The main problem is that we are still confused as to which parts of the human brain are responsible for language.

Thornhill: Questions on the functional design of language and on the origins of language are fundamentally different. What are your views on the psychological adaptations that are involved in cross-cultural changes in language?

Pinker: We certainly don't have an adaptation specifically to change language over historical time spans. What we have is an adaptation to acquire language, and this can then be mixed with changes in local circumstances, such as the requirement of inventing new words to describe objects of local interest, combined with mispronunciation and mishearings that spread from speaker to speaker and eventually become endemic to a language community. If you put these small day-to-day innovations into a statistical model like those used in population genetics and epidemiology, where mutations or diseases can start out in a single individual and then come to fixation, and iterate that over many innovations in vocabulary or pronunciation across the generations, then you would see how Shakespearean English turned into modern English.

Cronin: You gave a crystal-clear characterization of language as an adaptation, which seems to apply not only to native language but also to a second or third language acquired by a child. That ability is presumably not an adaptation because it would rarely have been required and so selection couldn't have acted on it. What are your views on this?

Pinker: I've had this discussion with John Tooby and Leda Cosmides, who have argued that acquiring multiple languages might indeed be an adaptation. There may have been situations in our ancestors' foraging lifestyle in which several groups with divergent languages were interacting, say for the purposes of exchange, or abduction, for that matter. Another possibility is that bilingualism is a by-product of the fact that learning a language involves some degree of separation between content and storage. We obviously don't have innate slots in the brain for every word we acquire. Instead, we have what is essentially a large bank of random access memory, and an ability to write words into it. If you're already equipped to learn some presumably large number of words in one language, that could give you the ability to use some of those slots for words in another language. Similarly, learning a grammar must involve writing down a set of rules in the brain, and a flexible amount of storage space for grammatical rules might give people an ability to write down the rules for another language.

Gaulin: You argued that there is no cost involved in the sharing of information. However, the situation where information is given but not received may imply that there is a cost. What is the evidence on conversational reciprocity?

Pinker: People are careful about who they talk to and what they say. Language is freely used in intimate settings such as among family members. However, outside the family you might choose to deploy this ability to share information only in situations where your faculty of social cognition allows you to predict that the cost of sharing information will be repaid through some kind of reciprocity with your conversational partner, either in kind, that is by other information conveyed by language, or in some other currency.

Cosmides: This is related to an interesting part of economic theory that involves rival and non-rival goods, which is a separate dimension from having or not having property rights. Language in this sense would be a non-rival good in that when you know something that somebody else also knows, it doesn't prevent you from using that knowledge. One of the strange properties of non-rival goods is that they are under-produced. Since ideas and knowledge are non-rival, this means they are under-produced as well. Moreover, the exchange of information is clearly not always cost free, as we all know when we can't get our research done because we're teaching too much.

Buss: I am interested in a common phenomenon, which is the derogation of competitors in the context of mating and the ways in which people use language to impugn the status, character and reputation of their rivals to make them less desirable to members of the opposite sex. I'm trying to understand how to conceptualize the nature of this. Clearly, it involves language and the theory of mind because you have to know the desires of the opposite sex and the exploitable qualities of the rival that you can use to tap into that desire. I have been thinking about the issue of domain specificity and domain generality in conjunction with the Dawkins & Krebs (1978) argument about animal signals as manipulations. You've also emphasized the communication function. Therefore, I'm wondering whether it is possible to conceptualize this as a set of different domain-specific mechanisms interfacing with language to produce this phenomenon, or can we conceptualize it as a derogation of competitors module that taps these different psychological capabilities?

Pinker: It is an empirical question, but I think it is unlikely that there is a derogation of competitors module. The derogation of competitors probably arises by an interaction of the person's romantic and sexual modules, which set up the goals of winning or keeping a romantic partner; their theory of mind module, which knows that the romantic interest will be weighing the desirability of oneself and one's rival; their biographical database, which records compromising information about people, especially rivals; their social or Machiavellian intelligence, which predicts that one's romantic interest will think less of the rival if she knows the compromising information; and their language faculty, which is used as a means to the end of conveying that information to her.

In general, though I talk about modules, I don't think we should take the module metaphor too seriously. A better metaphor may be the tissues, organs and systems of the body. The blood and the skin are specialized organs, but they don't function in isolation. And you can't draw sharp spatial boundaries around the circulatory system, the lymphatic system, the nervous system and so on; they are specialized in their structure, but they are integrated into a system. Behaviour, too, requires the co-ordination of many mental systems. We use modules as a shorthand for specialized, heterogeneous structure, but in actual behaviour there are complicated interactions among many of these specializations.

Buss: Would there be design circuits for tapping those different mechanisms?

Pinker: There would certainly have to be an interface, i.e. data conduits between the machinery of grammar, our knowledge about people and their beliefs and desires, our goals and so on. These information pathways could be adaptations for the specific purposes that language is most useful for. It's easy to use language to convey the idea 'John had an affair with Mary but wants you to think he didn't' but it's difficult to use language to convey sights, smells and motor sequences. Perhaps the difference comes from there being a high bandwidth data pathway between language centres and social centres, but no pathway of that kind between the language centres and centres for olfaction, motor control and early stages of visual processing.

References

Darwin C 1872 The expression of the emotions in man and animals. John Murray, London

Dawkins R, Krebs JR 1978 Animal signals: information or manipulation. In: Krebs JR, Davies NB (eds) Behavioural ecology: an evolutionary approach. Blackwell Scientific Publications, Oxford, p 282–309

Ohala JJ 1983 Cross-language use of pitch: an ethological view. Phonetica 40:1–18

Ohala JJ 1984 An ethological perspective on common cross-language utilization of F_o of voice. Phonetica 41:1–16

Cross-species comparisons

David F. Sherry

Department of Psychology, University of Western Ontario, London, Ontario, Canada N6A 5C2

Abstract. Cognitive and neural adaptations in animals have been analysed using the comparative method. Comparisons between closely related species that differ in a cognitive or neural character, and comparisons between distantly related species that share a cognitive or neural character, can be used to identify adaptations. Recent research has identified adaptive modifications of memory and the hippocampus that have evolved convergently in two clades of food-storing birds, the chickadees and tits (Paridae), and the jays and nutcrackers (Corvidae). Similar modifications of the hippocampus occur in other groups of animals, such as the cowbird brood parasites, in which there has been selection for spatial memory. Three general patterns that emerge from the comparative study of animal cognition provide a framework for research on human psychological adaptations: the existence of both specialized and general cognitive capacities; a clear relation between specialized capacities and specific selective pressures; and evolutionary change in the relative size of brain areas with cognitive functions.

1997 Characterizing human psychological adaptations. Wiley, Chichester (Ciba Foundation Symposium 208) p 181–194

Comparisons between species reveal psychological adaptations in two ways. Species that are closely related phylogenetically may differ in some character of interest, indicating that despite an evolutionary history that is largely shared, adaptation to different selective pressures has produced evolutionary divergence of the character in these species. Comparisons among species of food-storing birds, for example, have revealed a number of correlated behavioural, cognitive and neural adaptations. Nowhere in its extensive range from North Africa to Scandinavia and from the west coast of Ireland to the east coast of Japan does the Great tit (*Parus major*) store food. Nearly all other members of the avian family Paridae, to which the Great tit belongs, store food (Hampton & Sherry 1992, Sherry 1989, Slikas et al 1996). This divergence within the Paridae suggests that there exists, or once existed, some selective pressure not shared by the Great tit (and its non-storing close relatives *Parus caeruleus* and *Parus monticolus*) and most other tits and chickadees. The evolutionary question is, what difference in selective pressure is responsible for this divergence in food-storing behaviour between Great tits and other parids? As we shall see, selection for food storing has had a cascade of effects on memory and the brain in food-storing birds. Identifying the selective pressure responsible for food storing in these birds amounts to identifying a selective pressure that can cause evolutionary modification of cognitive function and its neural architecture.

A second way that comparisons between species can reveal psychological adaptations is the occurrence of a shared character in species that are only distantly related. Such species share little of their evolutionary history, but have at least one selective pressure in common, to which both have been exposed, and which has produced evolutionary convergence of the character in question. Chickadees, nuthatches and jays all store food, but belong to different avian clades (Sheldon & Gill 1997). Clades are monophyletic, that is, all members of a clade share a common ancestor and the clade contains all descendants of that common ancestor. The three clades to which chickadees, nuthatches and jays belong also contain species that do not store food: penduline tits in the 'chickadee' clade, gnatcatchers, thrushes and starlings in the 'nuthatch' clade, and vireos in the 'jay' clade. This phylogenetic pattern and the fact that only a handful of the world's other passerine species store food indicate that food storing is not an ancestral trait, shared by chickadees, nuthatches and jays because their most recent common ancestor stored food, but is instead a derived character that has arisen independently in each group. The evolutionary question in this case is, what shared selective pressure has produced the convergent evolution of food storing in chickadees, nuthatches and jays? Storing behaviour differs somewhat among these groups of birds, not surprisingly given their independent evolutionary histories. Nevertheless, all three groups possess long-term memory for scattered cache sites and a hippocampus that is much larger than that of closely related non-food-storing birds. These similarities indicate that selection for food storing may have modified memory in similar ways in these birds and produced similar anatomical changes in the hippocampus. Evolutionary change in cognition and the brain has proceeded independently in these three groups of birds but has arrived at similar end points in response to a shared selective pressure.

The remainder of this chapter will provide a brief overview of what is known about cognitive organization and the brain in food-storing birds (for more detail on the comparative method, see Harvey & Krebs 1990, Harvey & Pagel 1991). A question that must be addressed, however, is whether such a comparative evolutionary approach can answer questions about human psychological adaptations. There is, clearly, only one human species and no fossil record of hominid behaviour and cognitive processes. How, then, can the comparative approach contribute to the study of human adaptation? The comparative study of cognition in humans and primates is one approach (Hauser 1997, this volume). Another approach is to investigate in humans the patterns in cognitive processes and brain organization that have been revealed by the comparative study of animal cognition. These patterns are as follows:

(1)   animal cognition consists of a collection of functionally specialized capacities together with more general-purpose capacities;

(2)   specialized cognitive abilities are adaptations to specific selective pressures; and

(3)   the structure of the brain, in particular the relative size of its component parts, is readily modified by natural selection in response to selective pressures on cognitive capacities.

These three patterns are so widespread and appear with such regularity in the comparative study of animal cognition and the brain that they are, in all likelihood, true of human psychological adaptation as well.

Food-storing birds

Behaviour

Chickadees, nuthatches and jays make thousands of widely scattered food caches in a year and retrieve them several days to several months later by remembering the spatial locations of caches. Experimental studies in the field and in the laboratory have shown that cache sites are remembered, and are not placed in favourite sites, re-encountered by chance, marked with beacons or revisited in any ordered sequence. The birds use nearby landmarks to identify the locations of cache sites and may also use global orientation cues such as the sun compass to relocate cache sites (Wiltschko & Balda 1989). Recent research on memory and orientation in food-storing birds is reviewed by Kamil & Balda (1990), Sherry & Duff (1996), Shettleworth (1995) and Vander Wall (1990).

   Most, but not all, corvids and parids store food, and the species that exhibit the behaviour do so to different degrees. The natural distribution of food-storing behaviour within these two avian families has made possible the comparative investigation of cognitive capacities. The logic of the comparisons is that if long-term memory for large numbers of spatial locations is the result of a general enhancement of memory in food-storing species, then comparisons of storing and non-storing members of a family should reveal general superiority on memory tasks in the food-storing species. If, on the other hand, the cognitive mechanisms of food storing are narrowly specialized for the task of cache recovery, then food-storing species should not be general-purpose mnemonists, but domain-specific mnemonists.

   The empirical results present two different pictures. In the family Corvidae the species that store more food in the wild show more accurate cache recovery in the laboratory and a general superiority on spatial memory tasks (Olson 1991, Olson et al 1995). In the family Paridae, in which the comparisons between storers and non-storers must necessarily always involve non-storing tasks, the differences observed between species are often slight, and the ranking of species by their performance depends on exactly which task is being performed (Shettleworth 1995). Among the parids, differences in behaviour and memory are not well predicted by whether or not a particular species stores food; whereas among the corvids, differences in behaviour are well predicted by differences in food storing. This is probably because the corvids are a much more heterogeneous group than the parids. All parids in which food storing has been described are members of the same genus (*Parus*), while the corvid food storers that have been experimentally examined to date come from three different genera (*Nucifraga*, *Aphelocoma* and *Gymnorhinus*). It is possible, therefore, that the evolution of food storing has followed a different course in corvids than in parids, and that this history accounts for some of the differences observed between these two food-storing groups.

It appears that in corvids, selection on memory has resulted in a system that enables food storers to perform well on a variety of tasks. In parids the evidence suggests that selection on memory has resulted in a specialized system that confers little advantage on memory tasks that do not involve food storing. In both cases these changes can be identified with selection acting specifically on the cognitive components of food-storing behaviour.

The hippocampus

Lesions of the avian hippocampus abolish accurate cache recovery without disrupting either caching or search for caches (Sherry & Vaccarino 1989). Chickadees cache and search for caches normally but fail to find them. This deficit is quite specific: lesions of the avian hippocampus disrupt memory for spatial locations, not memory in general (Bingman et al 1995, Hampton & Shettleworth 1996, Sherry & Vaccarino 1989). Furthermore, comparisons between food-storing and non-storing species of passerine birds indicate dramatic specialization of the hippocampus in food-storing birds. The hippocampus is considerably larger in food-storing species, compared to non-storing species of the same body size (Sherry et al 1989, Krebs et al 1989). Comparisons within food-storing families show that greater food storing in the wild is correlated with a relatively larger hippocampus (Basil et al 1996, Hampton et al 1995, Healy & Krebs 1992).

These comparative studies have helped to reveal the selective pressures acting on cognitive organization and the brain of food-storing birds. Greater hippocampal size in food-storing parids and corvids compared to non-storing members of these families shows that change in the brain has occurred as a result of selection for the ability to retrieve scattered caches of food. Differences in body size, habitat, social structure and other variables are not correlated with the observed difference in hippocampal size (Sherry et al 1989, Krebs et al 1989). Variation in hippocampal size within the parids and corvids can be related quite directly to variation in the amount of food storing typical of each species.

The evolution of food storing

What was the evolutionary origin of food storing in the passerines? The parids probably first appeared during the rapid radiation of the passerines in the early Miocene, about 20 million years ago (Feduccia 1995). Given the basal position of the Great tit in the parid clade, its distribution and its non-storing status, it is likely that further speciation among the chickadees and tits in the late Miocene, approximately seven million years ago, involved radiation into northern temperate coniferous forests and the simultaneous evolution of food storing (Slikas et al 1996, Sheldon & Gill 1997). It is likely that selection for effective retrieval of cached food resulted in selection for long-term memory for large numbers of scattered spatial locations, which in turn resulted in modification of the hippocampus.

The evolution of food storing among the corvids presents a more complex picture, in part because the corvids are a more diverse group anatomically and behaviourally, and in part because of behavioural specializations in some corvids that result from co-evolved mutualisms with the pines, oaks and beeches, for which they are the primary agents of dispersal (Tomback & Linhart 1990). Nevertheless, it is clear that in some corvids, selection for effective cache retrieval, comparable to that proposed for parids, has resulted in evolutionary change in the cognitive and neural apparatus required for long-term memory of large numbers of scattered spatial locations.

Brood parasites

It is likely that many animals have specialized memory requirements. Brown-headed cowbirds (*Molothrus ater*) are brood parasites that lay their eggs in the nests of several hundred different host species. Females search for the nests of potential hosts but do not lay an egg in a host nest immediately upon discovering it. Females usually lay one egg near dawn each day in a nest they have previously located, and spend the remainder of the morning searching for host nests in which to lay eggs on subsequent days. Females parasitize up to 40 host nests in a breeding season and probably locate many more host nests than they eventually parasitize. Male brown-headed cowbirds do not assist females in searching for host nests. Female brown-headed cowbirds are thus exposed to selection for remembering the spatial locations of previously discovered host nests whereas males are not. One result of this sex-specific pattern of selection is a greater hippocampal size in female cowbirds compared to males, a sex difference that does not occur in other members of the icterid family to which cowbirds belong (Sherry et al 1993). Interestingly, not all subspecies of the brown-headed cowbird show this sex difference in hippocampal size, perhaps because of differences among subspecies in habitat, behaviour or recent evolutionary history (Uyehara et al 1997).

Cowbirds as a group show remarkable variation in their brood parasitic behaviour, as illustrated by three species found in South America. The bay-winged cowbird (*Molothrus badius*) is not a brood parasite at all. It breeds communally, usually building its own nest but sometimes appropriating the nests of other species (Ridgely & Tudor 1989). Male and female bay-winged cowbirds both exhibit parental care, and helpers at the nest have been reported (Fraga 1991). The screaming cowbird (*Molothrus rufoaxillaris*) is a specialist on one host only, the aforementioned bay-winged cowbird. Unlike brown-headed cowbirds, though, screaming cowbird males accompany females in searching for the nests of their hosts (Mason 1987). Finally, the shiny cowbird (*Molothrus bonariensis*) is a generalist brood parasite more similar in behaviour to the brown-headed cowbird, and females search for host nests unaccompanied by males. Reboreda et al (1996) found that the hippocampus is larger overall in the two brood parasitic species than in the non-parasite, and that a sex difference in hippocampal size, in favour of females, occurs in the generalist parasite but in neither of the other two species.

Lanyon (1992) proposed a phylogeny for the cowbirds, based on DNA sequences of the mitochondrial cytochrome c gene, in which non-parasitism was the ancestral condition and evolution of the group proceeded by exploitation of increasing numbers of hosts. The use of a greater number of hosts by cowbirds is thus correlated with an evolutionary increase in hippocampal size.

Evolutionary change in the hippocampus

Selection for spatial memory, or perhaps spatial orientation, can influence the avian hippocampus, whether the source of the selective pressure is search for food caches or search for potential host nests. Furthermore, a sex difference in the size of the hippocampus can arise if the sexes are exposed to different selective regimes. A sex difference in hippocampal size, in favour of males, occurs in some polygynous rodents but not in congeneric monogamous species because of selection for spatial ability in polygynous males that compete for mates by increasing the size of their home range (Jacobs et al 1990, Jacobs & Spencer 1994). As Lande (1980) has pointed out, if the genetic basis of a sex difference is autosomal, as is likely for complex polygenic traits such as the relative size of brain regions, more intense sex-specific selection must occur for sex differences to evolve, compared to the intensity of selection required for character divergence between species. Species do not exchange genes, and divergence between species can occur more rapidly and be maintained with less intense selection than is necessary to establish and maintain divergence between the sexes, which, of course, do share genes.

Implications for human psychological adaptations

Three general patterns in the evolution of animal cognition were listed at the beginning of this chapter: (1) the existence of both specialized and general-purpose cognitive capacities; (2) adaptation in specialized capacities in response to specific selective pressures; and (3) modification of brain areas associated with specialized capacities. For humans, there is reasonable support for the first assertion. Human memory consists of multiple neurally localized systems, some serving specialized functions, others of broad domain (Sherry & Schacter 1987, Schacter 1996). Language acquisition is a similar instance of a highly specialized cognitive system with its own neural hardware (Pinker 1997, this volume). Further comparative work on the neural localization of cognitive function in the brains of humans and other primates, perhaps using functional imaging techniques, is likely to provide new information relevant to the third assertion and elucidate the paths of evolutionary change in the human brain.

The area in which few unequivocal conclusions can be reached at present is the second assertion, the identification of selective forces that have shaped the evolution of human cognition. Sex differences in human spatial memory, for example, are commonly reported, usually showing better performance by men on tasks that involve mental rotation (Hampson 1995). In both men and women spatial ability can change with hormonal status (Kimura & Hampson 1994). A variety of interesting but still untested hypotheses have been proposed to account for this sex difference in humans, including sex differences in foraging, dispersal and sexual selection (e.g. Silverman & Eals 1992, Gaulin 1992). The major challenge that research on animal cognitive processes presents to the study of human psychological adaptations is to identify the as yet largely unknown selective pressures that have directed the evolution of human cognition.

Acknowledgements

I thank Sue Healey and Frank Gill for their helpful comments on the manuscript. Preparation of this article was supported by the Natural Sciences and Engineering Research Council of Canada.

References

Basil JA, Kamil AC, Balda RP, Fite KV 1996 Differences in hippocampal volume among food storing corvids. Brain Behav Evol 47:156–164

Bingman VP, Jones T-J, Strasser R, Gagliardo A, Ioalé P 1995 Homing pigeons, hippocampus and spatial cognition. In: Alleva E, Fasolo A, Lipp H-P, Nadel L, Ricceri L (eds) Behavioural brain research in naturalistic and semi-naturalistic settings. Kluwer Academic Publishers, Dordrecht (Proc NATO Advanced Study Institute) p 207–223

Feduccia A 1995 Explosive evolution in Tertiary birds and mammals. Science 267:637–638

Fraga RM 1991 The social system of a communal breeder, the bay-winged cowbird *Molothrus badius*. Ethology 89:195–210

Gaulin SJC 1992 Evolution of sex differences in spatial ability. Yearb Phys Anthropol 35:125–151

Hampson E 1995 Spatial cognition in humans: possible modulation by androgens and estrogens. J Psychiatry Neurosci 20:397–404

Hampton RR, Sherry DF 1992 Food storing by Mexican chickadees and bridled titmice. Auk 109:665–666

Hampton RR, Shettleworth SJ 1996 Hippocampal lesions impair memory for location but not color in passerine birds. Behav Neurosci 110:831–835

Hampton RR, Sherry DF, Shettleworth SJ, Khurgel M, Ivy G 1995 Hippocampal volume and food-storing behavior are related in parids. Brain Behav Evol 45:54–61

Harvey PH, Krebs JR 1990 Comparing brains. Science 249:140–146

Harvey PH, Pagel MD 1991 The comparative method in evolutionary biology. Oxford University Press, Oxford

Hauser MD 1997 Tinkering with minds from the past. In: Characterizing human psychological adaptations. Wiley, Chichester (Ciba Found Symp 208) p 95–131

Healy SD, Krebs JR 1992 Food storing and the hippocampus in corvids: amount and volume are correlated. Proc R Soc Lond Ser B 248:241–245

Jacobs LF, Spencer W 1994 Natural space-use patterns and hippocampal size in kangaroo rats. Brain Behav Evol 44:125–132

Jacobs LF, Gaulin SJC, Sherry DF, Hoffman GE 1990 Evolution of spatial cognition: sex-specific patterns of spatial behavior predict hippocampal size. Proc Natl Acad Sci USA 87:6349–6352

Kamil AC, Balda RP 1990 Spatial memory in seed-caching corvids. In: Bower GH (ed) The psychology of learning and motivation, vol 26. Academic Press, New York, p 1–25

Kimura D, Hampson E 1994 Cognitive pattern in men and women is influenced by fluctuations in sex hormones. Curr Direct Psychol Sci 3:57–61

Krebs JR, Sherry DF, Healy SD, Perry VH, Vaccarino AL 1989 Hippocampal specialization of food-storing birds. Proc Natl Acad Sci USA 86:1388–1392

Lande R 1980 Sexual dimorphism, sexual selection, and adaptation in polygenic characters. Evolution 34:292–305

Lanyon SM 1992 Interspecific brood parasitism in blackbirds (Icterinae): a phylogenetic perspective. Science 255:77–79

Mason P 1987 Pair formation in cowbirds: evidence found for screaming but not shiny cowbirds. Condor 89:349–356

Olson DJ 1991 Species differences in spatial memory among Clark's nutcrackers, scrub jays, and pigeons. J Exp Psychol Anim Behav Processes 17:363–376

Olson DJ, Kamil AC, Balda RP, Nims PJ 1995 Performance of four seed-caching corvid species in operant tests of nonspatial and spatial memory. J Comp Psychol 109:173–181

Pinker S 1997 Language as a psychological adaptation. In: Characterizing human psychological adaptations. Wiley, Chichester (Ciba Found Symp 208) p 162–180

Reboreda JC, Clayton NS, Kacelnik A 1996 Species and sex differences in hippocampus size in parasitic and non-parasitic cowbirds. NeuroReport 7:505–508

Ridgely RS, Tudor G 1989 The birds of South America. University of Texas Press, Austin, TX

Schacter DL 1996 Searching for memory. Basic Books, New York

Sheldon FH, Gill FB 1997 A reconsideration of songbird phylogeny, with emphasis on the evolution of titmice and their sylvioid relatives. Syst Biol 45:473–495

Sherry DF 1989 Food storing in the paridae. Wilson Bull 101:289–304

Sherry DF, Duff SJ 1996 Behavioural and neural bases of orientation in food-storing birds. J Exp Biol 199:165–172

Sherry DF, Schacter DL 1987 The evolution of multiple memory systems. Psychol Rev 94:439–454

Sherry DF, Vaccarino AJ 1989 Hippocampus and memory for food caches in black-capped chickadees. Behav Neurosci 103:308–318

Sherry DF, Vaccarino AJ, Buckenham K, Herz RS 1989 The hippocampal complex of food-storing birds. Brain Behav Evol 34:308–317

Sherry DF, Forbes MRL, Khurgel M, Ivy GO 1993 Females have a larger hippocampus than males in the brood-parasitic brown-headed cowbird. Proc Natl Acad Sci USA 90:7839–7843

Shettleworth SJ 1995 Comparative studies of memory in food storing birds. In: Alleva E, Fasolo A, Lipp H-P, Nadel L, Ricceri L (eds) Behavioural brain research in naturalistic and semi-naturalistic settings. Kluwer Academic Publishers, Dordrecht (Proc NATO Advanced Study Institute) p 159–192

Silverman I, Eals M 1992 Sex differences in spatial abilities: evolutionary theory and data. In: Barkow J, Cosmides L, Tooby J (eds) The adapted mind. Oxford University Press, Oxford, p 533–549

Slikas B, Sheldon FH, Gill FB 1996 Phylogeny of titmice (Paridae). I. Estimate of relationships among subgenera based on DNA–DNA hybridization. J Avian Biol 27:70–82

Tomback DF, Linhart YB 1990 The evolution of bird-dispersed pines. Evol Ecol 4:185–219

Uyehara JC, Grisham W, Arnold AP 1997 Sexual monomorphism in hippocampi in a subspecies of the brown-headed cowbird, submitted
Vander Wall SB 1990 Food hoarding in animals. University of Chicago Press, Chicago, IL
Wiltschko W, Balda RP 1989 Sun compass orientation in seed-caching scrub jays (*Aphelocoma coerulescens*). J Comp Physiol 164:717–721

DISCUSSION

Mealey: I would like to ask a few questions about male screaming cowbirds, which search for nests. Are they monogamous? Do they search separately from the female? And how do they ensure their paternity?

Sherry: The screaming cowbird, the specialist parasite, is monogamous and Mason (1987) observed pairs together at host nests. Mated males keep close to females and probably do not search separately from the female.

Kacelnik: And the fact that they are together all the time ensures the paternity of the male.

Mealey: This would imply that the females have a bigger hippocampus because they're the ones doing the searching and the males are just tagging along.

Sherry: Reboreda et al (1996) did not find any sex differences in relative hippocampal size in the screaming cowbird. They suggested that this is because the two sexes have similar spatial habits, i.e. they search for host nests together. In contrast, they did find a sex difference in the shiny cowbird, the generalist parasite, because in this species females search for host nests alone. This is what we had previously found in brown-headed cowbirds, which are also generalist parasites. Brown-headed cowbird females are reported to search for host nests alone, by walking on the forest floor, sitting in the canopy or flying into the understorey to flush incubating birds.

Kacelnik: We need to know exactly what is going on in terms of who remembers what and who is leading who. There are no data on this, although much of the current research is centred on finding out the behavioural details that underlie these different lifestyles. The same animals are now undergoing spatial tests in the laboratory to see whether there are sex differences in spatial problem solving in the species that show these sexual differences in behaviour with respect to space in the field.

Hauser: There are species in which overall brain size is the same but the hippocampus differs, so there must be some other piece of the brain that is larger in those other species. This may not tell us anything about specialization, but there must be some sort of a trade-off and a resultant cost.

Sherry: In the data that I showed, telencephalon size is measured excluding the hippocampus, so the same-sized brain does not necessarily have a larger hippocampus. It is often supposed that there is some constraint on overall brain size but this may not be true if building a bigger skull, at least in birds, poses no problem, so it's not clear whether there is a size cost. However, although the hippocampus may not be displacing other brain areas, it is nevertheless energetically costly, and it is probably fair to say there is an energetic trade-off when hippocampal cells are added.

Møller: There is an interesting story about the brain space in males in relation to song because the song centres in males regress after the breeding season, implying that there is some cost involved in maintaining this structure.

Kacelnik: We have found that the sexual differences disappear in winter, as have others (Smulders et al 1995, Barnea & Nottebohm 1994). Other groups have found that the food-storing species that have a larger hippocampus than those who don't store food don't show these differences at the times of the year when they are not actively storing food. We should also take into account that the hippocampus is only a small fraction of the entire telencephalon.

Sherry: The seasonal evidence comes from three sources: (1) Barnea & Nottebohm (1994) have found seasonal neurogenesis in the hippocampus of black-capped chickadees, with a maximum in autumn; (2) Smulders et al (1995), also working with chickadees, have found that the hippocampus is larger, relative to the rest of the brain, in autumn than at other times of year; and (3) Karin Petersen and I have collected data that show the hippocampus of food-storing, white-breasted nuthatches is larger in winter than in spring. Food storing is seasonal in chickadees and nuthatches, with a maximum in autumn and winter, and a minimum in spring and summer. These neural changes in the hippocampus may be related to the seasonal waxing and waning of food-storing behaviour.

Hauser: Some of those results are subject to a different interpretation because of the staining techniques, i.e. there may be seasonal variations in the ability of neurons to pick up the stain. Indeed, some of these data on bird song have been criticized by Gahr (1990) because the staining techniques themselves can be sensitive to the season. Therefore, seasonal changes in brain volume may have been overestimated.

Sherry: Gahr has suggested that if variation in cell activity affects sensitivity to a stain, Nissl stain for example, then changes in cell activity and phenotype can be mistaken for a change in the overall size of the nucleus. This is not a problem uniquely associated with season. Gahr's recommendation is that in addition to standard staining techniques, nuclei should be identified by some functional criterion, such as connectivity or a cytochemical marker (Gahr 1994).

Gaulin: I would like to add that meadow voles are seasonal breeders and that the sex differences in range size in this species disappear outside of the breeding season, as do sex differences in maze performance (Gaulin 1995).

Beecher: I would like to bring up the issue of special- versus general-purpose mechanisms. You began talking about specialists, i.e. caching birds, but then you mentioned a variety of other examples. You showed that there was an elaboration of a relatively general mechanism: a structure that we know works in birds, in mammals and probably in other vertebrates, and that is spatial memory. Therefore, we're simply observing a general-purpose hypertrophy in a variety of animals that have specialized spatial memory needs.

Sherry: We may be combining general, in the sense of phylogenetically widespread, and general, in the sense of general purpose. It is possible to have a mechanism that is highly specialized in function, but is phylogenetically widespread. I would still call that a specialized cognitive mechanism.

Beecher: I'm thinking more about the original arguments about scatter caching birds. Were these birds using memory mechanisms that are specialized to this task or were they simply extending mechanisms that they already had for basic memory operations? For example, Bob Reineke (1995) has compared two corvids, crows and grey jays. Grey jays are long-term food storers, storing hundreds of items that they will retrieve months later, in the middle of winter. Crows, on the other hand, just dabble, i.e. they place some food somewhere and they go back and get it the same day. He gave both species a difficult spatial task with over 1000 locations, and he tested them up to 30 days after caching. He found that the crows performed just as well as the grey jays, suggesting that to some extent they're using general-purpose mechanisms.

Sherry: It is a mixed picture. In other comparisons among corvids, those that cache the most, such as Clark's nutcracker and the pinyon jay, do better on spatial memory tasks than corvids that cache less, such as the scrub jay and Mexican jay (Olson et al 1995). On a test of memory for colours, using the same touch-screen procedure, no differences were found among these four species. This suggests that whatever spatial memory mechanism nutcrackers are using to find caches, it is general in the sense that it can also be used for spatial non-matching to sample on a touch screen. Other corvids can perform the task, but not nearly as well as nutcrackers, so the enhancement of spatial memory is not general across all corvids. Furthermore, nutcrackers do no better than other corvids on memory for the colours of stimuli. Among the chickadees and tits the differences observed between food-storing and non-food-storing species are less clear-cut, and seem to depend more on exactly how the memory tasks are set up (Shettleworth 1995).

Nisbett: Are there sex differences in the human hippocampus?

Sherry: Magnetic resonance imaging data that I am familiar with on the size of the human hippocampus show no sex differences, either for the whole hippocampus or for the left and right hippocampi taken separately (Jack et al 1989).

Daly: There is a recent paper by Tom James and Doreen Kimura, who work in David Sherry's department (James & Kimura 1997). They used the Silverman and Eals test, which involves seeing an array of objects in space and then switching pairs (see Silverman & Eals 1992), and they found that women did significantly better than men at remembering the initial whereabouts of objects that were switched, at detecting switches and at identifying which objects had been switched. James & Kimura also found, using the same subjects in the same experimental session, that the sex differences were wiped out if the arrays were similar and the movements were not pair switches but were movements of objects to positions that were not previously occupied. They're arguing that we don't yet know what it is that women are better at doing.

Tooby: Your claim that this doesn't involve spatial memory in the same way that the other task does is puzzling. Intrinsic to both tasks are spatial location tags.

Sherry: It's probably not worth debating about the appropriate definitions of 'spatial' and 'non-spatial' tasks. The definition determines what is a spatial task and what is not. I think there are important differences, nevertheless, between the task that Silverman and Eals work with and other tasks, such as mental rotation or orientation and navigation.

Gangestad: My understanding is that in humans the hippocampus has functions pertinent to memory tasks not clearly spatial in nature. There is some controversy about exactly what those functions are. Can you comment on hippocampal specializations in humans?

Sherry: The human hippocampus is involved in many kinds of cognitive processing that are not spatial in any obvious way. It is an interesting question whether the evolutionary origin of contemporary hippocampal function in humans is spatial in some way. People interested in the spatial function of the hippocampus might argue that it is because what other evolutionary origin is it likely to have had? The hippocampus clearly has important spatial functions in many other animals. Others might argue that olfactory processing is the ancestral function, or some much more general cognitive process.

Gangestad: The hippocampus seems to be involved in some configural memory tasks, although how it is involved remains debatable. Can you comment on the possibility that in humans selection generalized a capability that was specifically spatial beforehand?

Sherry: Yes, that is likely. The theoretical argument among hippocampal workers is whether there is a broader umbrella that describes hippocampal function, under which spatial memory and other proposed functions can be subsumed. There are several proposals for what this broader umbrella might be, including declarative memory and the formation of configural associations. But while acknowledging that there may be a more encompassing way of describing the many things that the hippocampus does, especially in humans, I am always drawn back to its spatial function, partly because of comparative data showing differences in the hippocampus among animals that differ in their use of space.

Scheib: When you study adaptations do you have to consider whether the present environment is somewhat different to that in which the animals evolved?

Sherry: The prevailing assumption in most animal work is that the contemporary environment is the environment in which behaviour makes adaptive sense. Therefore, when we're trying to explain song repertoires or food caching in birds, we look at the current mating system or the current diet in the current environment.

Scheib: But for example isn't the cuckoo's environment constantly changing, in the sense that it has a co-evolutionary arms race going on with its hosts?

Sherry: Yes, there are cases where the contemporary environment is known to be unlike the environment in which the animal spent most of its evolutionary history. Brown-headed cowbirds, for example, moved into eastern North America in historical times, probably moving along corridors opened by the westward expansion of grazing (Mayfield 1965). This has had the effect that the cowbird's potential hosts in the east, forest species or grassland edge species, are heavily hit by cowbird parasitism, while potential hosts in grasslands in the middle of the continent are adapted to cowbird parasitism and raise many fewer cowbird nestlings.

Dawkins: Some of the methodology that you use for birds might be applied to other species such as humans. In the case of birds you had some theoretical reason for looking at the hippocampus, but suppose you hadn't. I am trying to suggest a methodology for doing this more generally, which might be useful for humans as well. Suppose you applied the methods of numerical taxonomists, who are in the habit of measuring hundreds of things without any preconceptions. So, you measure hundreds of variables, including information about which are food storers, and input them all into a computer. You could then do a discriminant function analysis to find out which measurements would weight heavily in order to differentiate the food storers from the non-food storers. This would be a way of teasing out not just the measurements that you have a priori reason to suspect, such as hippocampal size, but also possibly many other things. If you could do this with primates you might then be able to work out something about the human natural environment that you wouldn't otherwise have known.

Mealey: Discriminant analysis can be used at a variety of levels, and not just for cross-species analysis. I hope to be looking at monozygotic twins in the future by categorizing them according to different outcomes, and then using discriminant analysis to look for the triggers, i.e. the facultative mechanisms and different environmental cues in their backgrounds that trigger different developmental outcomes. Another level at which this kind of analysis can be used is to look for facultative differences that occur between cultures. That is, the categorization of different cultures, where the running of a discriminant analysis on all the ecological and environmental variables may give some clues on the triggers that set one culture off on a different track from another. Therefore, this method can be used across individuals, across cultures and across species; we should earnestly start doing such analyses.

References

Barnea A, Nottebohm F 1994 Seasonal recruitment of hippocampal neurons in adult free-ranging black-capped chickadees. Proc Natl Acad Sci USA 91:11217–11221

Gahr M 1990 Delineation of a brain nucleus: comparisons of cytochemical, hodological, and cytoarchitectural views of the song control nucleus HVc of the adult canary. J Comp Neurol 194:30–36

Gahr M 1994 Brain structure: causes and consequences of brain sex. In: Short RV, Balaban E (eds) The differences between the sexes. Cambridge University Press, Cambridge, p 273–300

Gaulin SJ 1995 Does evolutionary theory predict sex differences in the brain? In: Gazzaniga M (ed) The cognitive neurosciences. MIT Press, Cambridge, MA, p 1211–1225

Jack CR Jr, Twomey CK, Zinsmeister AR, Sharbrough FW, Petersen RC, Cascino GD 1989 Anterior temporal lobes and hippocampal formations: normative volumetric measurements from MR images in young adults. Radiology 172:549–554

James TW, Kimura D 1997 Sex differences in remembering the locations of objects in an array: location shifts versus location exchanges. Evol Hum Behav, in press

Mason P 1987 Pair formation in cowbirds: evidence found for screaming but not shiny cowbirds. Condor 89:349–356

Mayfield H 1965 The brown-headed cowbird, with old and new hosts. Living Bird 4:13–28

Olson DJ, Kamil AC, Balda RP, Nims PJ 1995 Performance of four seed-caching corvid species in operant tests of nonspatial and spatial memory. J Comp Psychol 109:173–181

Reboreda JC, Clayton NS, Kacelnik A 1996 Species and sex differences in hippocampus size in parasitic and non-parasitic cowbirds. NeuroReport 7:505–508

Reineke R 1995 A comparison of cache retrieval between gray jays and crows. PhD thesis, University of Washington, Seattle, WA, USA

Shettleworth SJ 1995 Comparative studies of memory in food storing birds. In: Alleva E, Fasolo A, Lipp H-P, Nadel L, Ricceri L (eds) Behavioural brain research in naturalistic and semi-naturalistic settings. Kluwer Academic Publishers, Dordrecht (Proc NATO Advanced Study Institute) p 159–192

Silverman I, Eals M 1992 Sex differences in spatial abilities: evolutionary theory and data. In: Barkow J, Cosmides L, Tooby J (eds) The adapted mind. Oxford University Press, Oxford, p 533–549

Smulders TV, Sasson AD, DeVoogd TJ 1995 Seasonal variation in hippocampal volume in a food-storing bird, the black-capped chickadee. Neurobiology 27:15–25

Cross-cultural patterns and the search for evolved psychological mechanisms

Steven J. C. Gaulin

Department of Anthropology, University of Pittsburgh, Pittsburgh, PA 15260, USA

Abstract. Darwin's principle of evolution by natural selection provides a theoretical basis for functional analyses of behaviour. This approach is complementary to traditional psychology: ideas about what behaviour was designed to do suggest how it might be organized. The cross-cultural record, because it focuses on the broadest characterization of human behaviour, can guide our search for psychological adaptations. Both universals (behavioural traits that are invariantly expressed despite cultural diversity) and conditional universals (behaviours that vary predictably with some environmental parameter) deserve attention from evolutionary psychologists. These behaviours are not themselves adaptations but are markers for underlying psychological traits that have been favoured by selection because they produce adaptive behavioural output. One universal (reciprocity) and one conditional universal (matrilineal investment bias) are used to exemplify how behavioural universals can guide the search for psychological adaptations.

1997 Characterizing human psychological adaptations. Wiley, Chichester (Ciba Foundation Symposium 208) p 195–211

Evolutionary biologists spend more of their effort trying to understand why particular kinds of behaviour — such as sociality or nepotism — exist, than they do detailing the proximate mechanisms that underlie the ontogeny and expression of these traits. Psychology is the discipline that has focused on mechanistic explanations of behaviour. While a minority of evolutionary psychologists may begin to ask why environment A is correlated with behaviour 1, most will still want to know how the correlation comes about; that is, psychology would retain its traditional and, I argue, proper focus on psychological mechanisms. What would be different about an evolutionary psychology is that the mechanisms it studies would typically be seen as adaptations in the Darwinian sense. Thus, the work of biologists and psychologists would be complementary, focusing on different levels of causation.

Not all the traits of organisms are adaptations; but those that are typically exhibit clear design for a particular function and are widespread in the population of interest. The central contribution of the cross-cultural record to evolutionary psychology is that it represents the definitive anthropological source on the distribution of human behaviour patterns. From that record, two categories of behaviour patterns, universals and what Brown (1991) calls implicational (or conditional) universals, should be of special interest to evolutionary psychologists. This is because they constitute markers or flags for candidate psychological adaptations.

Human universals

A universal is, as the name implies, a trait that is present in all described human societies. Some examples are: language, gossip, the one-year-old's fear of strangers, marriage, the conceptualization of intention, sexual division of labour, reciprocity and music. 'Culture' is often the explanation for why people do things differently in different places. But Brown (1991) has argued that when we turn to universals it is not easy to understand how cultural processes on their own might have produced a single uniform outcome in all human societies. Instead, there must be some intrinsic behavioural bias, some aspect of a basic and species-wide human nature, that is shining through and influencing the outcome in the same direction in all cases. It is at the level of these behavioural biases that we can see the outcomes of adaptive evolution.

This is a persuasive view and Brown gives only one plausible alternative explanation for human universals: a purely cultural trait could become universal if it were both ancient and useful such that it had been transmitted traditionally among all peoples everywhere. His best example is the use of fire. It is not necessary to posit some shared psychological adaptation(s) that motivates and shapes our use of fire; a simpler argument is that hominids have been using fire for a long time, that all populations have been exposed to it and that exposed people recognize its utility.

We need to be careful to rule out universals such as the use of fire. If we are, we can use the remaining ones to guide our search for psychological adaptations. Let me be clear about what I mean by 'guide our search'. We have an overarching body of theory — the modern synthetic theory of evolution including notions of natural, kin and sexual selection, population genetics and their supporting modelling techniques — that tells us broadly what classes of outcomes to expect from Darwinian processes. We also have the cross-cultural record of what people do, the universals and conditional universals of human behaviour. But these universals are not themselves psychological adaptations. Instead, they are the behavioural output of underlying features of the human mind, psychological mechanisms that have been perpetuated by Darwinian processes because they produced adaptive behaviour, on average, in ancestral human environments. In this view, human behaviour is not directly selected, but the underlying psychological mechanisms cannot be shaped by selection except to the extent that they produce behavioural output that has fitness consequences.

Behavioural universals then are guideposts to evolved, i.e. adaptive, psychological mechanisms. But a guidepost is just a meaningless point in space unless we know our direction and goal. So data on universals need to be viewed in the context of an understanding of Darwinian processes (the 'direction') and some informed guess about the selective regime in ancestral environments where the psychological mechanisms were shaped (the 'goal'). In detailing this approach to the study of psychological adaptations I am only reporting what many of us are already doing, rather than advocating some new method. A few examples will suffice.

Reciprocity

Reciprocity is a human universal. It is a cornerstone of social relations everywhere. People act as if they have a special claim on those they have previously aided and strive to compensate those who have aided them. In this case, the recognition of the universal came first, but Trivers (1971) provided the relevant overarching theory, and went so far as to list several interconnected psychological adaptations which would be essential in the evolution of reciprocity: the ability (1) to discriminate conspecifics as unique individuals; (2) to recognize when cheating has occurred; and (3) to associate that cheating with a particular individual, so that (4) future aid could be withheld from non-reciprocators. We have a universal and we have a relevant theory. Do we find the psychological adaptations that would permit reciprocal altruism in humans?

Normal people have distinct brain mechanisms for discriminating new faces and for remembering ones they have seen before (Benton 1980, Johnson & Morton 1991). Even those with specific brain injuries that compromise face recognition (prosopagnosics) can discriminate many individuals based on voice cues alone, suggesting redundant systems of individual recognition (Restak 1994). Cosmides (1989) and Cosmides & Tooby (1992) have shown that while people find many classes of logical problems opaque they are hypersensitive to cases where an individual has taken a social benefit to which s/he is not entitled: that is, they seem to have a specialized cheater-detection algorithm. These systems of individual recognition and cheater detection are cognitively linked in a way that fits the expectations from theory. Mealey et al (1996) have shown that mere verbal reports of character affect the potency of facial memories; faces that are labelled as belonging to cheaters are more likely to be remembered than those otherwise labelled. (They also observed an interaction with status that might have been predicted from a knowledge that status differentials are universal.) Other experimental studies indicate that people are less likely to ask for help when they feel unable to reciprocate (Greenberg & Shapiro 1971), and that if asked for help people are more likely to comply when the cost is low, the perceived benefit is large and when the requester has the potential to offer valuable help in return (Yinon & Dovrat 1987). The environment in which these psychological adaptations evolved is assumed, on analogy with modern hunter–gatherers, to consist of small, relatively closed social groups, where individual recognition and monitoring would have been efficient, and collective discrimination such as ostracism against non-reciprocators could have been an effective deterrent to cheating.

There are many other universals that could and should be explored for what they suggest about underlying psychological mechanisms. Linguistic universals (Brown 1991, Pinker 1994) offer an especially rich and promising domain for future research.

Conditional universals

The previous examples are ones where a single pattern of response and hence a relatively obligate or non-conditional set of psychological adaptations was assembled by selection. This is presumably because the relevant variation in ancestral environments was minimal. Presumably cheaters undermined the benefits of group life everywhere they occurred and thus there was a universal advantage to identifying and discriminating against them. In contrast, conditional universals are expected where the optimal behaviours varied in parallel with some environmental parameter. Here the underlying psychological mechanism(s) is necessarily more complicated; it must be plumbed in such a way as to translate the relevant environmental variation into an appropriate shift in behavioural output. Any such evolved psychological mechanism would constitute what biologists call a facultative adaptation.

A simple example of a conditional universal is that, in polygynous societies, co-wives generally share a residence if the polygyny is sororal — two sisters married to the same man — whereas they have separate residences if it is not. As far as I am aware, no one has sought a specific psychological mechanism underlying this universal, and perhaps in this instance one is not necessary. At the theoretical level Hamilton (1964) has outlined why all conflicts of interest are expected to be reduced among close genetic relatives. Co-wives are inevitably in competition over the resources of their common husband, but losing some share of those resources to a sister entails a smaller net cost than losing them to a non-relative because in the former case any lost resources should augment the production or fitness of nieces and nephews. On this view, the patterns of residence are a result of kin-selected mechanisms that moderate the level of competition as a function of genetic relatedness. This of course begs the question of exactly what mechanisms orchestrate the recognition and benign treatment of kin, but this is a general problem and it demonstrates that even some conditional universals may be the result of relatively generalized psychological mechanisms.

There are many societies that have 'classificatory' kinship systems in which, for example, father and father's brother are called by the same term, or mother and mother's sister are indiscriminable. There are, however, no languages that obligatorily lump the terms for mother and father. This linguistic universal undoubtedly reflects a deep psychological distinction between these two kin categories. One evolutionary basis for that distinction probably lies in the fact that maternity is certain while paternity is not.

Inheritance rules

Following up on this distinction, Hartung's (1985) findings on the cross-cultural distribution of matrilineal inheritance present a more complicated case of a conditional universal. Patrilineal inheritance involves the transmission of property from father to son and it is a common practice. Matrilineal inheritance is not the obverse, transmission from mother to daughter, rather it is transmission from a man to his sisters' sons (Fig. 1). There are two kinds of property, real and movable, whose transmission is regularly recorded by anthropologists. Hartung examined only societies that transfer both types in the same way — being either fully patrilineal or fully matrilineal — based on coding by Murdock (1967). Societies also differ in the extent to which extramarital sex is tolerated. Hartung, following the lead of Alexander (1974) and others (see references in Hartung 1985), argued that as extramarital sex becomes more prevalent a typical man is less and less likely to be the progenitor of his wife's children; that is, his probability of paternity decreases. Hartung cross-tabulated Flinn's (1981) estimates of probability of paternity against Murdock's inheritance data and found that matrilineal inheritance was characteristic of societies with moderate to low paternity probability, whereas patrilineal inheritance was the norm in societies with high paternity probability (Table 1). The validity of this strong correlation is not in serious question.

The relevant theory and modelling in this case are moderately complicated. Sexual selection theory with its emphasis on sex differences in reproductive variance (e.g. Trivers 1972) is relevant to understanding why resources might be transmitted primarily to males. Hamiltonian theory again emphasizes the shared interests of close relatives. Finally, the differential effects of extramarital sex on paternal and maternal probabilities explain why men might find closer relatives among their sisters' children than among their wife's children (see Table 2).

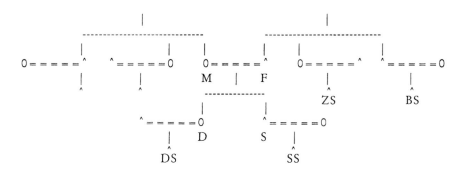

FIG. 1. Genealogy showing a married couple, their children, grandsons, siblings and nephews. Patrilineal inheritance flows from father (F) to son (S); matrilineal from F to sister's son (ZS). B, brother; D, daughter; M, mother. In compounds the first term is possessive, e.g. DS, daughter's son.

TABLE 1 Cross-tabulation of inheritance practice by probability of paternity

Inheritance practice	Probability of paternity	
	Moderate	High
Matrilineal	17	3
Patrilineal (sons only)	5	45

Data from Hartung (1985).

We have a reliable conditional universal and a body of related theory. The relevant ancestral environment would have been one in which males had resources to invest in immatures and paternity probability was variable. Neither of these conditions can be directly verified from the fossil or archaeological records but neither seem strongly at variance with our current picture of human evolution. Before we start the search for psychological mechanisms, are there any other lines of evidence that low paternity probability is the right explanation for matrilineal inheritance?

TABLE 2 Expected relatedness among first- and second-degree relatives

	Expected relatedness	Probability of paternity[a]						
		1.00	0.95	0.90	0.85	0.80	0.75	0.70
Females								
M–S(D)	1/2	0.500	0.500	0.500	0.500	0.500	0.500	0.500
MM–DS	1/4	0.250	0.250	0.250	0.250	0.250	0.250	0.250
FM–SS	$1/4p$	0.250	0.238	0.225	0.213	0.200	0.188	0.175
MZ–ZS	$1/8(1+p^2)$	0.250	0.238	0.226	0.215	0.205	0.195	0.186
FZ–BS	$1/8p(1+p^2)$	0.250	0.226	0.204	0.183	0.164	0.146	0.130
Males								
F–S(D)	$1/2p$	0.500	0.475	0.450	0.425	0.400	0.375	0.350
MF–DS	$1/4p$	0.250	0.238	0.225	0.213	0.200	0.188	0.175
FF–SS	$1/4p^2$	0.250	0.226	0.203	0.181	0.160	0.141	0.122
MB–ZS	$1/8(1+p^2)$	0.250	0.238	0.226	0.215	0.205	0.195	0.186
FB–BS	$1/8p(1+p^2)$	0.250	0.226	0.204	0.183	0.164	0.146	0.130

[a]Values given are P-values.
B, brother; D, daughter; F, father; M, mother; S, son; Z, sister. In compounds the first term is possessive, e.g. ZS, sister's son. Wherever S appears as the last term D could be accurately substituted.

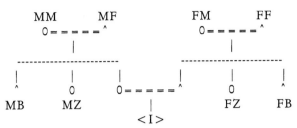

FIG. 2. Genealogy showing an informant <I> and his/her matrilateral and patrilateral grandparents (whose investment was examined by Euler & Weitzel 1996), and matrilateral and patrilateral aunts and uncles (whose investment was examined by Gaulin et al 1997). B, brother; F, father; M, mother; Z, sister. In compounds the first term is possessive, e.g. MB, mother's brother.

More subtle matrilineal bias

Hartung's models show that, even considering multi-generational effects, paternity probability must be slightly below 0.5 (0.46 actually) before a man would realize an evolutionary benefit by refocusing his investment on sisters' sons rather than wife's sons. Without making the calculations explicit he also noted that all of a man's relatives would benefit were he to invest matrilineally whenever paternity probability is less than 1.0 (Table 2). A further implication is that whereas fathers would still be expected to favour 'their' sons, even if only half were really their own, second-degree relatives — grandparents of both sexes, aunts and uncles — should always show an investment bias toward their matrilateral kin and the magnitude of this bias should grow larger as the probability of paternity decreases. Only recently has anyone tried to assess whether such a bias exists; two new studies assess the investment of the second-degree relatives labelled in Fig. 2.

In Germany Euler & Weitzel (1996) asked people to rate on a seven-point scale how much each of their grandparents had cared for them during childhood. They found a pronounced matrilateral bias, with maternal grandmothers being the most caring and paternal grandfathers being the least (Table 3). The results for paternal grandmothers and maternal grandfathers were intermediate but variable (note that both are expected to be related to the grandchild by 1/4p). Euler & Weitzel (1996) found that when each

TABLE 3   Mean grandparental investment

Grandparent	Expected relatedness	Co-resident	Separated	Widowed
MM	1/4	5.09, 633	5.06, 48	5.10, 602
FM	1/4p	4.20, 595	3.25, 36	4.41, 571
MF	1/4p	4.51, 551	2.06, 34	4.17, 517
FF	$1/4p^2$	3.80, 470	1.77, 30	3.89, 487

Values are: mean, N.
Data from Euler & Weitzel (1996).
FF, father's father; FM, father's mother; MF, mother's father; MM, mother's mother.

grandparental couple was co-resident mother's fathers were more caring than father's mothers; but if the analysis is restricted to widowed or divorced grandparents then the reverse is true, so co-residence with mother's mothers may increase mother's fathers involvement with the grandchild. Unfortunately, two additional factors — grandparental age and residential distance between the grandparent and grandchild — might also affect investment levels; but although these effects were explored, they were not fully controlled in this study. This is a concern because sex differences in both age at marriage and dispersal patterns could introduce systematic biases.

In the US Gaulin et al (1997) studied the investment of four other kinds of second-degree relatives: matrilateral and patrilateral aunts and uncles (see Fig. 2). In this study co-residence was naturally controlled because, for example, mother's brothers and sisters have typically dispersed from their natal homes by the time they have nieces and nephews in whom they might invest. Also, age and distance effects were statistically controlled. Informants were asked to rate, on a seven-point scale, how much concern each of these four relatives showed for them; those informants who had both a patrilateral and a matrilateral uncle (or both a patrilateral and a matrilateral aunt) were also asked to answer a forced-choice question about which uncle (or aunt) showed more concern. Both the rating-scale data and the forced-choice data indicate a significant matrilateral bias.

For example, 72% (98/137) of those with both matrilateral and patrilateral aunts said the former showed more concern ($\chi^2 = 25.4$, $P < 0.001$). Confounds were removed by regressing the informant's choice of which aunt showed more concern on the difference in the two aunts' ages and the difference in their residential distances from the informant. The y-intercept of the resulting regression gives the matrilateral bias that would exist if age and distance differences were zero. With these confounds thus removed there was still a significant matrilateral bias in the y-intercept ($t(130) = -5.09$, $P < 0.0001$). With respect to uncles 64% (83/130) of informants chose the matrilateral relative as more concerned ($\chi^2 = 9.97$, $P < 0.005$), and regression again yielded a significant matrilateral bias after removing the effects of age and distance differences ($t(119) = -5.09$, $P < 0.0001$).

For those subjects who had all four classes of relatives, patrilateral and matrilateral aunts and patrilateral and matrilateral uncles, the seven-point rating-scale data (Table 4) provide the most direct comparison with the grandparental data of Euler & Weitzel (1996) (Table 3). The adjusted means are based on repeated-measures ANCOVA of the rating scale data; sex of relative and laterality of relative were treated as within-subject factors and residential distance and age of the relative were treated as covariates. Based on that analysis aunts are significantly more solicitous than uncles ($F(1,66) = 23.20$, $P < 0.0001$) and matrilateral relatives are significantly more solicitous than patrilateral relatives ($F(1,66) = 5.31$, $P = 0.024$), after controlling for factors that might be associated with laterality.

In summary, the cross-cultural data indicate a strong association between formalized matrilineal inheritance and low paternity probability. And, as predicted from theory, there seems to be a weaker but more general matrilineal investment bias among more

TABLE 4   Mean, S.D. and adjusted mean investment of aunts and uncles

Aunts and uncles	Expected relatedness	Mean (S.D.)	Adjusted mean
MZ	$1/8(1 + p^2)$	4.75 (1.84)	4.71
FZ	$1/8p(1 + p^2)$	3.96 (1.87)	3.97
MB	$1/8(1 + p^2)$	3.65 (1.93)	3.65
FB	$1/8p(1 + p^2)$	3.28 (1.71)	3.31

$N = 73$. FB, father's brother; FZ, father's sister; MB, mother's brother; MZ, mother's sister.

distal relatives in societies where paternity probability is likely to be appreciably higher. The cross-cultural pattern is obviously conditional (facultative) in that inheritance norms vary systematically with sexual practices. But grandparents, aunts and uncles in Germany and the US are also presumably behaving in a conditional manner simply because a given individual is often, simultaneously, both a patrilateral and a matrilateral relative, e.g. one may have nieces through brothers and nieces through sisters. Although the studies discussed above are not designed in such a way as to permit these direct comparisons, their results strongly suggest that grandparents, aunts and uncles are investing differentially in relatives to whom they are related through females. If the paternal probability argument is correct, matrilineal investment biases are the behavioural output of facultative psychological mechanisms. What precisely are these mechanisms? Unfortunately, mostly indirect evidence is available at present.

Clues about psychological mechanisms

Whenever the probability of paternity is less than 1.00 women benefit if their sons invest matrilineally but their husband invests patrilineally; men of course benefit by avoiding any misdirection of their investment to non-relatives. Evidence from both North and Central America suggests that paternity probability is monitored by males and that systematic attempts are made to manage or manipulate its perception, most notably by wives, but also by other matrilateral relatives.

A woman's behaviour provides her husband with some cues to paternity, but even brief (and therefore easily concealed) acts can compromise paternity. In contrast, the offspring's phenotype presents durable cues to genetic similarity. For this reason, paternal care might be expected to be responsive to any perceived phenotypic similarity or dissimilarity between a man and his wife's children. Given the divergent selection pressures on men and women, mothers are expected to exaggerate paternal resemblance to their offspring whereas their husbands are expected to be sceptical.

Daly & Wilson (1982) reviewed tapes of spontaneous utterances at 111 North American births and surveyed the parents and other relatives of 526 Canadian infants via a mailed questionnaire. With respect to spontaneous utterances they found that paternal resemblance was remarked significantly more often than maternal resemblance and that this was due entirely to a strong bias in the remarks of mothers. Fathers tended to be sceptical in the face of mothers' assertions of paternal resemblance and mothers tended to make such claims more in the father's presence than in his absence. In their mail survey some indications of paternal scepticism were marked. For example, third-party allegations of paternal resemblance were significantly more likely to be doubted by fathers and accepted by mothers than were allegations of maternal resemblance to be doubted by mothers and accepted by fathers. Maternal grandmothers were significantly more likely to respond to the mail survey than were paternal grandmothers, perhaps signalling greater interest in the infant. Mothers, but not other respondents, were significantly more likely to claim paternal resemblance for first-born infants than for infants of later birth orders.

Based on face-to-face interviews in southeastern Mexico, Regalski & Gaulin (1993) found that paternal resemblance was alleged much more frequently than maternal resemblance and that mothers and their relatives remarked such resemblance more often than did fathers and their relatives. In addition, allegations of paternal resemblance were significantly more likely for low birth-order children and when the mateship was new. Using a multivariate approach the best predictors of resemblance claims were duration of marriage and laterality of the speaker. There were indications both that fathers were systematically sceptical about claims of paternal resemblance, (i.e. fathers were much more likely to express doubt about claims of paternal resemblance than about claims of maternal resemblance), and that mothers' and fathers' claims agreed more often than would be expected by chance, suggesting that a mother's claims can affect her mate's perceptions.

Post hoc interpretations of these data are possible. For example, perhaps these patterns are due to some sort of male bias in Anglo and Hispanic culture. Perhaps declaring paternal resemblance is just the 'right thing' to do in these cultures. This perspective may deserve serious consideration if it can explain why matrilateral relatives are more compelled to do the right thing than are patrilateral relatives, why everybody seems to forget the right thing for marriages of longer duration or for infants of higher birth order, and why patrilateral relatives forget faster than matrilateral relatives.

Mothers, fathers and their relatives are all players in this evolutionary game, but so are infants themselves. An infant could benefit through being perceived as phenotypically similar to a male (carrying the argument to its logical conclusion this is true whether or not that male was her father) if such resemblance triggered an increase in investment from that male. A recent study shows an effect that is coincident with this idea. Christenfeld & Hill (1995) found that people performed randomly at matching facial photographs of children to facial photographs of their parents, with the exception that fathers and their one-year-old progeny were recognized as related at significantly better than chance levels. Such a result could be a consequence of biased parental gene expression (e.g. genomic imprinting) or of a specialized psychological mechanism that allows especially accurate assessment of father/infant similarities.

Before we over-interpret this finding it is important to note that one other careful study (Nesse et al 1990) found the opposite effect: children were more accurately matched to their mothers than to their fathers. There were significant methodological differences between these studies, however, one being that the age of the child was an experimental variable in Christenfeld & Hill's (1995) study, whereas it was a random variable in the design of Nesse et al (1990). This could well be an important difference for two reasons. First, the superior matching of fathers was limited to father/infant pairs and absent in father/10-year-old and father/20-year-old pairs, and thus could have been missed by Nesse et al (1990), who used a pooled sample of children ranging in age from six months to 18 years. Second, a male would benefit from the earliest possible recognition of non-paternity so as to minimize the amount of wasted investment; this suggests that any specialized paternity discrimination mechanism should be primed for peak sensitivity early in the investment cycle. If an offspring passed the test of phenotypic resemblance at that stage there would be little advantage in continuing to monitor its paternal resemblance. Given the discrepant findings and the centrality of the issue of phenotype matching, this is clearly an area where there is much work to be done.

Individual children who show little resemblance to their fathers may be rejected; a related solution would be to terminate one's involvement with a family when there were too many — or too many risks of — unrelated children. In this light Betzig (1989) studied the factors that precipitate divorce among the 186 societies of the Standard Cross-Cultural Sample. She found that adultery was the most common cause, being mentioned for 88 of the 186 societies; moreover, it was significantly more common than any other single cause except sterility (75 societies). In the case of adultery she noted a strong sex bias: either partner could be divorced for adultery in 25 societies, only wives could be divorced for the offence in an additional 54, but there were just two societies where adultery was a justification for divorce of husbands only (in the remaining seven cases the party was unspecified). Several of the less common causes of divorce are also relevant to paternity. Elopement with a lover was mentioned as grounds of divorce in 38 societies: in 11 either party could be divorced, in 24 wives only and in three husbands only. Lack of virginity is grounds for divorce in six societies; in all six only the wife can be divorced. These grounds for terminating a marital arrangement, their predominance and especially their distribution by sex are precisely in line with the idea that perceptions of paternity confidence are important motivators of behaviour.

There is too little overlap between Hartung's and Betzig's samples to offer a definitive answer, but the available data are at least suggestive of less intense guarding of paternity in matrilineal societies. Of the 186 societies surveyed by Betzig, 11 are high paternity patrilineal societies and nine of these (82%) allow divorce on grounds of wife's adultery. In contrast, of the four societies in Betzig's sample that are low paternity matrilineal societies only two (50%) allow divorce for the wife's adultery; one of the two societies that does not is among the few that permit divorce for husband's adultery only.

Matrilateral investment biases, rejection of unrelated children and divorce of unfaithful wives in some sense close the barn door after the horse is gone. There may be predispositions that can minimize the chance that a man will face unrelated children in his family and Daly et al (1982) have argued that male sexual jealousy is one such predisposition. Their cross-cultural, legal and historical reviews suggest that unauthorized sexual contact with a married woman is a crime, that the victim is her husband, that his jealousy is regarded as normal and often excuses what would otherwise be recognized as a crime of violence on his part. Male sexual jealousy is often a precipitating factor in homicides in both western and non-western societies (see also Daly & Wilson 1988), and Daly et al (1982) specifically argue that male sexual jealousy, with its associated use of repressive violence, is an adaptation to threats to paternity.

Conclusion

One obligate universal and one set of interrelated facultative universals have been discussed. For the case of reciprocity, considerable work has exposed a functionally integrated group of psychological traits that could underlie the universality of social exchange. For the case of matrilineal investment biases, several lines of evidence suggest that selection has operated on psychological traits related to the perception, monitoring and control of paternity; but beyond male sexual jealousy (Daly et al 1982) and the possibility of specialized father/infant phenotype matching (Christenfeld & Hill 1995) research in this area has yet to reach the level of psychological mechanisms. The same is true for a great many universals. Evolutionary psychologists should find plenty of work for the next century.

References

Alexander RD 1974 The evolution of social behavior. Annu Rev Ecol Syst 5:325–383

Benton A 1980 The neuropsychology of face recognition. Am Psychol 35:176–186

Betzig L 1989 Causes of conjugal dissolution. Curr Anthropol 30:654–676

Brown D 1991 Human universals. McGraw-Hill, New York

Christenfeld NJS, Hill EA 1995 Whose baby are you? Nature 378:669

Cosmides L 1989 The logic of social exchange: has natural selection shaped how humans reason? Studies with the Wason selection task. Cognition 31:187–276

Cosmides L, Tooby J 1992 Cognitive adaptations for social exchange. In: Barkow J, Cosmides L, Tooby J (eds) The adapted mind. Oxford University Press, New York, p 163–228

Daly M, Wilson MI 1982 Whom are newborn babies said to resemble? Ethol Sociobiol 3:69–78

Daly M, Wilson MI 1988 Homicide. Aldine de Gruyter, Hawthorne, NY

Daly M, Wilson MI, Weghorst S 1982 Male sexual jealousy. Ethol Sociobiol 3:11–27

Euler H, Weitzel B 1996 Discriminative grandparental solicitude as a reproductive strategy. Hum Nat 7:39–59

Flinn M 1981 Uterine and agnatic kinship variability and associated cousin marriage preferences: and evolutionary biological analysis. In: Alexander R, Tinkle D (eds) Natural selection and social behavior. Chiron, New York, p 439–475

Gaulin S, McBurney D, Brakeman-Wartell S 1997 Matrilateral biases in the investment of aunts and uncles: a consequence and measure of paternity uncertainty. Hum Nat 8:139–151

Greenberg M, Shapiro S 1971 Indebtedness: an adverse aspect of asking for and receiving help. Sociometry 34:290–301

Hamilton WD 1964 The genetical evolution of social behavior. J Theor Biol 7:1–52

Hartung J 1985 Matrilineal inheritance: new theory and analysis. Behav Brain Sci 8:661–668

Johnson M, Morton J 1991 Biology and cognitive development: the case of face recognition. Blackwell Science, Oxford

Mealey L, Daood C, Krage M 1996 Enhanced memory for faces of cheaters. Ethol Sociobiol 17:119–128

Murdock G 1967 The ethnographic atlas. University of Pittsburgh, Pittsburgh, PA

Nesse R, Silverman A, Bortz A 1990 Sex differences in ability to recognize family resemblance. Ethol Sociobiol 11:11–21

Pinker S 1994 The language instinct. Harper Collins, New York

Regalski JM, Gaulin SJC 1993 Whom are Mexican infants said to resemble? Monitoring and fostering paternal confidence in the Yucatan. Ethol Sociobiol 14:97–113

Restak R 1994 The modular brain. Macmillan, New York

Trivers RL 1971 The evolution of reciprocal altruism. Q Rev Biol 46:35–57

Trivers RL 1972 Parental investment and sexual selection. In: Campbell B (ed) Sexual selection and the descent of man, 1891–1971. Aldine, New York

Yinon Y, Dovrat M 1987 The reciprocity-arousing potential of the requester's occupation, its status and the cost and urgency of the request as determinants of helping behavior. J Appl Soc Psychol 17:429–435

DISCUSSION

Sperber: I would like to raise a methodological question regarding the causal correlation between matrilineal inheritance, on the one hand, and paternity doubts, on the other. You presented the case as if the preference for matrilineal inheritance would emerge in societies where doubts about paternity were greater. However, in the anthropological and historical record, kinship structures and related patterns of inheritance seem to be much more stable than sexual mores and attending questions about paternity. It makes sense, therefore, to think of a causal correlation operating in the direction opposite to the one you envisage. If the inheritance system is matrilineal, then a man knows that his heirs will be his sister's children rather than those of his wife. It may matter relatively less to him, then, if his wife's children are biologically his own, especially if the counterpart of greater paternity doubts is a greater chance of having children with other men's wives.

Gaulin: Yes, that's possible. My aim is to get a grip on some kind of psychological mechanism that can not only explain those extreme cases of matrilineal investment bias that we call matrilineal inheritance, but that could also explain the patterns that I showed for Europe and North America — hardly matrilineal societies — where there is still a significant matrilineal investment bias. You could say that they are totally different phenomena, but the explanation that you're offering on reversing the correlation (such that matrilineality drives a lack of concern for paternity) does not sufficiently explain the observed bias in European and American grandparents', aunts' and uncles' investments.

Wilson: In light of this argument, you would want to look in matrilineal societies at the sexual habits of females and the psychological reactions of males. I'm not an expert on this literature, but it seems to me that ethnographic accounts indicate females have opportunities to have extra-pair copulations (EPCs) and males are sexually jealous. And in ethnographic studies of some Caribbean peoples who are relatively matrilineal men target their gift giving and investment in children who they believe are theirs.

Gaulin: Clearly, there are more opportunities for EPCs in those societies.

Wilson: But there are no data on the paternity.

Gaulin: There are data on sexual practices.

Thornhill: What one would anticipate the adaptation to do is to track information about EPC rates, rather than paternity, in the culture in which an individual grows up.

Daly: There's an unfortunate tendency in the non-human behavioural ecology literature to use the terms 'confidence' and 'certainty' of paternity to refer to a population parameter: one minus the cuckoldry rate. 'Confidence' of paternity is variable in a given society: some males have good reason to doubt their paternity and some have good reason to be confident. I wish it hadn't become the practice to use the phrases 'confidence' and 'certainty' when one meant paternity probability. The population incidence of cuckoldry might be one input to each individual's confidence and certainty, but there are other idiosyncratic inputs.

Tooby: The wider question is, what can one use the immensely rich data provided by the human ethnographic record for? Everything known about what humans have done in other places and times can be used to inform research into the species-typical design of the human psychological architecture. All of these different cultures provide natural experiments showing how this design responds to different circumstances, and so information about them constitutes a remarkable resource. On the other hand, you cannot control natural experiments, so there are the obvious correlational criticisms that arise from this. For this reason, we cannot attain tight descriptions of our architecture until we apply experimental and neuroscience methods as well. Nevertheless, the ethnographic record can be a productive tool.

However, some of the phrases you used in your talk disturbed me a little, such as counterposing as alternative hypotheses 'is there a mental mechanism for such and such?' or 'is it just a good idea?'. Mental mechanisms always necessarily underlie every behaviour. Presumably, as Don Symons argues, the real questions are over whether the mechanisms involved are domain general or specialized in their function, and over whether the mechanisms were designed or shaped by selection to produce the behaviour in question or whether behaviour is just a by-product of their designs. In the case of kinship the claim that people might be choosing to invest in their sisters' descendants because they recognize that it is a good idea to use this better fitness pathway is nonsensical because fitness is not the kind of detectable entity that could become the evolved object of a motivational system. For practices to spread because of their utility, the human mind has to be able to perceive and value the utility they produce, and humans cannot experience 'fitness'. Instead, one would expect that the post-Pleistocene human response to various categories of kin is the result of naturally selected psychological adaptations designed to solve kin-related problems as they manifested themselves in hunter–gatherer contexts. After all, so-called kinship systems are not the object of selection themselves, since they are the interactive product of members of the social group, who are to some degree in conflict with each other. I would expect that the enormous profusion of human kinship systems are the result of a complex set of evolved kin-oriented motivational and inferential mechanisms operating in different social ecologies, and that if we work backwards from these systems we might be able to determine the computational designs of these specializations. One would expect one or more specialized representational spaces for assigning individuals to kin categories, special motivational operators to translate suffering of self and other in terms of these categories, prepared attention to informative, ecologically valid cues and so on.

So, I would be wary of too easily accepting the notion that ancient and long-standing behavioural practices have spread solely because of general learning mechanisms. For example, I would even question your paradigm example — that the use of fire spread simply because it is a 'good idea'. Humans have been using fire for at least 500 000 years, and perhaps as long as 2×10^6 years, a period that is sufficient to have shaped specialized mechanisms. When they can afford to, humans go out of their way to replace cheap and safe means of heating, cooking and lighting with fires and other flame-based tools such as candles and lanterns, they prefer foods with fire-based flavours, such as the smell of wood smoke in the distance, they go through a particular age in childhood where they are particularly fascinated by fires and fire making, and they sometimes exhibit a particular fire-oriented pathology, pyromania. No family I know of gathers around the microwave on special occasions just to savour all of the special associations created by decades of eating microwaved food. All of these things suggest that even the use of fire by humans may be partially shaped by evolved psychological design features specialized for fire.

Miller: I have a technical question about Hartung's analysis. The higher the paternity uncertainty, the more likely it is that your putative sister is actually your half-sister. This half-sister effect is so severe that you would have to have paternity certainty lower than about 30% for matrilineal inheritance to make sense. What man would put up with a co-habitating woman who gave lower than 30% paternity certainty?

Gaulin: Hartung came up with a number of 0.268. He also found that it was higher for multigenerational effects but that it still did not exceed 0.5. His argument is that matrilineal inheritance isn't a male strategy, he calls it a grand-maternal strategy. Males only prefer matrilineal inheritance when is P is below 0.268, but all his relatives, with the exception of his wife, prefer it anytime P is less than one. Therefore, the argument is that it is a compromise.

Maynard Smith: Is it totally absurd to suppose that humans have a motivation to increase their fitness and act accordingly?

Tooby: The reason why we and others have argued that humans cannot have evolved a motivation specifically to increase their fitness is because fitness is itself inherently a future consequence that cannot be directly observed. Because fitness is invisible and cannot be tracked by the organism, adaptations cannot evolve to pursue it. What can become the object of motivations are cues that are perceptually detectable to the organism, that stably recur across many generations and that reliably predict the decision that increases fitness. So, we eat food not because we are motivated to increase our fitness, but because foods have cues that once predicted that eating them would increase fitness, and we evolved adaptations to appreciate and be guided by those cues. The same is true in all areas: not fitness itself but fitness-predicting cues become the objects of motivation in the areas of breathing, kinship, sex, aggression, danger and so on.

Wilson: In a matrilineal society, where there is fishing, for example, or warfare with male absences for extended periods, is there any information about the probability of male mortality, and about conflict between a man's sister and the wife? And how much say does the male have in any conflict? The reason I ask is that if she's half sib to him, it is probable that her husband may have disappeared (died at sea or in war) and the material possessions constituting the inheritance are presumably from her mother. Therefore, she might be inclined to think 'that's mine'.

Gaulin: You have raised two questions. Peter Ellison, in a comment to Hartung's original paper, argued that instead of being a response to low paternity matrilineal inheritance was a response to high rates of male lineage extinction, as might occur with dangerous activities such as fishing and warfare (Ellison 1985). Hartung agreed that Ellison might have been correct. Second, there will inevitably be a conflict between the wife and the sister because patrilineal inheritance is a good strategy for the male unless P is low. Therefore, if there is any way that a wife can get her husband to divert some of his resources in her children's direction then she is doing well. The sister of course has cultural norms on her side and is going to tug the male in the other direction.

Daly: One of the complications with cross-cultural studies is that to do quantitative analyses you inevitably rely on these typological characterizations of societies as matrilineal, patrilineal and so on. Some of the more interesting questions for psychologists seeking the psychological underpinnings of these things are to know more about who exactly within these societies is defying the rules, how often are they really followed, etc.

Pinker: We haven't talked much about psychological mechanisms, which presumably we would all experience as emotions, because we don't literally calculate degrees of relatedness or think about our nieces and nephews in terms of genetic overlap. I'm wondering whether the proximate psychological mechanism boils down to the sense of loyalty and solidarity that you have to your blood family versus your spouse. What goes through a person's mind when he contemplates whom he should leave his assets to? First, you leave your inheritance to a generation later than yours; that is, you don't usually leave money to your parents or your siblings because they're going to die before or around the same time as you. Therefore, the question is whether you leave it to your nieces and nephews, or to your own children. Consequently, the most important variable is: do I feel more loyalty and solidarity to this woman, or to my parents? It is analogous to the feelings one must sort through in any ambiguous romantic relationship. For example, one asks, how many times have I dated this woman, how serious is the relationship and should I spend Thanksgiving and Christmas with her and her parents or with my parents? Surely these deliberations are influenced by my hunches of whether she is not sleeping with anyone else, and whether I'm not sleeping with anyone else. I suspect that these kinds of thoughts are the proximate psychological mechanism, and everything else that you have talked about is a consequence of the emotional seesaw between solidarity with wife and solidarity with blood family.

Gaulin: This is the argument I had in mind when I mentioned sexual jealousy as the only thing close to a psychological mechanism that anybody has talked about in this domain.

Pinker: But sexual jealousy is not the only variable. It would only be one emotion feeding into the degree of commitment I feel toward my wife versus toward my blood family.

Reference

Ellison P 1985 Lineal inheritance and lineal extinction. Behav Brain Sci 8:672

Evolutionary psychology and genetic variation: non-adaptive, fitness-related and adaptive

Steven W. Gangestad

Department of Psychology, University of New Mexico, Albuquerque, NM 87131, USA

Abstract. Behavioural variation across individuals can be substantial. A broad generalization emerging from three decades of behavioural genetic studies is that most psychological individual differences have moderate broad heritabilities (30–60%). There are at least three possible scenarios for this genetic variation. First, it may be adaptively neutral and not subject to selection. Second, it may be related to fitness despite selection. Third, it may be maintained by selection for alternative adaptations. Some authors favour the first of these possibilities, but the latter two cannot be ruled out. First, temporally varying selection pressures (e.g. pathogens) can maintain fitness-related genetic variance in a population despite current selection pressures. Moreover, direct and indirect evidence on humans support the notion that some phenotypic variance is fitness related. Second, while adaptive alternatives are unlikely to be found at a level of highly complex design, frequency-dependent selection can maintain variation at finer, quantitative levels. One potential example is discussed. Because of their particular relevance to evolutionary psychology, fitness-related and adaptive genetic variance deserve further attention.

1997 Characterizing human psychological adaptations. Wiley, Chichester (Ciba Foundation Symposium 208) p 212–230

Human evolutionary psychology is the study of our species' psychological adaptations evolved through natural selection. As exemplified by this symposium, two features pervade its practice. First, its workers typically embrace a notion of universal human nature, a single, species-typical collection of adaptations. Second, the working assumption is thoroughly adaptationist: until demonstrated otherwise, the features of complex human psychological processes are assumed to exist because they solved adaptive problems posed by environments recurrent along the path of our evolution. Of course, no serious student of modern biology would deny a role for non-adaptive forces, such as developmental and genetic constraints, in evolution. As noted by Mayr (1983), however, these roles can only be appreciated after an adaptationist perspective has been applied; that something isn't adaptive can be inferred only by knowing what is or would have been.

Individual differences have provoked discussion about the generality of universal design and the limits of adaptation (e.g. Tooby & Cosmides 1990, Wilson 1994). Behavioural variation across individuals can be substantial. For instance, in two studies the coefficients of variation (S.D. over mean) of men's number of intrasexual fights and number of sex partners were 1.7 and 0.9, respectively (B. F. Furlow, T. Armijo-Prewitt & S. W. Gangestad, unpublished data 1996, Thornhill & Gangestad 1994)—with some having over five times the mean. Variations merely due to differences in experience can readily be incorporated into an adaptationist framework, of course; universal, adaptive psychological design specifying behavioural expression contingent on experience produces variation in behaviour, not despite adaptive design but rather because of it. As it turns out, however, three decades of behavioural genetic studies point to a striking generalization: nearly all behavioural traits show moderate broad heritability (0.3–0.6; e.g. Bouchard 1994). For instance, both aggressivity and sexual variety have heritabilities in this range, and thus men's number of fights and number of sexual partners may have genetic coefficients of variation greater than 0.5—much higher than most morphological traits (the genetic coefficient of variation of height, for instance, is about 0.04).

There are at least three possible scenarios for this variation. First, it may be adaptively neutral and not subject to selection. Second, it may be related to fitness despite selection. Third, it may be maintained by selection for alternative adaptations. Because of their particular relevance for evolutionary psychology, I focus on fitness-related and alternative adaptive variations.

The causes of non-adaptive (neutral and fitness-related) genetic variance

The accumulation and maintenance of non-adaptive variation in natural populations is caused by a variety of factors.

Mutations accumulate due to weak stabilizing selection

On continuous traits that have single, stable optima in a population, variance removal through stabilizing selection and variance introduction through mutation balance out at an equilibrium variance around an adaptive norm (Lande 1975). Stabilizing selection at individual loci is generally weak, though mutation–selection equilibria can tolerate a modest but non-negligible amount of fitness-related variation between individuals (Rice 1988). Mutation–selection balance undeniably occurs but probably does not account for all genetic variance on fitness traits in natural populations (e.g. Houle et al 1994).

Some personality variants may be maintained by this simple process. Perhaps the extremes of distributions of the Big Five (see Bouchard 1994) are selected against through stabilizing selection, such that, for instance, optimal emotional arousal to negative events is in the intermediate range.

Trade-offs between fitness components create negative covariances between them

From a life history perspective, important 'traits' include decision rules about how to allocate effort. Because decisions are constrained by a budget (a total resource base of energy, time, etc.), expenditures on one feature limit expenditures on others. Thus, allocation of effort to early reproduction limits resources that can be dedicated to somatic growth, costing future reproduction and longevity. Similarly, the more a male exerts mating effort, the less he can invest in offspring.

Genes that affect these sorts of decisions exhibit antagonistic pleiotropism, increasing one fitness component (e.g. early age-specific reproduction) at the expense of others (e.g. later age-specific reproduction). Within some range, then, variation in fitness components may exert little effect on fitness, resulting in genetic variation in them (e.g. Mousseau & Roff 1987). On average, negative genetic covariance between fitness components is expected, through it may be quite weak (Charlesworth 1990). Direct evidence for antagonistic pleiotropy is rare, but life-history trade-offs are inevitable and thus so too is antagonistic pleiotropy.

Impulsivity vs. constraint (heritability about 0.5) can be thought of as a life history dimension, the discount rate of time, such that impulsive individuals value the future less than constrained individuals. Perhaps variability on this trait is partly maintained by antagonistic pleiotropism.

Genetic constraints limit the effectiveness of selection against non-adaptive phenotypes

A classic instance of genetic constraint is heterozygote superiority, i.e. greater fitness of heterozygotes than homozygotes. Due to sexual recombination, heterozygosity cannot breed true and, hence, in these instances neither can fitness. A second form of genetic constraint is antagonistic pleiotropy, caused not by budget constraints (above) but by the fact that, given the genetic background against which it's expressed, an allele has multiple phenotypic effects.

An intriguing form of pleiotropy was proposed by Tooby & Cosmides (1990). They suggested that, because many pathogens parasitize their hosts by adapting to particular protein structures, selection favours rare molecular structures (e.g. proteins), all else equal. All else is equal if alternative molecular structures serve identical functions. Hence, frequency-dependent selection yields immense variation at the molecular level, while maintaining uniformity of function at a molar (e.g. complex behavioural) level. As it turns out, however, not all can be precisely equal at a molar level and thus, Tooby & Cosmides suggest, molecular variation creates some 'parametric' variation at a molar level, including behavioural trait variation. This trait variation is unlikely to be completely neutral (indeed, mutation–selection balance itself creates non-neutral trait variance). Thus, alleles will have pleiotropic effects on fitness: one effect on pathogen resistance due to relative prevalence, another on behaviour. An allele that has slightly deleterious effects on behaviour will be rarer at equilibrium than one that, all else equal, has slightly advantageous effects on behaviour.

While mutation–selection balance or pleiotropic effects due to budget constraints cannot be entirely overcome by selection, in principle genetic constraints can. Genetic expression, such as pleiotropy, is partly due to the genetic background against which a gene appears. Thus, costs to gene expression can in principle be overcome through alteration of the genetic background (e.g. selection for modifier genes that prohibit pleiotropic expression of genes). Relatedly, benefits of heterozygosity (e.g. functional redundancy of alternative alleles with the benefits of molecular diversity) can potentially be obtained through other means (e.g. selection for functional redundancy of structurally different alleles at different loci) without the costs of heterozygote superiority. Cases in which selection on the genetic background is itself constrained certainly exist and, thus, variation sometimes truly is due to genetic constraints (see below on temporally varying selection).

Temporally varying selection can promote genetic variation

Temporally changing selection in populations with overlapping generations results in genetic variability because selection does not work against specific alleles long enough to drive or keep them out of the population. Clearly, many aspects of human environments do not change at rates sufficiently fast to maintain genetic variability (e.g. climatic change may be slow). As emphasized in the work of Hamilton (1982, Hamilton & Zuk 1982), selective pressures that can change rapidly are pathogens, who themselves evolve to adapt better to 'us', requiring us to change if we are to remain adapted to them.

Temporally changing selection pressure maintains not only trait variance but also fitness-related trait variance. Alleles in the population include those that are currently favoured as well as ones that were favoured in the recent past. Naturally, fitness characters affected most by the changing selection pressures should have the greatest genetic variance and, hence, pathogen resistance might be an important source of fitness-related genetic variance. Variability in pathogen resistance can affect psychological characters in at least three ways.

First, pathogen attack can itself affect the development of neural structures.

Second, less resistant individuals may avoid disease by allocating more effort to the immune system, and incur the costs of having fewer resources available for other phenotypic features. Thus, given the great amount of effort dedicated to both the immune system and brain growth in early childhood, increases in one may occur at the expense of the other. Similarly, increased allocation of effort to immune system components may sacrifice somatic growth and size, which may indirectly affect psychological characters. As discussed below, selection for costly, honest signals plays into these latter processes and increases the phenotypic effects of fitness-related variation.

Third, variation in fitness markers may evoke differential treatment from others, which then affects phenotypic characters. Differential treatment by peers and potential mates (discussed below) are obvious examples. Parental treatment may be another. Interestingly, parental affection received, assessed in a variety of ways, appears to have a heritability of about 50% (here, construing affection received as an 'extended phenotype') (Plomin & Bergeman 1991).

Temporally changing selection also increases genetic variation by affecting other processes. As noted above, selection for genetic modifiers can eliminate genetic constraints by altering the genetic background. In a changing environment, this process may not occur. Genetic modifiers get selected along with favoured genes. If selection on favoured genes is not consistent, neither is selection for genetic modifiers. Thus, genetic variability due to genetic constraints is more likely to accumulate when selection pressures are variable. Pleiotropic effects of genes selected for rareness at a molecular level due to pathogens, as suggested by Tooby & Cosmides (1990) and discussed earlier, might be particularly expected.

Summary

In all likelihood, multiple processes contribute to non-adaptive variation (though their relative contributions to genetic variability in natural populations remain unknown). Most, such as mutation–selection balance, are expected to create only modest genotypic fitness differences between individuals (see Rice 1988). An exception is temporally changing variation, which may cause appreciable fitness differences across individuals. At a molecular (e.g. protein) level, most genetic variation is probably near-neutral, a small fraction related to fitness, but that need not be true of genetic variance underlying behavioural traits. Because fitness-related variance is of greater relevance than near-neutral variance to evolutionary psychology, I turn to discuss evidence for its existence and implications.

Evidence for and implications of fitness-related variation

The study of fluctuating asymmetry

Biologists can study some aspects of maladaptation by examining their outcomes. One such outcome is fluctuating asymmetry, absolute deviation from symmetry on bilateral traits that, on average in the population, are symmetrical. Asymmetry on these traits is thought to result from imprecise expression of design due to developmental stress to which the organism is not adapted. It increases in populations exposed to pollutants and pathogens and, within populations, has been found to covary with developmental health and viability (see Møller & Swaddle 1997). Clearly, fluctuating asymmetry does not reflect all aspects of maladaptation in organisms, only those that affect expression of developmental design. Because pathogens and other newly introduced environmental features are important sources of developmental stress in natural populations, however, fluctuating asymmetry may reflect much maladaptation due to temporally changing selection pressures. Though the proximate causes of fluctuating asymmetry may largely be environmental (e.g. pathogens), fluctuating asymmetry generally has moderate heritability (perhaps partly due to variation in pathogen resistance; Møller & Thornhill 1997). The measure of human fluctuating asymmetry most widely used in research (asymmetry of ears, elbows, wrists, hands, fingers, ankles and feet, or some combination thereof, summed into a single index) has about 30% narrow heritability, and 50% when measurement error is subtracted out (Livshits & Kobylianski 1989).

Fluctuating asymmetry and atypical brain asymmetries

Does fitness-related variation affect human psychological adaptations and its substrates? One intriguing study in this regard is that of Thoma (1996), who found that external body fluctuating asymmetry (using the measure just described) strongly predicts atypical brain asymmetry (positively) as well as size of the corpus callosum, the structure that links the two hemispheres (negatively, as measured by magnetic resonance imaging). These findings suggest that variation in general developmental imprecision affects even gross anatomical brain structure. Other studies indicate that developmental imprecision has functional consequences. In addition to atypical anatomical brain structure, individuals with high fluctuating asymmetry possess atypical lateralization of functions — e.g. language functions are overrepresentively localized in the right hemisphere and mechanisms that decode facial expression are overrepresentively localized in the left (Yeo et al 1997). Atypical organization of cognitive modules might be expected to be associated with performance deficits, though no study has addressed that point directly. Relatedly, however, fluctuating asymmetry negatively predicts fluid intelligence in college students (Furlow et al 1997). Moreover, a variety of neurodevelopmental disorders (e.g. schizophrenia, autism, dyslexia and attention deficit disorder) are associated with signs of more general developmental imprecision (see Yeo & Gangestad 1993 for a brief review). These disorders, which together affect a significant portion of the US population (perhaps more than 10%) and have substantial heritabilities, may be the extreme outcomes resulting from a continuum of fitness-related genetic variation due to temporally varying selection pressures.

If fitness-related variation related to developmental imprecision affects the psychological adaptations of over 10% of western populations in obvious ways, it is worth considering the possibility that few brains and the collections of psychological adaptations they represent are truly untainted by this variation. Participants in Thoma's study, after all, were 'normals', mostly university students and hospital employees, yet their brains exhibit significant variation. Pervasive variation may be further evidence that temporally varying selection, a particularly potent cause of fitness-related variation, has played a significant role in human history, though firm conclusions in this regard must await fuller documentation of the functional impact of neurodevelopmental imprecision.

Fitness-related variation has probably been a recurrent feature of human
populations and thus humans should have evolved adaptations in response to it

Perhaps the most important implications of fitness-related variation for evolutionary psychology follow from the fact that, if genotypic fitness differences have been a recurrent feature of human populations, humans should have evolved adaptations in response to them. Two major sorts of adaptations that could have evolved are allocation decisions due to selection for costly advertisements of fitness and phenotype-contingent strategies.

Allocation decisions.   Sexual selection should favour those who can discern and choose mates who are phenotypically or genetically most fit, thereby benefiting their offspring directly through material benefits or indirectly through genetic effects. These selection pressures in turn appear to favour the evolution of costly traits that honestly advertise fitness to mates or intrasexual competitors (Hamilton & Zuk 1982). Evolved costly traits can be thought of as evolved allocation decisions: what is selected is allocation of effort to their development. They evolve as signals of mate quality, then, precisely because of the allocation trade-offs discussed earlier. Individuals who are less fit because of, say, lower pathogen resistance, will have fewer resources to devote to their development because of the resources they must dedicate to active immune functions. Thus, allocation of effort to development of costly traits pays only for the more fit, which makes them honest signals (Grafen 1990). For similar reasons, it may pay more fit individuals to allocate more effort to potentially costly intrasexual competition and characters useful for intrasexual competition. Sexual selection operates more strongly on males' costly traits than females' and therefore men's evolved allocation decisions are expected to be more affected by fitness-related genetic variance than women's.

   If fluctuating asymmetry relates to ability to evolve costly sexually selected traits, then men's symmetry should predict their sexual success as well as characters that signal fitness. Evidence supports these predictions. Men's fluctuating asymmetry negatively predicts their number of sex partners (Thornhill & Gangestad 1994), extra-pair copulations (EPCs), times selected as an EPC partner (Gangestad & Thornhill 1997a) and a purported phenomenon of cryptic female choice, female orgasm (Thornhill et al 1995). Even when body mass has been controlled (which may itself negatively covary with fluctuating asymmetry, e.g. Manning 1995), men's fluctuating asymmetry also negatively predicts measures that may tap willingness or ability to engage in potentially costly intrasexual competition: frequency of intrasexual fights, particularly over status and respect (B. F. Furlow, T. Armijo-Prewitt & S. W. Gangestad, unpublished data 1996); and judged ability to physically protect a mate (Gangestad & Thornhill 1997b). What phenotypic traits mediate these relations is not fully known, but they appear to include metabolically costly traits (e.g. body size, muscularity; Gangestad & Thornhill 1997b, Manning 1995).

Phenotype-contingent strategies.   The effects of fitness-related variation may not stop with allocation decisions to costly traits and resultant sexual success. Other adaptive choices may be contingent on intersexual success with women and success at intrasexual competition with men. Thus, while less likely to engage in risky intrasexual competition with men, high fluctuating asymmetry men appear to invest more heavily in a mate—spending more time with her, sexualizing other women less and being more honest with her (Gangestad & Thornhill 1997b). Moreover, they advertise these benefits they provide to potential mates and intrasexual competitors. Single men interviewed for a potential 'lunch date' by an attractive female confederate were asked to tell both the woman and another male interviewed for the date why they should be chosen. Relatively asymmetrical men tended to be less likely to state that they were better than the other guy, and instead tended to focus on just how nice they were (J. A. Simpson, S. W. Gangestad & N. Christensen, unpublished data 1996).

Of course, if fluctuating asymmetry is heritable then so too must be these behavioural dimensions. These heritabilities will be instances of 'reactive heritability', heritability due to universal evolved adaptations of selecting a behavioural strategy on the basis of one's own heritable phenotype (Tooby & Cosmides 1990).

Adaptive variation through frequency-dependent selection

Thus far, I have focused on genetic variation due to the limitations of natural selection to remove variation. One potentially important cause of variation due to natural selection itself is negative frequency-dependent selection. When the fitnesses of genotypes decrease with their frequency, selection can maintain a mixture of genotypes. Because individuals often compete most directly with others who pursue a behavioural strategy similar to their own (e.g. occupy a similar feeding or social niche) frequency-dependent selection for rare behavioural phenotypes may be common (see Figueredo & King 1996 for an extended argument).

Is it reasonable to expect adaptive behavioural variation?

Tooby & Cosmides (1990) argued that frequency-dependent selection on behavioural phenotypes probably accounts for their genetic variation only rarely due to genetic constraints. Complex psychological adaptations require the co-ordination of multiple genes specifying different aspects of the developmental programme that constructs the adaptation. Alternative designs within a population thus require alternative suites of genes co-ordinated in a way that construct an adaptation. Sexual recombination breaks up alternative functional suites of genes, however, resulting in non-functional, uncoordinated suites of genes in progeny. Thus, alternative designs cannot be the direct product of alternative suites of genetic differences but instead should be the result of environmentally contingent expression of alternative but universal developmental designs.

Wilson (1994) qualified these claims in a number of ways. First, while uniformity probably exists at a general level of overall design, non-trivial variations at finer, quantitative scales (the levels measured by most personality trait measures) may be maintained by adaptive selection on behavioural phenotypes. Tooby & Cosmides (1990) argued that genetic variation on behavioural traits is akin to non-adaptive parametric variation in the lung or heart capacity around an adaptive norm. Yet, as noted at the outset, even in 'normal' populations the genetic coefficients of variation on specific behavioural dimensions can be much greater than those of morphological characters. These behavioural variations would be akin to variations in gross morphology only if, say, adults with two or three times the mean heart or lung capacity in a population were not uncommon.

Second, frequency-dependent selection for alternatives on such traits does not require a lack of intermediate, maladapted forms. For instance, frequency-dependent selection can operate on a continuous polygenic variable, the extremes of which represent two adaptive alternatives. Selection will maintain variance despite the fact that most progeny end up somewhere in the middle. On a somewhat more complex scale, frequency-dependent selection in a two-dimensional space defined by two variables, A and B, for two combinations when rare — A and not-B, and B and not-A — can maintain variance on both dimensions despite the fact that many offspring end up with combinations not favoured at any frequency.

Third, the costs of maladaptive progeny can be eliminated by selection on the genetic background of complex polymorphisms. Just as genetic modifiers can adaptively alter pleiotropic expression of genes (as discussed earlier), modifiers can adaptively alter expression of a polymorphism. Threshold effects on complex polygenic continua, which serve to 'cut out' intermediate forms, appear to account for the genetic basis of a wide range of complex adaptive polymorphisms in non-human species (Roff 1996).

Are there any empirical examples?

Though mentioning candidates for adaptive human polymorphisms humans, Wilson (1994) cited only suggestive evidence for them. Jeff Simpson and I have discussed an example that, while not demonstrated beyond doubt, is intriguing (Gangestad & Simpson 1993).

Individuals differ in the extent to which they can exercise deliberate control over their expressive behaviour. While some can feign realistic portrayals of emotional expression, others have great difficulty at being able to hide their true emotional states. This individual variation is one component of the Self-Monitoring Scale (Snyder 1974) and accounts for much of its heritability. Using a twin methodology, J. A. Simpson and I developed a longer scale that tapped this genetic variance. As far as I know, this measure, Genic Expressive Control, has one of the highest heritabilities (0.76 in a cross-validation sample) of any personality measure (as it should, given it was designed to have high heritability, unlike other measures). The item that covaries most highly with the rest of the measure is simply 'I would probably make a good actor'.

Is this variation simply non-adaptive variation around an adaptive mean? Or, might it be adaptive variation? Frank (1988) has proposed a model for the latter. He suggested that signs of inability to feign expression of emotion and commitment can have benefits due to increased opportunities for social exchange. When a population in which few individuals can feign emotion, however, ability to do so may be advantaged due to others' inattentiveness to this ability. As the ability increases in the population, however, so does vigilance, resulting in frequency-dependent selection on ability and inability to feign emotion.

If genetic variance in expressive control is non-adaptive variation around an adaptive mean due to accumulated small effects of many genes, it should be near-normally distributed around a single mode. If, on the other hand, it represents alternative strategies, selection on its genetic background could have produced a threshold effect, in which case it could be bimodally distributed. Of course, variation on Genic Expressive Control is not entirely genetic variation (despite the name); it has about one-quarter non-genetic variation (including measurement error). Averaging monozygotic twins' scores on the measure increases genetic variance to about 85%. Gangestad & Simpson (1990) plotted the distribution of these scores and, indeed, they were bimodally distributed. Other quantitative analyses also suggest that this variation is expressed as a threshold effect (Gangestad & Snyder 1985).

While these results hardly demonstrate beyond doubt that variation in expressive control has been maintained by frequency-dependent selection, they are intriguing and suggest that adaptive selection on behavioural traits deserves serious attention (for another example see Mealey 1995).

Conclusions

Both the genetic and environmental components of individual differences must be incorporated into an evolutionary framework. Genetic variations seem likely to be partly fitness related or adaptive in nature and these sorts of genetic variation deserve further attention. Though gains have been made, progress thus far has been stymied by the fact that existing personality measures do not adequately tap meaningful behavioural variations from an evolutionary theoretical framework. Thus, the development of new measures of individual differences remains an important task for evolutionary psychologists.

References

Bouchard T J 1994 Genes, environment, and personality. Science 264:1700–1701
Charlesworth B 1990 Optimization models, quantitative genetics, and mutation. Evolution 44:520–538
Figueredo A J, King J E 1996 The evolution of individual differences. WCPA Observer 2:1–4
Frank R 1988 Passions within reason. Norton, New York

Furlow BF, Armijo-Prewitt T, Gangestad SW, Thornhill R 1997 Fluctuating asymmetry and psychometric intelligence. Proc R Soc Lond B 264:1–8

Gangestad SW, Simpson JA 1990 Toward an evolutionary history of female sociosexual variation. J Pers 58:69–96

Gangestad SW, Simpson JA 1993 Development of a scale measuring genetic variation related to expressive control. J Pers 61:133–158

Gangestad S, Snyder M 1985 'To carve nature at its joints': on the existence of discrete classes in personality. Psychol Rev 92:317–349

Gangestad SW, Thornhill R 1997a The evolutionary psychology of extrapair sex: the role of fluctuating asymmetry. Evol Hum Behav 18:69–88

Gangestad SW, Thornhill R 1997b Human sexual selection and developmental stability. In: Simpson JA, Kenrick DT (eds) Evolutionary social psychology. Lawrence Erlbaum Associates Inc., Hillsdale, NJ, p 169–195

Grafen A 1990 Biological signals as handicaps. J Theor Biol 144:517–546

Hamilton WD 1982 Pathogens as causes of genetic diversity in their host populations. In: Anderson RM, May RM (eds) Population biology of infectious diseases. Springer-Verlag, New York, p 269–296

Hamilton WD, Zuk M 1982 Heritable true fitness and bright birds: a role for parasites? Science 218:384–387

Houle D, Hughes KA, Hoffmaster DK, Ihara J 1994 The effect of mutation on quantitative traits. I. Variance and covariance of life history traits. Genetics 138:773–785

Lande R 1975 The maintenance of genetic variability by mutation in a polygenic character with linked loci. Genet Res 26:221–234

Livshits G, Kobylianski E 1989 Study of genetic variance in the fluctuating asymmetry of anthropometric traits. Ann Hum Biol 116:121–129

Manning JT 1995 Fluctuating asymmetry and body weight in men and women: implications for sexual selection. Ethol Sociobiol 16:145–153

Mayr E 1983 How to carry out the adaptationist programme? Am Nat 121:324–334

Mealey L 1995 The sociobiology of sociopathy: an integrated evolutionary model. Behav Brain Sci 18:523–599

Møller AP, Thornhill R 1997 The heritability of fluctuating asymmetry. J Evol Biol 10:1–16

Møller AP, Swaddle JP 1997 Asymmetry, developmental stability and evolution. Oxford University Press, Oxford

Mousseau TA, Roff DA 1987 Natural selection and the heritability of fitness components. Heredity 69:181–197

Plomin R, Bergeman CS 1991 The nature of nurture: genetic influences on 'environmental' measures. Behav Brain Sci 14:373–427

Rice WR 1988 Heritable variation in fitness as a prerequisite for adaptive female choice: the effect of mutation–selection balance. Evolution 42:817–820

Roff DA 1996 The evolution of threshold traits in animals. Q Rev Biol 71:3–35

Snyder M 1974 Self-monitoring of expressive behavior. J Pers Soc Psychol 30:526–537

Thoma R 1996 Developmental instability, handedness, and brain lateralization: MEG and MRI correlates. MSc thesis. University of New Mexico, Albuquerque, NM, USA

Thornhill R, Gangestad SW 1994 Fluctuating asymmetry and human sexual behavior. Psychol Sci 5:297–302

Thornhill R, Gangestad SW, Comer R 1995 Human female orgasm and mate fluctuating asymmetry. Anim Behav 50:1601–1615

Tooby J, Cosmides L 1990 On the universality of human nature and the uniqueness of the individual: the role of genetics and adaptation. J Pers 58:17–67

Wilson DS 1994 Adaptive genetic variation and human evolutionary psychology. Ethol Sociobiol 15:219–235

Yeo RA, Gangestad SW 1993 Developmental origins of human hand preference. Genetica 89:291–296
Yeo RA, Gangestad SW, Thoma R, Shaw P, Repa K 1997 Developmental instability and cognitive lateralization. Neuropsychol, in press

DISCUSSION

Mealey: In your interview data of males with high fluctuating asymmetry versus low fluctuating asymmetry there is, apparently, substantial variation in the way that different males portray themselves, even at any given level of fluctuating asymmetry. For the males who have high fluctuating asymmetry, yet who still present themselves as 'super-macho', do you have any data on how females respond? Have you given the video tapes to raters to see who gets rejected or accepted?. And do the ones who are lying get rejected more often?

Gangestad: That's a good question, and this research is in progress. We have video tapes of men interviewed in this study, which we're now using as stimulus materials in studies of how people perceive and respond to men's behaviour. One of the questions we've asked our perceivers is, how much do you think this person is bluffing? And yes, it's possible that men who have high fluctuating asymmetries who nevertheless are self-promoting will be seen as bluffing and accordingly unattractive.

Cosmides: Talking about the heritability of personality traits is tricky because one doesn't actually know what exactly is causing the variation. For example, one could imagine a mechanism regulating the circumstances under which one chooses to use aggression to get what one wants. Imagine it is uniform across people and there is no heritable variation in it. Nevertheless, the inputs to that mechanism could be how mesomorphic you are (i.e. how muscular and physically formidable you are), and there could be heritable variation in those inputs. If you were to do a personality test for aggressiveness, you would find heritable variation in the propensity to be aggressive. But this would not be because of genetic variation in the mechanism regulating when you're aggressive. It would be because of genetic variation in one of the inputs to that mechanism.

Gangestad: That's a good point, but its implications for the issues I addressed are limited. I addressed evolutionary accounts of what has allowed variation in personality to persist, which is entirely different from addressing the mechanisms that underlie the variation. From a quantitative genetic perspective, there is clearly genetic variation in aggressiveness. You suggest that the genetic effect is indirect and I agree that it may well be. Even if genetic variation affects the phenotype indirectly, however, can one ask questions about evolutionary processes that account for the variation? Thus, if aggressiveness is heritable and has implications for fitness, we can ask why the heritable variation persists, and whether the genes affect aggressiveness directly or indirectly through its effects on, for instance, the mesomorphic phenotype.

Tooby: This is a critical and interesting question because this symposium is on characterizing psychological adaptations and this addresses exactly the issue of how best to characterize what's going on. One might think from selection pressures that there are rules such as 'be aggressive when you're in a position to make it pay and don't when it's not'. In one sense that might be a complex interesting adaptation and it may be relatively uniform. There is a tendency when talking about psychology to talk about dispositions and tendencies but that's a crude way of characterizing psychological adaptations. I agree with your answer that in an abstract way anything heritable which changes the character is a genetic factor regulating that character. However, from a procedural point of view of how to conceptualize adaptations these questions about the dissection of characterizations are critical. These general personality measures are gateways to something interesting, but in the guise of 'dispositions' and 'tendencies' they obscure the complexity and block the precise characterization of the underlying machinery. To say someone is more aggressive or less aggressive does not give a description of the cognitive programme that organizes and regulates aggressive behaviour. Descriptively, it turns a complex machine into a dimension.

Gangestad: I don't disagree with these points. Again, my interest here is in why genetic variation is present. In the case of fluctuating asymmetry, there is some source of maladaptive variation, which could be mutation, temporally variable selection or spatially variable selection. Consistent with what you're suggesting, the variation appears to be characterized in terms of imprecise expression of adaptations and its consequences.

Tooby: Genetic variation detected and expressed at the psychological level could be there because of compromises with other non-psychological selection pressures, such as parasite-driven pressure for disease resistance through biochemical individuality. This intense selection pressure should pump genetic diversity into all systems that can be infected, which is virtually everything. I am unaware of any evidence that shows there is either more or less parasite-driven diversification in the genes underlying psychological characteristics compared to other types of characteristics.

Gangestad: Neither am I. The genetic coefficients of variation of many behavioural phenotypes appear to be greater than those of non-psychological traits, even though behavioural phenotypes may have lower heritabilities. This suggests that there is tremendous potential for selection on the genetic variation (e.g. Houle 1992) underlying behavioural phenotypes. One has to ask why this genetic variation persists.

Nesse: How are we going to figure this out? I suspect that within the next 10 years we will have specific genes that map onto many of these personality traits. The first of these — the serotonin transporter — was identified recently (Lesch et al 1996). There are two variants of this: 50% of us have one variant and 50% have the other. There are also recent data on aggression in mice showing that only three or four loci account for a high proportion of the variance. Therefore, there may be only a few genes involved. It seems to me that the design of studies we should be thinking of is to look for people who vary at these specific loci and look at their behavioural characteristics.

Gangestad: One cautionary note is that although genes have been found associated with phenotypes, such as schizophrenia and bipolar disorder, these studies have failed to be replicated. Gene mapping could lead to an understanding of the genetic basis of some behavioural traits but we should also be aware that many traits may be indirectly influenced by pathogen susceptibility or developmental stresses due to mutations widely distributed across the genome. In these cases, the genetic basis of traits may not be illuminated by molecular studies.

Sherry: I would like to point out that almost on an annual basis the newspapers report, 'Scientists find gene for learning', which sounds foolish. It isn't, and the scientists in question are not foolish. What they have usually found is a single gene substitution that disrupts learning. In *Drosophila* most of these have something to do with the cAMP cascade. There is no such thing as the gene for learning, but the action of many genes affects learning and modifying them can disrupt learning.

Miller: One of the things that best catalysed the study of domain-specific intelligence was Spearman's discovery of 'g', the factor that intercorrelates between many different kinds of intelligence and which we now measure with IQ tests. Could fluctuating asymmetry play a similar role in characterizing what's heritable? For example, if we're trying to estimate heritabilities for height or IQ does it make sense to partial out the heritability of fluctuating asymmetry, so that we can obtain a more specific measure of the genes for particular traits?

Gangestad: Possibly. We found that fluctuating asymmetry correlates, for instance, with the number of sexual partners, number of fights and intelligence, variables that don't correlate with each other. Thus, if you partialled out fluctuating asymmetry you would see slightly negative correlations. One construal of these findings is that the developmental imprecision underlying fluctuating asymmetry influences metabolic efficiency, and thereby the overall expendable energy budget (Manning 1995). On top of this variation are variations in life history decisions about how to allocate expendable energy, which results in the slightly negative partial correlations between behavioural phenotypes such as fights and intelligence. Life history theory is a case where people have expected negative covariations but then have had a difficult time finding them. One reason may be because variation in developmental imprecision masks them.

Thornhill: One way to think about this adaptation is by considering a test for its existence. Suppose you found that almost everyone has the ability to assess their own asymmetry and adjust their behaviour (such as in men's attitude about fighting or pursuing females) accordingly. If there existed such an information-processing adaptation then this would be consistent with heritable fluctuating asymmetry and species-typical, or non-heritable, psychological adaptations.

Tooby: This could be addressed easily by showing subjects mirror or camera distortions of themselves to see if that has an immediate effect on their choices in these areas.

Daly: But fluctuating asymmetry may be the scientist's window on something that the organism is monitoring in some other way.

Thornhill: Plastic surgery is often used to correct asymmetry, and about 99% of the patients report profound positive effects on the quality of their lives, which may suggest that some of these modifications are being assessed and lead to individuals adjusting their behaviour accordingly.

Bouchard: I have one comment that disagrees with this. One of the arguments made against my study of monozygotic twins reared apart is that monozygotic twins look alike (Bouchard et al 1990). This in turn is said to cause them to have similar personalities. The literature on physical attractiveness is enormous, but the correlations between physical attractiveness, personality traits and IQ are zero. If there's no relationship between the feature and the trait it's not a causal factor. I believe your data but I'm just saying that you've got to be careful when looking at the actual evidence.

Cosmides: You talked about pathogenic theories and in particular Hamilton's theory (Hamilton & Zuk 1982). I would like to mention that Hamilton's pathogenic theory is different from that of Tooby (1992) because Hamilton's theory is based on long-lived pathogens and looks at heritable variation in disease resistance, whereas Tooby's is based on short-lived pathogens and looks at forces maintaining genetic variation. Some of this variation will show up as non-additive genetic variation; when this is the case, picking a symmetrical/asymmetrical mate will not necessarily predict something about the future with respect to disease resistance. The opposite assumption holds for Hamilton's theory.

Møller: I've reviewed this literature quite recently. There are data on parasite loads and asymmetry for about 35 different organisms (Møller 1996a) and most of the results agree with Steven Gangestad's. There are also a couple of studies showing a relationship between asymmetry and resistance (Møller 1996b, Polak 1993).

Cosmides: But non-additive genetic variation will not necessarily transfer benefit to your offspring.

Gangestad: Why are you suggesting that it's not additive?

Cosmides: It is non-additive in so far as you're getting the biochemical individuality by just shuffling the genes. Much of the individuality is as a result of the shuffling rather than in the selective retention of parasite resistance.

Bouchard: In principle it could work that way, but it is an empirical question.

Cosmides: I agree. I am fascinated by the problems and the pulling apart of these different hypotheses, especially those on social dominance and the finding that symmetrical men are more socially dominant, for example.

Gangestad: There are certain relationships between fluctuating asymmetry and personality traits. For example, symmetrical males are relatively more resistant to the control of others. In our college sample they also got into more fights, and particularly fights concerning status. Social dominance is a broadly imprecise term, however, and we've not found symmetrical men to be dominant in all ways. Evidence suggests that they do not, for instance, more often assume leadership roles or seek social attention, two major facets of social dominance.

Cosmides: I'm interested in disentangling certain hypotheses. For example, I have to admit to my bias that, as a female, I always find the good genes hypotheses not credible.

For various reasons they can be theoretically problematic, excepting possibly cases of pathogen resistance (e.g. what counts as 'good genes' is, in many cases, exactly what would make a good investor). You can avoid being controlled, for example, in two ways: using your brain or using your brawn. Sometimes I feel that people underappreciate investment in protection, i.e. being able to protect somebody. It is possible that there are facultative variations in mate preference, depending on the society. One way of finding out exactly what women are choosing on the basis of is to look at populations in which being able to fight is important versus not important, because my intuition is that this may make a difference.

Nisbett: As far as I know, there might be reason enough to assume that your phenomena might look different in different cultures. D. Cohen (personal communication 1996) has performed a simple experiment in which he insults a man and the man does one of two things: he behaves in a meek fashion or he responds aggressively. He then asked women who observed the event to choose which man they would like to date, and he found that women from the northern USA prefer the wimp, whereas women from the southern USA prefer the tough guy.

Gangestad: I would like to address a few points. First, when we showed video tapes of men to women and asked them to rate them on the basis of attractiveness as a long-term mate versus a brief sexual partner, we found a difference. Preliminary results, at least, indicate that more symmetrical men are particularly preferred as short-term partners, suggesting that the benefits they provide can be had in a short-term relationship. Second, non-additive genetic variance can be examined empirically, and in fact the measure of fluctuating asymmetry we have used has 30–50% additive genetic variance as well as non-additive genetic variance. Third, if it is related to fitness, men who have higher fitness also have more partners and ancestrally perhaps more offspring. If so, good genes sexual selection operated, perhaps in addition to other kinds of sexual selection. Of course, symmetry need not be related to fitness, because there must be some cost to maintaining it.

Tooby: Why do you say there must be a cost to symmetry?

Gangestad: In the sense that there is a cost to all adaptations, including those that underlie developmental precision and resultant symmetry. Nonetheless, I'd assume that those who incur this cost are probably those best able to do so, and hence those who have highest fitness. Thus, I'd assume that genetic variance underlying fluctuating asymmetry is probably largely fitness-related genetic variation.

Maynard Smith: I am rather concerned about the population genetics of this approach. Maynard Smith's law says that the broad-sense heritability of everything is 0.5, but additive heritability of fitness is not 0.5. If the additive heritability of asymmetry is near 0.5 then the correlation between asymmetry and fitness has to be low. Therefore, let's not assume symmetry equals fitness or good genes, because it doesn't.

Gangestad: It's clearly not necessary to equate asymmetry with fitness to assume that developmental imprecision contributes some variance to fitness.

Dawkins: But your correlations are extremely low.

Gangestad: So the question might be, what correlation is necessary to fuel sexual selection? Rice (1988) published a model suggesting that mutation–selection balance alone could account for up to 20% genetic coefficient of variation in fitness. He concluded that this was enough for good genes sexual selection, although he may be wrong.

Maynard Smith: Sexual selection is not impossible even if there is no genetic variance.

Gangestad: But the question is not whether sexual selection is operating, but whether it is fuelled by some genetic variation in fitness.

Thornhill: You can't determine whether sexual selection was based on good genes historically by examining the heritability of some trait. If there has been effective good gene sexual selection in evolutionary history there must exist adaptations for assessing genetic quality of mates. Sexual selection in humans and in many other species is based on symmetry or developmental health, and sexual selection in humans is also based on other health markers. It is reasonable and appropriate to conclude that health was heritable enough in the past to generate good gene sexual selection for the evolution of the adaptations that are assessed in potential mates.

Gangestad: Until we know more about the heritability of symmetry and the strength of relations between symmetry and fitness components, it remains debatable whether good genes are the main benefits that females receive by mating with symmetrical males. There are some theoretical reasons to expect that the main benefit is not going to involve an expensive investment on the part of men that's going to take away from their mating effort. A study of romantically involved couples found that more symmetrical males were less honest and less giving of time to their mates. Out of the 11 facets of relationship investment that we examined, the one that symmetrical men provided more of was physical protection. Possibly that is a major benefit contributing to symmetrical men's attractiveness.

Maynard Smith: Some sort of mathematical modelling has to be done here. I'm uncomfortable about much of what I've heard in this discussion. There seems to be a lot of confusion about what the mechanisms are. I am also puzzled by the assertion that additive heritability of asymmetry is so high. I worked on asymmetry many years ago in *Drosophila* when it wasn't fashionable. Although it certainly had a genetic basis, it was almost entirely a direct measure of degree of inbreeding.

Daly: What do we know about heritability estimates for asymmetry across a large group of organisms?

Maynard Smith: The total heritability will be high, and the additive heritability will be low.

Møller: There have been 32 different studies of this, and the average heritability is about 15%. However, I'm a little suspicious of some of the laboratory studies because they were not carried out in stressful environments. For this reason there may be relatively low levels of asymmetry, and therefore we may not be able to pick up correlations between parents and offspring.

Gangestad: Most of these studies only used one trait's asymmetry. Aggregated data based on the asymmetry of multiple traits give a better estimate of general developmental imprecision and therefore the heritability of aggregated data is generally higher. For instance, Livshits & Kobyliansky (1989) found higher heritability of an aggregated measure of asymmetry in humans—about 30%, 50% if measured error is subtracted.

Maynard Smith: It doesn't follow that because you measure 10 traits you get a higher heritability. You're just introducing measurement error.

Gigerenzer: I would like to follow up on this concern about mathematical modelling and the use of correlation coefficients in this area. Ernest Hilgard, an experimental psychologist, said a long time ago that correlation is an instrument of the devil, but don't take this personally! The point is that any modelling of the dependency between two variables, for instance body asymmetry and brain asymmetry, assumes a theory about what the relationship actually is. The Pearson correlations reported in your research assume a linear relationship; that is, that some increment here with some probability results in an increment there. The data on the correlation between body asymmetry and brain asymmetry when incorporated into a scatter diagram are not indicative of a linear relationship. Instead, there is a cloud within which some 80% of the points lie, and where there is hardly any covariation; and it is only outside of this cloud that a linear relationship starts. Therefore, the underlying mechanism does not seem to produce an increment here and an increment there, rather it produces a threshold, i.e. there is only a linear relationship for extreme values above the threshold. Consequently, it might not be useful to use correlations as models of dependency, but rather to look for threshold models that have other kinds of underlying mechanisms.

Gangestad: These points are well taken. However, with only 28 data points it's difficult to discern the difference between a linear relationship and the sort of relationship you are suggesting. Moreover, I have no idea why one should expect an elbow-shaped relationship, and until I come across some strong evidence for it or a theoretical reason to expect it, I'm reluctant to believe that the relationship is truly elbow shaped. I agree that correlations do not, in many cases, give us as good evidence about causality as experimental analyses, but at the moment this is what we're stuck with. Generally speaking, in those sciences that depend on correlations, what one would hope for is an integrated network of findings that all make sense. If we only had a correlation with atypical asymmetries I might be worried, but we've got one for the corpus callosal area, atypical functional lateralization and intelligence. Finally, we have done randomization tests to confirm that there is a linear trend.

Kacelnik: In my opinion, this problem is being exaggerated. The correlations don't have a huge difference in the distribution of residuals in one direction or another. They were not perfect, but they seem reasonably normally distributed around the correlations.

References

Bouchard TJ, Lykken DT, McGue M, Segal NL, Tellegen A 1990 Sources of human psychological differences: the Minnesota study of twins reared apart. Science 250:223–228

Hamilton WD, Zuk M 1982 Heritable true fitness and bright birds: a role for parasites? Science 218:384–387

Houle D 1992 Comparing evolvability and variability of quantitative traits. Genetics 130:195–204

Lesch K-P, Bengel D, Heils A et al 1996 Association of anxiety-related traits with a polymorphism in the serotonin transporter gene regulatory region. Science 274:1527–1531

Livshits G, Kobylianski E 1989 Study of genetic variance in the fluctuating asymmetry of anthropometric traits. Ann Hum Biol 16:121–129

Manning JT 1995 Fluctuating asymmetry and body weight in men and women: implications for sexual selection. Ethol Sociobiol 16:145–153

Møller AP 1996a Parasitism and developmental instability of hosts — a review. Oikos 77:189–196

Møller AP 1996b Sexual selection, viability selection and developmental stability in the domestic fly, *Musca domestica*. Evolution 50:746–752

Polak M 1993 Parasites increase fluctuating asymmetry of male *Drosophila — Nigro spiracula —* implications for sexual selection. Genetica 89:255–265

Rice WR 1988 Heritable variation in fitness as a prerequisite for adaptive female choice: the effect of mutation–selection balance. Evolution 42:817–820

Tooby J 1982 Pathogens, polymorphism and the evolution of sex. J Theor Biol 97:557–576

Evolution and human choice over time

Alan R. Rogers

Department of Anthropology, University of Utah, Salt Lake City, UT 84112, USA

Abstract. This chapter reviews previous work on an evolutionary model describing the effect of time delays on human preferences. The model explains why the long-term real interest rate is usually near 3% and why rates of crime and driving accidents are highest among young adults. It does not succeed in explaining the phenomenon of preference reversal. The chapter reports new results on uncertainty and on a more comprehensive model allowing consumption to have simultaneous effects on mortality and fertility.

1997 Characterizing human psychological adaptations. Wiley, Chichester (Ciba Foundation Symposium 208) p 231–252

Every investment decision is a choice involving costs and benefits that accrue at different times. We give money to the bank today in the expectation that the bank will return that money to us with interest at some future date. We give money to the store today in return for a washing machine that (we hope) will provide service across a wide range of future dates. At first sight, it seems unlikely that evolution can tell us anything at all about such choices. Our ancestors did not have banks and did not buy washing machines. Most investment opportunities that are open to us were closed to them. It is difficult to see how selection could have shaped intertemporal preferences.

But ours was not the first species to make investments. Every squirrel or bird who stores nuts for the winter is sacrificing immediate consumption in return for consumption at a later date. Nests and burrows can also represent investments: the energy spent building one is retrieved by sheltering in it later on. Thus, choice over time has an ancient history. It is possible that such choices influenced the evolution of human brains. Let us therefore ask what human choice over time would look like if it had been shaped by natural selection.

The connection between evolution and economics

In Fig. 1 the horizontal axis measures the quantity of some good — say bread — that is available for your consumption today, and the vertical axis measures the quantity that you expect at some future date, which I will call 'tomorrow.' The dotted lines describe your preferences using a sort of contour map. On an ordinary contour map, the lines connect points of equal altitude. Here, the dotted lines connect points to which you are

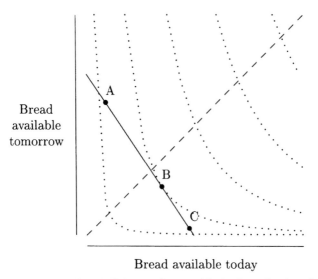

Bread
available
tomorrow

Bread available today

FIG. 1.   Irving Fisher's indifference diagram. The horizontal and vertical axes show the bread
available for consumption today and tomorrow. The dotted lines connect the combinations of
equal utility and provide a contour map of the utility function. The solid *budget line* connects
points that can be reached by lending and borrowing.

indifferent. In the economic jargon, they connect points of equal *utility*. For example,
you might be just willing to exchange the combination {10 ounces of bread today, 10
tomorrow} for the combination {11 today, eight tomorrow}. If so, then these two
combinations provide equal utility and would lie on the same contour line. For
consistency with what follows, I will use the term *utility contour* to refer to the
contour lines of the utility function. (In the literature, these are called *indifference curves*.)

When the utility contour is steep, one is willing to sacrifice a lot of tomorrow's bread
in exchange for a small increase in today's supply. This can be expressed by saying that
one's *marginal rate of time preference* is high.

By borrowing one can increase today's bread consumption at the expense of
tomorrow's. By lending one does the opposite. For example, the individual at A in
the diagram could move to points B or C by borrowing, whereas the individual at C
could move to points B or A by lending. The solid line along which this movement
occurs is called the *budget line* and has a slope that is determined by the interest rate. (We
are ignoring here the difference between lending and borrowing rates of interest.) If the
interest rate is high, one must give up a lot of tomorrow's bread in order to gain a little
today, so the budget line is steep.

Although the individuals at A and C may wish to be in the graph's upper right
corner, their options are limited. In this model world, they can change consumption
only by lending and borrowing, and this only moves them back and forth along the

budget line. Given this constraint, the best course of action is to lend or borrow until reaching point B, for B gives greater utility than any other point on the budget line.

Notice that the utility contour passing through B has the same slope as the budget line. This will always be true of the optimal point unless the optimum occurs at one of the two ends of the budget line. This is the result that I have been driving at and is so important that I will set it off for emphasis:

Principle 1. In a credit market, equilibrium occurs at a point where the slope of the budget line equals that of a utility contour (Fisher 1930).

This principle underlies a great deal of economic analysis.

So much for economics. Let us now switch gears and talk about evolution. Where economists talk of maximizing utility, evolutionists talk of maximizing *Darwinian fitness*, or just fitness for short. The fitness of an individual measures his or her genetic contribution to future generations. If a gene encourages its carriers to behave in ways that increase fitness, then that gene will tend to increase in frequency within the population. Thus, the population will eventually be dominated by the behaviours that impart greatest fitness.

With this in mind, let us give Fig. 1 a different interpretation. Suppose that one's fitness depends somehow on the consumption of bread today and tomorrow, and that the figure's dotted lines connect combinations that impart equal fitness. Then the dotted lines are *fitness contours*, and together they provide a contour map of fitness. If bread is good for you, then fitness is highest at the upper right corner of the diagram and lowest at the lower left. Suppose now that it is somehow possible to exchange bread today for bread tomorrow. For example, in gaining bread today you may deplete resources that are needed to gain bread tomorrow. Then your options are constrained and will fall along some line with negative slope such as the solid line in Fig. 1. (There is no reason to think that this new constraint line would be straight, but let us not quibble about details.) Selection will favour behaviours that choose the highest fitness among those fitnesses that are feasible. If we are constrained to stay on the solid line, then fitness is maximized at point B. As before, the equilibrium is reached at a point where two slopes are equal. The conclusion is:

Principle 2. Evolutionary equilibrium occurs at a point where the slope of the constraint curve equals that of the fitness contour.

We see again that the equilibrium lies at a point where two curves have equal slope. This result underlies much of the theoretical work that has been done in evolutionary ecology.

The processes that lead to principles 1 and 2 are entirely different. One involves lending and borrowing; the other involves changes of gene frequencies. Yet the results themselves are similar. I want to argue now that these results are connected by something deeper than analogy. They are connected by a third equilibrium principle, which was first described by Hansson & Stuart (1990). The new equilibrium principle states that:

Principle 3.   At evolutionary equilibrium, the contour map of the fitness function is identical to that of the utility function. In other words, the contour lines of the two functions have equal slope at every point.

To see why this is true, consider Fig. 2, which shows what would happen if the result were false. If the two maps differ, there must be some place where a fitness contour crosses a utility contour. The figure shows such a place, with U indicating a utility contour and F a fitness contour. Consider points X and Y. X has higher utility than Y, so we would take X rather than Y if given a choice. Yet Y imparts higher fitness than X, so our preference for X is harmful. A mutation that reversed the preference between X and Y would be favoured by selection. It follows that the preferences shown in Fig. 2 cannot represent an evolutionary equilibrium.

The new equilibrium principle connects evolution to economics. At evolutionary equilibrium, the slope of the utility contour must now equal that of the fitness contour as well as that of the budget line. To apply this principle to choice over time, we must first ask how such choices affect Darwinian fitness.

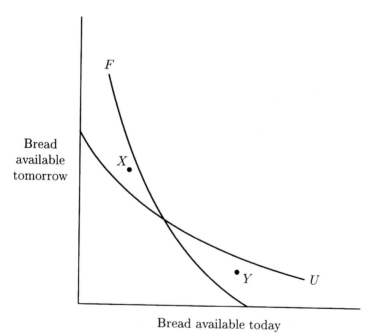

FIG. 2.   This figure depicts a fitness contour (F) crossing a utility contour (U).

How intertemporal choices affect fitness

How does natural selection weigh costs and benefits that arrive at different times? There are several factors to consider. First, nothing can affect fitness except by affecting survival and fertility. At a minimum, there will be effects at two different times. To relate these to fitness, we will need to ask several apparently unrelated questions:

(1) *How fast is the population growing?* This question is relevant because the contribution of each new child to the gene pool is proportional to the reciprocal of population size (Hamilton 1966). Thus,

Principle 4.   Evolution favours early reproduction in growing populations but late reproduction in shrinking ones.

(2) *Whose survival or fertility was affected, and how are these individuals related to the individual whose choice we are studying?* After menopause, women affect the gene pool only by the care they provide for children and grandchildren. To understand the effects on fitness of benefits that arrive late in life, we must account for parental (and grandparental) care. To a fair approximation (Hamilton 1964):

Principle 5.   Benefits to relatives are weighted by the coefficient of relationship, which equals 1/2 for the parent-offspring relationship, 1/4 for grandparent-grandchild, and so on.

(3) *How old was the individual whose survival or fertility was affected?* Age matters for two reasons. First:

Principle 6.   When a gene is expressed at a single age, the force of selection upon it is proportional to the probability that its bearer will live long enough to express it.

Age enters into the analysis in a second way for genes that affect age-specific survival: each death deprives the world of all of the children that might have been produced in the remainder of that individual's life. This remainder is likely to be larger for a 20-year old than for a 40-year old. In general:

Principle 7.   A change in the probability of death at age x has an effect on fitness that is proportional to $v(x)$, the reproductive value at that age.

The reproductive value (Fisher 1958) counts the individual's expected future contributions to the gene pool and discounts these according to changes in population size (see above).

Putting it all together

We are now in a position to calculate fitness contours. To move along a fitness contour, one must find a way to change 'bread today' and 'bread tomorrow' without affecting fitness. This involves adding up the fertility and survival effects of the two age-specific changes in bread consumption, discounting according to all the factors just listed (possibly allowing tomorrow's bread to be enjoyed by a child or a grandchild) and ensuring that the result equals zero — a zero change in fitness.

This programme faces an immediate obstacle: we don't know how consumption of bread (or any other commodity) affects fitness. Here, I will summarize earlier work (Rogers 1994) in which I took the coward's way out and ignored consumption altogether. Instead of dealing with consumption, I ask a simpler question: suppose that one individual (called the 'donor') performs some action that slightly changes her survival probability at age x. After a delay of length τ, this causes a change in the survival probability of an individual called the 'recipient' when the recipient's age is y. What condition do these changes have to satisfy in order to ensure that the donor's fitness in unaffected? The answer to this question will describe one of the fitness contours in Fig. 3, where fitness is shown on survival axes. Later work (A. R. Rogers, unpublished data 1996) has shown that substantially the same answers are reached by a less timid analysis that incorporates consumption and allows it to affect fertility as well as survival. Thus, we will not be led astray by the simpler analysis described here.

I will also suppress all calculations and cut straight to the answer: The slope of the fitness contour is

$$-\frac{v(x)}{re^{-\rho\tau}v(y)} \tag{1}$$

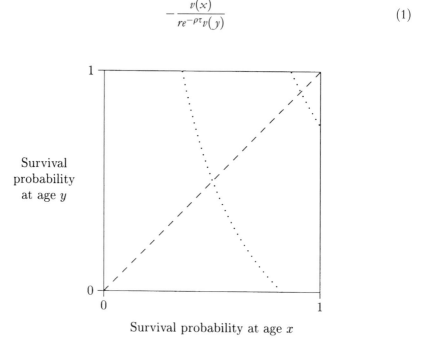

FIG. 3.   Fitness contours on survival axes.

where $v(x)$ and $v(y)$ are the reproductive values of donor and recipient, r is the coefficient of relationship between them and ρ is the rate of population growth (Rogers 1994).

To gain some intuition into this formula, you should ask yourself, when is its value high (implying that the future is discounted steeply)?. The fitness contour has a steep slope if the donor's reproductive value is high, the recipient's is low, the rate of population growth is high, or the relatedness between the parties is low. All of this makes intuitive sense. The fitness contours in Fig. 3 were drawn using this formula, using human demographic data for the reproductive values, setting $\rho = 0$, and assuming that the recipient is the donor's daughter so that $r = 1/2$.

The interest rate

Let us now apply this machinery to a real problem: the long-term interest rate. Figure 4 shows British interest rates over the period from 1727 to 1900. These data reflect several effects.

First, interest rates climb during wars because military spending forces governments to borrow money. To attract the huge loans that are needed, governments offer high rates of interest. To compete for capital, other borrowers must also offer high rates. Thus, all interest rates rise. In Fig. 4, the high rates before and after the year 1800 reflect the American revolution and the Napoleonic wars. The other upswings are associated with less costly wars.

Second, lenders demand high interest rates during periods of rising prices, since they anticipate being repaid with pounds that are worth less than the pounds they are lending. The lower panel in Fig. 4 shows an index that summarizes prices. High interest rates during the Napoleonic wars may reflect the rising prices of that period, and low rates during the 1890s may reflect falling prices.

But I wish to draw your attention not to the wars and depressions but to the periods that intervene. In these intervening periods, prices were relatively constant. The money interest rate should therefore equal the *real interest rate* — the rate that would obtain in loans involving consumption goods rather than money. Throughout this period, it appears that the long-term real rate of interest was close to 3%. This is the fact that I hope to explain.

Since we are interested here in the long-term interest rate, let us consider an investment that matures after a period of one generation, or T years and is enjoyed by the daughter of the investor so that $r = 1/2$. After one generation, the daughter's age will on average equal the age of the mother when she made the investment. Thus, $x = y$ and $v(x) = v(y)$. The slope of a fitness contour (formula 1) becomes $-2e^{\rho T}$.

At equilibrium, the slope of the fitness contour must equal that of the utility contour, which must in turn equal that of the budget line. The slope of the budget line is $-e^{iT}$, where i is the real interest rate. It equals the slope of the fitness contour only if:

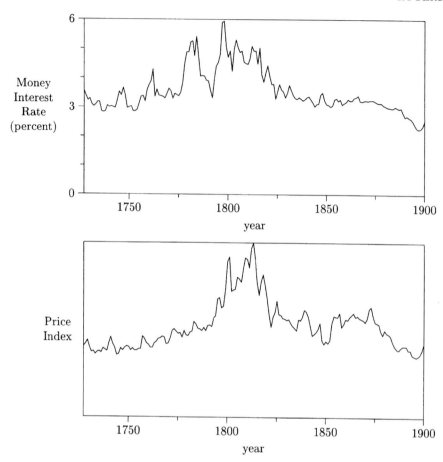

FIG. 4.   Interest rates and prices. Upper panel shows the annual money interest rate calculated from British bond prices (Homer & Sylla 1991). Lower panel shows an index of prices as described by Shiller & Siegel (1977).

$$i = (\ln 2)/T + \rho. \qquad (2)$$

According to the model, this interest rate should hold if human preferences are at evolutionary equilibrium.

Before discussing what this calculation has ignored, let us leap ahead and ask whether it makes quantitative sense. The relevant rate of population growth is not the current one, but some sort of average over evolutionary history. I will assume that this average is near zero, since sustained rapid demographic growth has probably characterized only the last few centuries of human history. In human populations, the generation time is usually a little less than 30 years, say 28. With this figure, equation 2 says that the long-term real rate of interest should be 2.5% per year — not far below the value of 3% that came from the data in Fig. 4. It is remarkable that an analysis so simple should lead so nearly to the correct answer.

Of the issues ignored here, two deserve mention. First, there is the time–productivity of investments. If one could invest money in a tree farm and realize a sure return of 10% per year, then banks paying less than 10% interest would lose customers. The interest rate is therefore tied to the time–productivity of investments. To see how this relates to the argument above, see my earlier paper on this topic (Rogers 1994).

Second, there is the issue of uncertainty, which can enter in two ways. One may be uncertain whether a delayed benefit will in fact arrive or uncertain about who will enjoy it. The second form of uncertainty is the subject of the section that follows. The first form is simpler to add into the model. Suppose that there is a constant hazard γ of some event that will prevent the delayed benefit from arriving. To incorporate this into the model, we must multiply the denominator of formula 1 by the probability, $e^{-\gamma\tau}$, that a benefit delayed by τ years will really arrive. This leads to

$$i = (\ln2)/T + \gamma + \rho. \tag{3}$$

The fact that i is close to 3% (rather than 2.5%) suggests that people expect a hazard of roughly 0.5% per year. In other words, the interest rate suggests that people are only about 95% confident that a benefit delayed by 10 years will actually arrive.

Uncertainty about recipients

The analysis above assumes that the recipient of any benefit is known with certainty at the time an investment is made. Let us now relax this assumption and allow the recipient to be drawn from the pool of people who are alive at the time the benefit arrives. In order to do calculations, we must specify the rule that is used for choosing among these potential recipients. I have used a simple rule: the benefit is allocated so as to maximize its discounted value to the donor's fitness. I rule out the possibility of distributing the benefit among several recipients.

Results will be described in terms of an evolutionary discount function, which is defined as follows. Suppose that I undertake some action that results in a delayed survival benefit to some individual determined by the rule above. If the benefit increases this individual's survival by an amount ΔP, then I will be indifferent between this benefit and an immediate benefit to myself of magnitude:

$$\Delta P \exp \left[- \int_x^y \lambda(x, z)dz \right]. \tag{4}$$

To be more precise, my fitness would be unaffected by the choice between these alternatives. The algorithm for computing λ is given elsewhere (Rogers 1994).

Figure 5 presents the evolutionary discount function implied by one set of demographic data. In the figure, 'age at investment' refers to the age at which a decision is made between an immediate and a delayed benefit. Ages beyond the age at investment are 'future ages'. Thus, the stars show the discount function for investments undertaken at age 20.

Suppose that a 20-year-old woman has been offered some benefit that will arrive at age 40. Since she is female and is now of age 20, the starred curve in the upper panel of Fig. 5 applies. It indicates that the average discount rates within the four five-year intervals spanning ages 20–40 are 0.059, 0.050, 0.012 and 0.007, respectively. The average of these is 0.032, and this implies that the future benefit should be discounted by a factor of $\exp[-20 \times 0.032] = 0.529$. The 20-year old, therefore, should value this delayed benefit at only about half of its nominal value.

Had the delay been shorter — say five years — this female should have discounted at an average rate of 0.059, nearly twice the 20-year rate. Thus, young adults should discount short delays at higher rates than long delays. This is certainly consistent with the conventional wisdom about young adults, who are said to worry their parents by sowing wild oats, burning the candle at both ends and living as though there were no tomorrow.

To compare this result with equation 2, we need to consider a delay of one generation, or about 30 years. The data in Fig. 5 imply that a 20-year-old female should discount this delay at average rate 0.026, or 2.6%, an answer that hardly differs from equation 2. On the other hand, a 30-year-old female should discount a 30-year delay a little more slowly, at 1.6%.

When the curve in Fig. 5 is averaged over long delays, the resulting average rate of discount is either just at or a little below the dotted line, which shows the prediction of equation 2. This more complicated model is thus in good agreement with the earlier analysis of interest rates.

The results in Fig. 5 depend on a particular set of demographic parameters. Such results should ideally be calculated from the demography of our ancestors, but that is of course unknown. This ignorance can't be helped. The best we can do is to assess the resulting error. If it were true that slightly different demographic assumptions imply markedly different answers, then Fig. 5 would be of little interest.

Figure 6 shows the effect of different assumptions about mortality. The upward-pointing triangles show a discount function reflecting extremely high mortality — so high that the average female lives only to age 20. The downward-pointing triangles show the effect of the opposite assumption — now the average female lives to age 80. The circles show the implication of an even more extreme assumption about

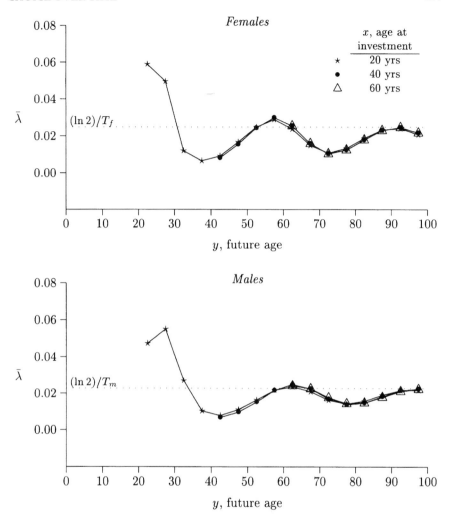

FIG. 5. Evolutionary discount function for 19th century Utah. $\bar{\lambda}$ is the average evolutionary discount rate within a five-year age interval. The dotted lines show the rate predicted by equation 2. See Rogers (1994) for data sources.

mortality — that no one dies until age 70, at which age everyone dies. This gives a rectangular survival curve that differs not only in magnitude but also in shape from realistic curves. But the three curves lie nearly atop one another, so even these gross changes in mortality rate have almost no effect on the answer. This insensitivity is surprising. Intuition suggests that discounting should be rapid when death rates are high, yet Fig. 6 shows that this intuition is false. Presumably, the small effect of

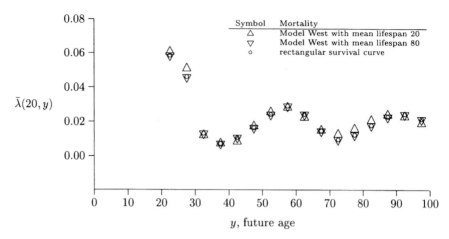

FIG. 6. The effect of mortality level on the evolutionary discount function. These evolutionary discount functions show the effect of three grossly different mortality schedules: two members of the 'Model West' family of mortality schedules (Coale & Demeny 1983) and a rectangular schedule in which all mortality occurs at age 70. $\bar{\lambda}$ is the average evolutionary discount rate within a five-year age interval.

mortality has something to do with the model's assumption of zero population growth, which requires changes in fertility to offset any change in mortality. Apparently, these opposing effects balance almost exactly.

To study the sensitivity of results to fertility data, I have avoided data from modern populations in which contraceptive use has distorted the fertility schedule. I have relied instead on 'natural fertility' populations, which may use contraceptives to space births but do not use them to achieve a target family size. The world over, natural fertility populations have fertility schedules with a characteristic shape. Because natural fertility schedules vary so little from population to population, it seems likely that this pattern has characterized our ancestors well into the past. Figure 7 shows the effect of various fertility schedules on the female evolutionary discount function. The three curves are similar in shape, implying high rates of time discount among young adults followed by damped oscillations that approach a long-term rate of about 2%. The main differences are in the timing of the peak. The Libyan and the Utahan fertility data imply that discounting should peak in the late teens or early 20s, while the Taiwan and Standard Natural Fertility schedules generate a peak in the late 20s. These differences mirror differences in the fertility schedules themselves (not shown).

Figure 8 shows the effect of various schedules of paternity. The discount functions of females are insensitive to assumptions about paternity, but the discount functions of males are more sensitive. All of the paternity schedules except that of 19th century Utah imply that male discount rates should peak much later than the female curves, at about age 40. This reflects the later age at which paternity rates peak in these populations. If

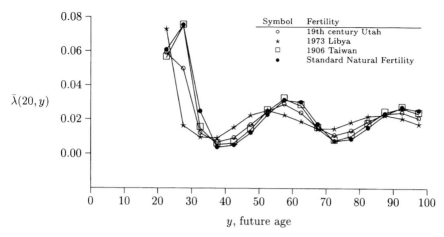

FIG. 7. The effect of fertility schedule on the evolutionary discount function of females is shown. $\bar{\lambda}$ is the average evolutionary discount rate within a five-year age interval.

our ancestry were characterized by demographic data such as these, then male rates of time preference should peak in middle age. On the other hand, the Utahan data imply a peak at age 30, not far from the peak for females. Without knowing more about the paternity schedules that characterized human evolutionary history, it is difficult to choose between these alternatives. Consequently, the model provides no clear prediction concerning the age at which male rates of time preference should peak. On the other hand, the male discount functions are broadly similar. All imply high marginal rates of time preference by young adults and lower rates among their elders. All converge toward roughly 2% in old age.

In more recent work (A. R. Rogers, unpublished data 1996), I have extended the model described here to deal with effects on consumption, which in turn affect mortality and/or survival. To allow for a diminishing marginal effect of consumption, I assume that the fertility of an individual who consumes κ units of resource is proportional to κ^α and that the age-specific survival of this individual is proportional to κ^β. The results turn out to depend on a single parameter, $\gamma \equiv \alpha/(\alpha + \beta)$, which measures the importance of marginal fertility relative to marginal survival. Figure 9 shows two results from this model along with the 20-year old from Fig. 5. When $\gamma = 0$, so that consumption's effect is entirely through survival, the curve for the new model lies on top of that for the old. This is remarkable since in the new model decisions are between consumption rather than survival alternatives and the relationship between consumption and survival is non-linear. Clearly, the old model lost little by ignoring consumption. When $\gamma = 1$ so that consumption's effect is through fertility, the discount function is similar in shape but its peak is shifted 20 years to the right. This predicts high rates of time preference among 40-year olds for decisions affecting fertility. I am still searching for a way to test this prediction.

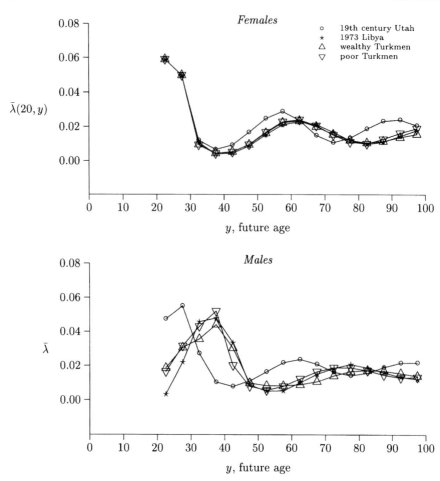

FIG. 8. The effect of paternity schedule on the evolutionary discount function is shown. $\bar{\lambda}$ is the average evolutionary discount rate within a five-year age interval.

To gain some intuition about the shape of the discount function, you should consider a hypothetical woman (Ego), who has a daughter at age 25, a granddaughter at age 50 and a great granddaughter at age 75. The benefits that Ego provides for herself and for these descendants increase her own fitness by an amount proportional to $rv(x)$, where r the coefficient of relationship, and $v(x)$ is the reproductive value of the recipient. The reproductive value curves, discounted by the relatedness r of each individual to Ego, are shown in Fig. 10 in log scale. At each age, Ego directs benefits to the individual whose curve is highest. She will therefore keep for herself any benefits that arrive between ages 20–31. Since her own

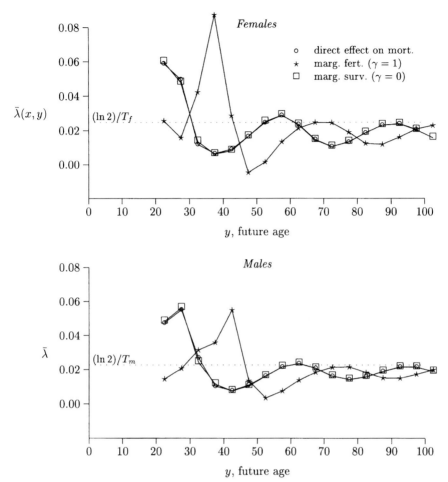

FIG. 9. The evolutionary discount function for effects on consumption. This figure is based on male and female fertility data of 19th century Mormons (El-Faedy & Bean 1987) and on the Model West life table (Coale & Demeny 1983) (mortality level 12, mean life span 47.5 for females and 44.5 for males). $\bar{\lambda}$ is the average evolutionary discount rate within a five-year age interval.

reproductive value is falling, these benefits are worth more the earlier they arrive, and her age-specific discount rates are therefore high. Benefits that arrive after age 31, on the other hand, will go to her daughter. Since the daughter's reproductive value is rising at first, these benefits are worth more if delayed, and the mother's discount rate is negative. As the daughter's reproductive value begins to fall, the process repeats itself, giving rise to oscillations similar to those seen in the other figures.

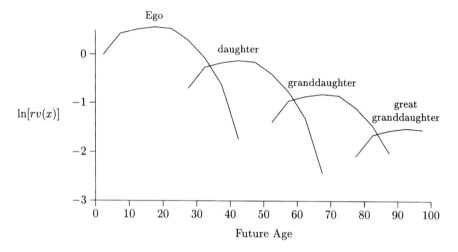

FIG. 10.   Values to Ego of self and descendants. This shows the log of contributions to Ego's reproductive success from herself, a child, a grandchild and a great grandchild, assuming that individuals of each generation are born 25 years apart.

Conclusions

The model described above seems to account for the fact that the long-term real interest rate is usually close to 3% and for the widespread perception that young adults do not worry much about their futures. This latter perception can be bolstered with data on the age pattern of crime and of driving accidents.

   Time preference is likely to be reflected in crime rates because crime has immediate benefits but delayed costs. If young adults discount the future steeply, they should also have high crime rates. This is exactly what is found: age-specific rates of many crimes peak among young adults. Gove (1985) observes that 'virtually all forms of deviance that involve substantial risk and/or physically demanding behaviour occur mainly among young persons, and the rates of such deviance decline sharply by the late twenties and early thirties'. This pattern is not an artefact of modern US culture, for it has been found wherever it has been sought (Hirschi & Gottfredson 1983, Daly & Wilson 1990). Although crime categories differ in the age where rates peak, 'no significant crime category deviates from the pattern of an early peak and a subsequent decline' (Wilson & Herrnstein 1985). Figure 11 illustrates the age pattern in homicide rates and relates it to predicted discount functions. Homicide rates are highest among young adults — precisely the ages at which steep discounting is predicted. Several authors argue that no explanation yet advanced is capable of accounting for the effect of age on crime (Gove 1985, Hirschi & Gottfredson 1983, Wilson & Herrnstein 1985, Wilson & Daly 1985). The model developed here may provide part of the answer. The similar pattern that is exhibited in rates of driving accidents (Wilson & Daly 1985) may have the same explanation.

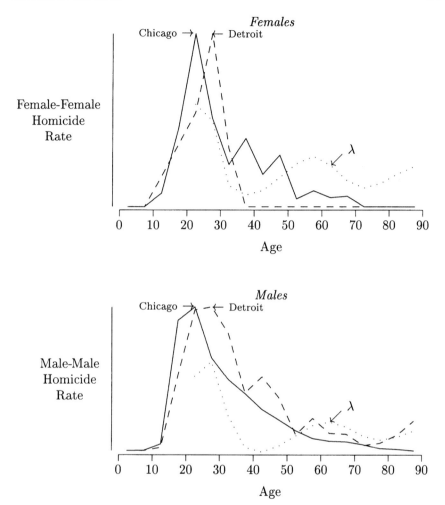

FIG. 11. Homicide rates. Rates per capita of homicide committed by females against females and males against males. Dotted lines show the evolutionary discount rate function from Fig. 5. Homicide data are those in Fig. 1 of Daly & Wilson (1990).

There is, however, a phenomenon that the model described above does not explain: reversal of preference. Reversals of preference occur when an individual changes his/her mind about the relative value of two alternative rewards as the time of their arrival approaches. For example, I may plan on Monday to diet on Tuesday, but when Tuesday arrives I may succumb instead to the temptation of a chocolate cake. Preference reversals are difficult to demonstrate in laboratory experiments with humans, but no one who has ever undertaken a diet can doubt that the effect is real.

Yet as I have shown elsewhere (Rogers 1994), preference reversals are not predicted by my own model.

To account for preference reversals, many authors advocate a model assuming that value is a hyperbolic function of time delay. Rachlin & Raineri (1992) provide a lucid account of how this model generates reversals of preference. The hyperbolic model implies that marginal rates of time preference will decline with longer and longer delays. Experiments that exhibit such declining rates have been offered as support for the hyperbolic model (Thaler 1981). Yet a glance at Fig. 5 will show that my own model also predicts declining rates in experiments with subjects of college age. Thus, the widely cited evidence for declining rates does not help separate the two models.

Nonetheless, the case for preference reversals is secure, and these are not explained by the evolutionary model advanced here. On the other hand, the hyperbolic model does not account for the long-term interest rate or for the evidence that young adults discount the future steeply. Each model succeeds where the other fails, and neither explains all the facts on the table.

Acknowledgements

This chapter was prepared during a sabbatical leave at the Research Centre of King's College, Cambridge.

References

Coale A J, Demeny P 1983 Regional model life tables and stable populations, 2nd edn. Academic Press, New York
Daly M, Wilson M 1990 Killing the competition: female/female and male/male homicide. Hum Nat 1:81–107
El-Faedy M A, Bean LL 1987 Differential paternity in Libya. J Biosoc Sci 19:395–403
Fisher I 1930 The theory of interest. Kelley & Millman, New York
Fisher RA 1958 The genetical theory of natural selection, 2nd edn. Dover, New York
Gove WR 1985 The effect of age and gender on deviant behavior: a biopsychosocial perspective. In: Rossi AS (ed) Gender and the life course. Aldine de Gruyter, Hawthorne, NY, p 115–144
Hamilton WD 1964 The genetical evolution of social behavior. J Theor Biol 7:1–52
Hamilton WD 1966 The moulding of senescence by natural selection. J Theor Biol 12:12–45
Hansson I, Stuart C 1990 Malthusian selection of preferences. Am Econ Rev 80:529–544
Hirschi T, Gottfredson M 1983 Age and the explanation of crime. Am J Sociol 89:552–584
Homer S, Sylla R 1991 A history of interest rates, 3rd edn. Rutgers University Press, New Brunswick, N J
Rachlin H, Raineri A 1992 Irrationality, impulsiveness, and selfishness as discount reversal effects. In: Loewenstein G, Elster J (eds) Choice over time. Russell Sage Foundation, New York, p 93–118
Rogers AR 1994 Evolution of time preference by natural selection. Am Econ Rev 84:460–481
Shiller R J, Siegel J J 1977 The Gibson paradox and historical movements in real interest rates. J Pol Econ 85:891–907
Thaler R 1981 Some empirical evidence on dynamic inconsistency. Econ Lett 8:201–207
Wilson J Q, Herrnstein R J 1985 Crime and human nature. Simon & Schuster, New York

Wilson M, Daly M 1985 Competitiveness, risk taking, and violence: the young male syndrome. Ethol Sociobiol 6:59–73

DISCUSSION

Tooby: There are a number of assumptions in your life history model, and I was wondering which of them are the most significant in driving your conclusions. For example, modelling senescence is essentially the same problem of discounting the future. The traditional argument made by George Williams (1957) is that there is always some independent probability of mortality, i.e. you might not be around tomorrow to collect a benefit deferred from today, and so in a trade-off between vigour or benefit now and later, there will be a preference for today over tomorrow that is shaped by this independent mortality. Also, as resulting selection leads to designs that are vigorous during youth and more damaged in old age, it seems as if the discounting rate should increase with age and approaching death. And yet, it seems as if youthful selection of high risk behaviours are the mirror image of what such discount models would lead you to expect. So, in your model, why exactly is there a sharp peak in discounting early in childhood?

Rogers: Because that is the age when women's reproductive value is declining fastest.

Tooby: But a feedback relationship is also present, and that itself is subject to selection.

Rogers: It's always the case with the life history models that everything depends on everything else.

Tooby: What is the argument for males?

Kacelnik: The predictions are based on current factors. The data on the incidence of homicides can be tested more directly by having an empirical discounting factor, and by forgetting that it is not exponential, to see whether age-specific discounting factors vary with the evolutionary model. Cropper et al (1991) found that the discounting factors increase with age, which is what you would expect because if you are old and you are going to die soon, delayed benefits matter less.

Rogers: I have some data relevant to that point. In a telephone survey, subjects were told to imagine that they had just won a prize and were asked to choose between an immediate reward or a larger delayed reward. From their responses, I was able to calculate the marginal rates of time preference (MRTPs) shown in Fig. 1. The shape of the curves shown there are in general agreement with my model: the MRTP is highest among young adults and wobbles down to a low value in middle age. The three curves were calculated from questions concerning delays of three months, one year and three years. As the model predicts, these differing delays have their largest effect on young adults. However, the model does not predict rates this high and does not predict the rising rates in old age. These rising rates, however, are in good agreement with Alex Kacelnik's suggestion. Bear in mind, however, that these results are preliminary. I have yet to show that these patterns are statistically significant.

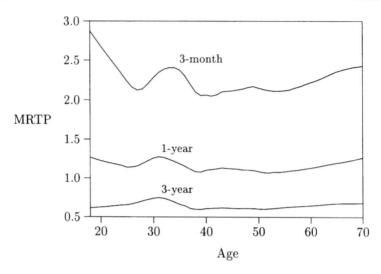

FIG. 1.   Estimates of the marginal rate of time preference (MRTP). Estimates were obtained from hypothetical questions in a telephone survey administered to 310 subjects by the Division of Survey Research of the University of Utah. The estimates were smoothed using the Lowess algorithm with each point-estimate using 30% of the data.

Daly: I'm sure this is a domain within which there are large framing effects. If you ask people to choose between outcomes with an invariant versus probablistic number of people surviving, for example, you get one answer if it's according to six deaths with a likelihood of 0.5 versus three with a likelihood of 1.0; whereas 600 versus 300 gives you a completely different answer (Wang 1996). The comparison of relatives versus non-relatives also gives different answers, which may be perfectly reasonable, but I worry about the validity of hypothetical questions.

Rogers: The literature on hypothetical questions is large. There was one study in which discount rates were calculated from hypothetical questions, and also from real behaviour, and the correlation between the two was zero. Therefore, there is plenty of reason for concern about these data. If you discard hypothetical questions then you're left with several other kinds of information about discounting. Most of the experiments have been done with children, although some have been done with young adults. The conclusions with young adults are always that discount rates are high. There are other ways of estimating discount rates, based on how much you save. These have been done mainly with elderly people, and in these studies the discount rates are low. Therefore, you could argue that this supports my model, but the methods are completely different so no one should have much faith in this inference. Having discarded the hypothetical questions, and having decided that we cannot compare these lines of evidence, the only thing that we're left with is preference reversals, e.g. we've all experienced not wanting to have any dessert because we want to lose weight and then the day after tomorrow we see a cake that just looks so good that we eat it. But this is only a thought experiment — an appeal to personal experience and intuition. The solid evidence in this field is limited.

Nisbett: There is a huge range of curves that all have the same function. For example, if we look at age and sex in relation to almost any dependent variable we get the same kind of picture: nobody does anything until their late teens and then they start doing things. Men do many more things than women, and then everybody declines from early adulthood until there is total inertia. It's true for playing jazz, writing novels, the onset of schizophrenia and having orgasms! These can't all be due to discount rates.

Miller: There is a large volume of economic evidence that may be relevant. If the real interest rate is an average across age-specific discount rates you could make the prediction that at any age where your discount rate is higher than the real interest rate you would borrow money, and when the reverse is true you would invest money. Investment and borrowing profiles as a function of age probably provide strong support for what you've suggested, i.e. young people borrow a lot, people aged 30–50 invest a lot and then it reverses again.

Rogers: That's correct, but there's a confounding factor: the amount of money that you have, and the amount of money that you expect to have. In Irving Fisher's diagram (Fig. 1 in my manuscript) young people are in the upper left: they don't have much now but they expect to have a lot in the future. This makes them less inclined to invest money and more inclined to borrow it. This is a second reason to expect saving to increase with age. I don't know how to separate the two.

Gigerenzer: Demographic data suggest that prior to the 19th century 20-year olds had different certainties about their life expectancies. For instance, life insurance companies in the 17th and 18th centuries did not discount, i.e. short-term life insurance cost the same amount for 25- or 75-year olds, which may have been because of the unpredictability of diseases and other risks (Daston 1987). Would these different data affect your model?

Rogers: My chapter considers the effect of mortality rates and concludes that you can have phenomenal changes in mortality rates without any substantial effects on the conclusion. Therefore, I am tempted to say no.

Pinker: One of the reasons for discounting the future is that you might not be here tomorrow; therefore, ordinarily you would expect that older people would discount the future more steeply. However, it's reversed if you transfer your resources to your children, so that instead of spending it now because you yourself may not be around tomorrow, you don't spend it now because there may be, in effect, two or three of yourselves around tomorrow.

Rogers: There may be one of yourself tomorrow that has a higher reproductive potential than you do now, and that future self may be better able to use resources. I don't think of it in terms of declining survivorship, I think of it in terms of declining reproductive potential.

Pinker: Wilson and Daly argued in their book *Homicide* that males are more reckless than females, and that young males are the most reckless of all because they're playing for the high stakes of being able to reproduce, which has a high variance in males (Daly & Wilson 1988). In your analysis, the reason the discounting rate peaks at a young age is that anything you accumulate might be transferred to progeny. But if the variance of whether or not you have progeny is high, young males may not expect to have any kin to transfer their wealth to when they die.

Rogers: That possibility is incorporated in the model. Males may be in a sort of 'winner takes all' market, but this is not included in the model.

Kacelnik: I've been mulling over the problem of discarding the empirical data based on hypothetical questions. The problem is that if we hypothesize that the incidence of crime of any kind is related to discount factors as measured by a variety of techniques, some of which are sensitive to framing, then we can't test it. We should look at whether the distortion produced by framing interacts with age because, for example, if (as in the Cropper et al 1991 example) we ask 'if we don't take action then people are going to die', as opposed to asking 'if we take action we're going to save lives', we may get different discount rates, but we may observe that the variation due to framing does or doesn't interact with age. The way in which Bateson & Kacelnik (1996) did the experiments with the starlings is also hypothetical, in the sense that it's not related to fitness because it is performed in the lab, but they get consistent results. Therefore, we shouldn't worry too much until we show real interactions.

References

Bateson M, Kacelnik A 1996 Rate currencies and the foraging starling: the fallacy of the averages revisited. Behav Ecol 7:341–352

Cropper ML, Aydede SK, Portney PR 1991 Discounting human lives. Am J Agric Econom 73:1410–1415

Daly M, Wilson MI 1988 Homicide. Aldine de Gruyter, Hawthorne, NY

Daston LJ 1987 The domestication of risk: mathematical probability and insurance 1650–1830. In: Krüger L, Daston LJ, Heidelberger M (eds) The probabilistic revolution, vol 1: ideas in history. MIT Press, Cambridge, MA, p 237–260

Wang XT 1996 Domain-specific rationality in human choices: violations of utility axioms and social contexts. Cognition 60:31–63

Williams GC 1957 Pleitropy, natural selection and the evolution of senescence. Evolution 11:398–411

Relationship-specific social psychological adaptations

Margo Wilson and Martin Daly

Department of Psychology, McMaster University, Hamilton, Ontario, Canada L8S 4K1

Abstract. Mainstream social psychology has sought parsimonious explanations of broad general applicability, and has in practice focused on stranger interactions. An evolutionary perspective, however, justifies predicting a rich diversity of relationship-specific social psychological adaptations. The demands of motherhood, fatherhood, mateship, sibship and other relationships are qualitatively distinct, and so, it appears, are the psychophysiological mechanisms that have evolved to deal with them. One window on what distinguishes social relationships is provided by the substance and epidemiology of interpersonal conflicts. Homicides exhibit victim–killer relationship-specific patterns in context, motive and demography, which we discovered only because we adopted an evolutionary psychological perspective. We offer a number of specific hypotheses about possible human psychological adaptations that have yet to be assessed, but are readily derived from consideration of the particular qualities of specific relationship types.

1997 Characterizing human psychological adaptations. Wiley, Chichester (Ciba Foundation Symposium 208) p 253–268

Psychologists have paid surprisingly little attention to social relationships. For decades, experimental social psychologists studied interactions with strangers, often imaginary strangers minimally described, on the presumption that basic processes — such as liking, conforming, etc. — are best illuminated in controlled, decontextualized studies. Recently, this restrictive focus has been somewhat alleviated as the study of 'close relationships' has become a significant subfield, but research under this rubric remains unduly narrow. Berscheid's (1994) extensive review, for example, is concerned primarily with dating and marital relationships and secondarily with same-sex friendships, making no reference to rivalries, patron–client relationships or any sort of familial relationship whatsoever.

With backgrounds in non-human animal behaviour, our bias has been to seek ecologically valid windows on real world behaviour. Homicides have provided us with an especially interesting 'assay' of interpersonal conflict, permitting tests of various hypotheses about the substantive issues of central importance in different relationship types (Daly & Wilson 1988, Wilson & Daly 1996). This window proves cloudy, however, when one wishes to characterize putative psychological adaptations in detail.

An evolutionary adaptationist perspective suggests that several distinct types of interpersonal relationships are sufficiently ancient and important to have affected the functional organization of our evolved social psyche. Sexual partnership, friendship, parenthood and so forth are qualitatively distinct kinds of close relationship differing in many specific ways. The attributes of an ideal mate, say, are quite different from those of an ideal sibling or friend, and it is clear that the mind processes information about these different sorts of intimates in different, specialized ways (Symons 1995). The things that require attention differ between relationship types, and so do affect and phenomenology. What constitutes a betrayal of one type of relationship might be irrelevant to another. Even the several fundamental sorts of close genetic relationship require distinct analyses.

Kinship psychology

Inclusive fitness theory suggests that evolved social psyches will respond discriminatively to cues indicative of degrees of genetic relatedness. Although there is intriguing cross-cultural diversity in human kinship systems, diversity of evolutionary psychological interest in itself, there are also some telling universals (Daly et al 1997). Parent–offspring relationships are the fundamental building blocks of ego-centred kindred terminologies in all societies, such that terminology implies genealogy, and people everywhere conceive of kin relations as arrayed along a 'close–distant' dimension. This closeness is always positively (albeit imperfectly) correlated with both genetic relatedness and actual co-operativeness. Diverse homicide analyses, for example, indicate that relatedness reduces risk, *ceteris paribus* (Daly & Wilson 1988). Some recent studies have belatedly introduced relatedness into the social psychological literature on the determinants of altruism, and the results have been highly supportive of selection-minded models (Burnstein et al 1994, Petrinovich et al 1993, Wang 1996; see Fig. 1). Whether the human mind contains procedures for computing something like relatedness coefficients and whether the same estimates are used in distinct domains demand investigation. For example, Wedekind et al (1995) have provided evidence that olfactorily mediated detection of shared major histocompatibility complex alleles dampens human sexual attraction and thereby facilitates outbreeding, but whether the same cue affects discriminative nepotism is unknown.

There is an a priori reason to suspect that kin classification might employ specialized mental processes. Unlike most other categorizations, kinship cannot be represented as a nested hierarchy because of sexual reproduction. Our ancestors double at each generation, making 'family' an altogether different computational problem. Nevertheless, people often seem determined to bend their conceptualization of family links to fit the inappropriate nested hierarchy model, placing themselves and others in named families (usually patrilineages). Why do people resort to this inappropriate mental model? D. Jones (unpublished work 1996) has argued that

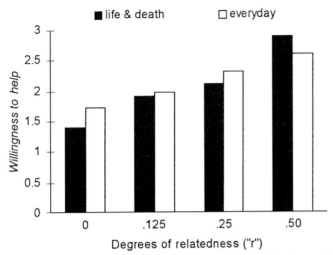

FIG. 1.   Effect of relatedness on undergraduates' professed inclination to help a same-sex person, depicted as sibling, parent, nephew/niece, cousin or acquaintance, in two hypothetical situations (Burnstein et al 1994). Relatedness was significantly more predictive of choice of beneficiary in the 'life-or-death' situation; sex and nationality (US vs. Japan) of subjects had no effects.

selection can favour a relatively undifferentiated kin group identity if mutually enforced group norms are incorporated into a variant of Hamilton's theory. Perhaps so, but even the most extreme systems of unilineal descent reckoning do not suppress bilateral affiliation and sympathy (Fortes 1969), so one may question whether clan memberships and patrinyms are really such powerful elicitors of familial feeling as they sometimes appear. Another issue for psychological research.

Genealogical kinship is not simply unidimensional. It encompasses several qualitatively distinct categories of relationship. The business of being a mother is quite different from that of being a father or an offspring or a sibling. There is no question that *Homo sapiens* possesses specific psychophysiological adaptations for dealing with the peculiar demands of motherhood and offspringhood, and we probably have others for dealing with the challenges of being fathers and siblings. Psychological adaptations for grandparenthood and perhaps for other distal kin relationships remain plausible, too.

Motherhood

The most intimate of mammalian social relationships and the one with the largest inventory of special-purpose anatomical, physiological and psychological machinery is that between mother and young. But the task demands of motherhood are a good deal more complex than even a consideration of the component demands of conceiving, gestating and raising a baby would imply. Because offspring are not equally capable of converting parental nurture into parental fitness gains, selection for subtle discrimination in allocating maternal effort is intense, and the evolved motivational mechanisms regulating maternal investment decisions are complexly contingent on variable attributes of the young, of the material and social situation, and of the mother herself, including life span developmental changes from youth to menopause.

Adaptive allocation of maternal investments is an especially subtle problem because of the active intervention of other parties with conflicting interests, particularly the offspring themselves. Parent–offspring conflict (Trivers 1974) is endemic to sexually reproducing species because of a certain asymmetry of relationship. Mother is equally related to her offspring, but each offspring is more closely related to self than to sib, so each offspring is selected to covet a little more from mother than would be optimal for her own fitness. This conflict accounts for the otherwise puzzling existence of seemingly maladaptive aspects of mother–young interaction, including weaning conflict, tantrums and the dangerously high levels of allocrine substances of fetal origin, such as human chorionic gonadotrophin and human placental lactogen, in the blood of pregnant women (Haig 1993).

Social psychologists and motivation theorists have largely ignored maternal inclinations, but paediatricians have done a lot of research on the immediate postpartum period. This literature implicitly treats maternal attachment as unidimensional, but an evolutionary perspective suggests instead that it is likely to involve at least three separable processes proceeding over different time courses: (1) assessment of the quality of the child and the situation; (2) rapid discriminative attachment to the baby as an individual; and (3) more gradual deepening of individualized love (Wilson & Daly 1994).

The first hypothesized process functions as an immediate postpartum assessment of the reproductive episode's prospects. Adaptationist thinking suggests that present material and social circumstances, the baby's phenotypic quality and the mother's residual reproductive value will all be determinants of a 'decision' whether to follow through or opt out. The hypothesized effect of a mother's reproductive value is supported by age-specific infanticide rates (Fig. 2) and by the finding of Jones et al (1980) that responsiveness to one's infant increased with maternal age, independent of any effects of marital status, socioeconomic status or postpartum contact. As for infant phenotype, deformities and signs of low viability raise the risk of maternal abandonment in non-state societies and elicit shocked rejection even in modern hospital settings. Postpartum maternal responsiveness also varies with more subtle cues of infantile quality and health. In this context, it is noteworthy that healthy human neonates exhibit a precocious social responsiveness that superficially mimics that of an older infant, including eye contact and selective attention to maternal speech, behaviours readily interpreted as infantile tactics for advertising quality and eliciting maternal commitment during this risky assessment phase.

Many new mothers experience a brief period of the 'blues', and some a more debilitating postpartum depression associated with concerns about their ability to

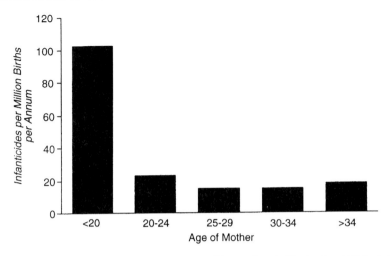

FIG. 2. Rates of infanticide ($N = 227$) perpetrated by mothers in England and Wales in 1977–1990 as a function of maternal age. Homicide data provided by the Home Office; age-specific births from the Office of Population Censuses and Surveys.

cope with the baby. Such depression is apparently especially likely when the mother is young, single or otherwise lacking social support, and when the infant is unhealthy. Postpartum depression is typically discussed as pathology, and in extreme cases it surely is, but these correlates suggest that it must be understood as an extreme form of an adaptively modulated emotional response.

If postpartum evaluation justifies further investment, the mother must then develop an individualized commitment such that she is emotionally prepared to invest heavily in her baby's welfare without being vulnerable to parasitism by unrelated children. This is the adaptive rationale for our second proposed maternal-bonding process. In social mammals with precociously mobile young and synchronous births, this discriminative attachment is urgent, and special-purpose neurochemical mechanisms for attaining it within a few hours of birth have been discovered in sheep (Kendrick et al 1992). In the human case, this individuation occurs at a more leisurely pace: mothers commonly report developing a feeling that their baby is uniquely wonderful over several days.

Mothers are highly sensitive to their newborns' distinctive features, recognizing them by voice, smell or touch after scant exposure, and these abilities are often portrayed as specific adaptations for discriminative bonding. This argument is weak. Evidence that discriminative abilities are actually heightened in new mothers is lacking, nor is it likely that postpartum infant recognition errors were an important selective force in human evolution. (And our more general ability to discriminate faces visually would solve the problem if they were.) Special offspring recognition adaptations exist where they are needed; for example, such abilities vary among closely related species of swallows according to whether the breeding ecology is such as to expose parents to the risk of misdirected parental investment (Beecher 1990).

The third predictable process of maternal attachment is much more gradual. A mother's love may be expected to grow with the child's increasing reproductive value, especially over the first few years as the probability that the child will survive to reproduce rises steeply. The relentless decline in filicide rates even into the teens (Daly & Wilson 1988) supports this expectation, as does the longitudinal study of Fleming et al (1990) in which positive maternal affect towards baby grew monotonically over months, while the same measures held relatively steady with respect to self and husband.

Information that mothers garner from monitoring offspring quality should affect the depth and time-course of their love and commitment, especially while mortality risk remains high. In many societies newborns are not immediately named or officially acknowledged by the community, a practice more or less explicitly linked to their uncertain future; naming bestows personhood and facilitates the individuation of affection. The delay in recognizing infants' personhood facilitates difficult divestment decisions, lessening emotional pain should the infant die. Since maternal effort is not to be squandered, chronic changes in infantile responsiveness and robustness due to malnutrition, dehydration and pathogens may be expected to diminish maternal inclinations, despite the sickly infant's greater need. In an observational study of low birthweight twins, Mann (1992) found that the healthier twin was indeed more effective in eliciting maternal responsiveness; factors such as duration of postpartum separation and infant smiling could not account for the mother's differential treatment.

Fatherhood

Much of the above theorizing applies to fathers as well, but there are crucial differences. Both parents are selected to assess offspring quality and need, but the father lacks the special avenues of communication (and manipulation) between mother and fetus or nursling. Both parents are also likely to respond to cues of the fitness value of available alternatives to present parental investment, but here it is even clearer that the specifics are different: the chronic chance that extra-pair mating effort might enhance male fitness is a selection pressure against intensive commitment to fatherhood. Finally, both parents are selected to discriminate with respect to available cues that the offspring is indeed one's own, but the relevant information sources are again distinct. Paternity can be misattributed, so putative fathers must rely on information about possible infidelity and/or familial resemblances. One implication is that fathers' affection may be influenced by their children's resemblance to themselves, whereas little or no such effect would be expected in mothers. This hypothesis awaits a good test, but there is evidence that all interested parties pay much more attention to babies' resemblances on the paternal than on the maternal side, and that mothers and their kin actively promote perceptions of paternal resemblance (Daly & Wilson 1982, Regalski & Gaulin 1993).

Some male songbirds adjust paternal investment adaptively in response to cues of paternity probability (e.g. Davies 1992). Male mice time the interval since they last ejaculated into a female, and become non-infanticidal weeks later when resultant offspring might be born (Perrigo & vom Saal 1994). Behavioural ecologists call these 'paternity (un)certainty' (or 'paternity confidence') adaptations with little risk of being misunderstood as implying awareness, but that implication is often drawn when *H. sapiens* is under consideration. It should not be. In principle, human paternity confidence adaptations may function in isolation from articulatable beliefs. The affection of adoptive fathers may be more strongly affected by the adoptee's resemblance to self than the affection of adoptive mothers, for example, simply because such resemblance is a cue to which the paternal psyche has evolved sensitivity.

Sibship

Sibling relations also warrant scrutiny from a selectionist perspective. Sisterhood is, of course, at the heart of the analysis of Hamilton (1964) of the evolution of altruism in social insects, but sibling relations are prominent in the sociality of diploid creatures too, especially in studies of 'helpers at the nest' (Stacey & Koenig 1990). Siblings are natural allies by virtue of relatedness, but they are also competitors for parental resources. Little wonder, then, that sibling relationships are often 'ambivalent'.

The costs that a toddler is willing to impose on its infant sibling in competition for maternal investment may vary adaptively in relation to nutritional status, the birth interval, the social situation and perhaps even the phenotypic quality of each youngster. This could be a profitable area in which to seek hitherto unsuspected psychological adaptations. It is also plausible that siblings are sensitive to paternity cues indicating full vs. half sibship. In Belding's ground squirrels, Holmes & Sherman (1982) found that full sisters are more co-operative neighbours than half sisters, even when both sister types are littermates so that familiarity is controlled. Could there be analogous discrimination in humans? Data on women's reproductive careers in contemporary foraging peoples (e.g. Hill & Hurtado 1996) imply that the distinction between full and half sibship may have been selectively significant in our ancestors, so toddlers may possess psychological adaptations for adjusting the intensity of competitive tactics towards newborn siblings in relation to either phenotypic cues or direct evidence of male turnover.

Such modulated sib competition adaptations could even be operative *in utero*. Live cells of fetal origin persist in maternal circulation for years after a birth, and may be active agents in genetic conflicts, and Haig (1993) has convincingly interpreted pregnancy pre-eclampsia as a by-product of maternal–fetal conflict over resource transfer. In light of these considerations, it is interesting that the incidence of pre-eclampsia in second and later pregnancies is elevated after a change of paternity (Trupin et al 1996). This may reflect escalated fetal conflict tactics, perhaps because the fetus can somehow assess its relatedness to its predecessor or, more plausibly, because the vestigial cells of the predecessor assess their relatedness to cells of the new fetus entering maternal circulation, and manipulate the mother to withold resources more if it is only a half sib, inspiring a compensatory response from the fetus.

Birth position within a sibship is a crucial aspect of one's social ecology. The eldest child commonly enjoys a parental priority that influences life options and personality development (Sulloway 1997). Subsequent children are apt to be less familially oriented and more inclined to forge reciprocal ties with non-relatives, but a lastborn child is in some ways more like a firstborn than a middleborn (Salmon 1997). Thus, more than merely seeking psychological adaptations for sibling interactions, we may need to consider even finer intrafamilial relationship categories.

Marital and affinal relationships

People refer to 'relatives' by marriage as well as by blood, and there is a certain logic to this conflation: in both cases, relatives have a commonality of interest grounded in the fact that both gain fitness from the reproduction of their common kin. A longstanding mateship becomes increasingly like a genetic relationship because as children are produced and mature the exigencies and resource allocations that would be ideal for promoting the fitness of the two parties are increasingly consonant. Indeed, if mating partners are faithfully monogamous and their efforts are channelled predominantly into reproduction rather than collateral nepotism, their commonality of interest — and hence of perspective — may become near-total (Alexander 1987). This would seem to explain why established couples may become more solidary than even the closest genetic kin. Interests in their separate kindreds remain a potential source of conflict in any couple, but these 'in-law problems' may be more acutely felt when there are no children to cement the marital relationship and the broader alliance between the two parties' families of origin that the marriage represents.

An important difference between mateship and genetic kinship is that the former is more readily and irredeemably betrayed. The correlation between mates' expected fitnesses can be abolished if either engages in extrapair mating effort, and should a cuckolded husband unwittingly invest in a rival's young, the very acts that promote his wife's fitness are positively damaging to his own. These considerations apparently explain why suspected or actual infidelity is a uniquely potent source of marital conflict and violence (Daly & Wilson 1988, Wilson & Daly 1996).

Step-relationship is like cuckoldry in that the child is a potential vehicle of fitness for one marriage partner but not the other. It is different, however, in that this asymmetry is out in the open, and has ideally entered into the negotiation of entitlements and reciprocities in the remarriage. Nevertheless, the presence of stepchildren is an important risk factor for marital disruption and violence, and the stepchildren themselves incur greatly elevated risk of violence too (Daly & Wilson 1996; Fig. 3). Step-parenthood itself is evidently the relevant risk factor, rather than some correlate or 'confound', reinforcing the point that parental psychology is designed to channel affection and investment preferentially toward own offspring.

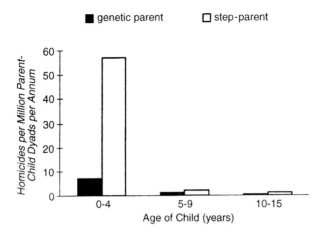

FIG. 3.   Rates of filicide perpetrated by genetic parents ($N = 726$) and stepparents ($N = 141$ victims) in England and Wales in 1977–1990 as a function of victim's age. Homicide data provided by the Home Office; population-at-large estimates derived from census data and Clarke (1989).

Non-familial relationships

We have concentrated on family relationships both because of the ready applicability of selectionist ideas and because psychologists have neglected them. But non-kin relationships may have a natural taxonomy and a set of distinct social psychological competences too. One might distinguish co-operation based on individual reciprocity versus co-operation based on shared group identity versus indifference versus enmity, for example.

Fiske (1991) and Haslam (1994) maintain that relationships themselves are not discretely categorical, but that people generate them from four discrete mental models, called 'communal sharing', 'authority ranking', 'equality matching' and 'market pricing'. We suggest that these models are better described as alternative solutions to co-ordination problems than as a complete 'mental grammar' of social relationships. Whether combinations of these models can generate all qualitatively distinct non-kin relationships we cannot say, but they surely fail to encompass all the relationship-specific psychological adaptations of kin relationships as described above.

Kinship is evidently the dominant mental model for helpful social interactions (Bailey 1988), and hence it provides an almost inescapable metaphor for establishing, describing or explaining co-operative relationships with non-relatives. Indeed, Alexander (1974) suggested that the cognitive abilities necessary for non-nepotistic reciprocal altruism must have evolved previously in the context of nepotistic sociality. Turning such conjectures into empirically testable propositions is a challenge for evolutionary social psychologists.

Acknowledgements

Our homicide research has been supported by the Social Sciences and Humanities Research Council of Canada, the Natural Sciences and Engineering Research Council of Canada and the Harry Frank Guggenheim Foundation.

References

Alexander RD 1974 The evolution of social behavior. Ann Rev Ecol Syst 5:325–383

Alexander RD 1987 The biology of moral systems. Aldine de Gruyter, Hawthorne, NY

Bailey KG 1988 Psychological kinship: implications for the helping professions. Psychotherapy 25:132–141

Beecher MD 1990 The evolution of parent–offspring recognition in swallows. In: Dewsbury DA (ed) Contemporary issues in comparative psychology. Sinauer, Sunderland, MA, p 360–380

Berscheid E 1994 Interpersonal relationships. Annu Rev Psychol 45:79–129

Burnstein E, Crandall C, Kitayama S 1994 Some neo-Darwinian decision rules for altruism: weighing cues for inclusive fitness as a function of the biological importance of the decision. J Pers Soc Psychol 67:773–789

Clarke L 1989 Children's changing circumstances: recent trends and future prospects. Ctr Pop Studies, London School of Hygiene and Tropical Medicine Res Paper, London, p 89–94

Daly M, Wilson MI 1982 Whom are newborn babies said to resemble? Ethol Sociobiol 3:69–78

Daly M, Wilson MI 1988 Homicide. Aldine de Gruyter, Hawthorne, NY

Daly M, Wilson MI 1996 Violence against stepchildren. Curr Dir Psychol Sci 5:77–81

Daly M, Salmon C, Wilson M 1997 Kinship: the conceptual hole in psychological studies of social cognition and close relationships. In: Simpson J, Kenrick D (eds) Evolutionary social psychology. Lawrence Erlbaum Associates Inc., Englewood Cliffs, NJ, p 265–296

Davies NB 1992 Dunnock behaviour and social evolution. Oxford University Press, New York

Fiske AP 1991 Structures of social life: the four elementary forms of social relationships. Free Press, New York

Fleming AS, Ruble DN, Flett GL, van Wagner V 1990 Adjustment in first-time mothers: changes in mood and mood content during the early postpartum months. Dev Psychol 26:137–143

Fortes M 1969 Kinship and the social order. Aldine de Gruyter, Chicago, IL

Haig D 1993 Genetic conflicts in human pregnancy. Q Rev Biol 68:495–532

Hamilton WD 1964 The genetical evolution of social behaviour. J Theor Biol 7:1–52

Haslam N 1994 Categories of social relationship. Cognition 53:59–90

Hill K, Hurtado AM 1996 Ache life history. Aldine de Gruyter, Hawthorne, NY

Holmes WG, Sherman PW 1982 The ontogeny of kin recognition in two species of ground squirrels. Am Zool 22:491–517

Jones FA, Green V, Krauss DR 1980 Maternal responsiveness of primiparous mothers during the postpartum period: age differences. Pediatrics 65:579–584

Kendrick KM, Levy F, Keverne EB 1992 Changes in the sensory processing of olfactory signals induced by birth in sheep. Science 256:833–836

Mann J 1992 Nurturance or negligence: maternal psychology and behavioral preference among preterm twins. In: Barkow J, Cosmides L, Tooby J (eds) The adapted mind. Oxford University Press, New York, p 367–390

Perrigo G, vom Saal FS 1994 Behavioural cycles and the neural timing of infanticide and parental behaviour in male house mice. In: Parmigiani S, vom Saal FS (eds) Infanticide and parental care. Harwood Academic, Chur, p 365–396

Petrinovich L, O'Neill P, Jorgensen M 1993 An empirical study of moral intuitions: toward an evolutionary ethics. J Pers Soc Psychol 65:467–478

Regalski JM, Gaulin SJC 1993 Whom are Mexican infants said to resemble? Monitoring and fostering paternal confidence in the Yucatan. Ethol Sociobiol 14:97–113

Salmon CA 1997 Effects of birth order and sex on familial sentiment and action. PhD thesis, McMaster University, Hamilton, Canada

Stacey PB, Koenig WD 1990 Cooperative breeding in birds. Cambridge University Press, Cambridge

Sulloway F 1997 Born to rebel. Pantheon, New York

Symons D 1995 Beauty is in the adaptations of the beholder: the evolutionary psychology of female sexual attractiveness In: Abramson PR, Pinkerton SD (eds) Sexual nature/sexual culture. University of Chicago Press, Chicago, IL, p 80–118

Trivers RL 1974 Parent–offspring conflict. Am Zool 14:249–264

Trupin LS, Simon LP, Eskenazi B 1996 Change in paternity: a risk factor for preeclampsia in multiparas. Epidemiology 7:240–244

Wang XT 1996 Domain-specific rationality in human choices: violations of utility axioms and social contexts. Cognition 60:31–63

Wedekind C, Seebeck T, Paepke AJ 1995 MHC-dependent mate preferences in humans. Proc R Soc Lond Ser B 260:245–249

Wilson MI, Daly M 1994 The psychology of parenting in evolutionary perspective and the case of human filicide. In: Parmigiani S, vom Saal FS (eds) Infanticide and parental care. Harwood Academic, Chur, p 73–104

Wilson MI, Daly M 1996 Male sexual proprietariness and violence against wives. Curr Dir Psychol Sci 5:2–7

DISCUSSION

Beecher: It would be interesting to look at olfactory signatures. Dick Porter has done quite a bit of work showing that mothers can recognize the smell of their children even when they are only one day old (Porter 1991).

Wilson: Mothers, fathers and just about anyone else can also recognize children on the basis of touch or on their voices. It's clear that the human species is good at individual recognition, but it's not clear whether mothers have a special ability at this time. It is interesting to speculate whether non-obvious cues, such as olfactory mechanisms, might impact on bonding. Martin Daly and I are interested in this issue because of the differential risks of children being killed or abused by step-parents versus genetic parents. There are mechanisms whereby individual children are identified but we know little about them.

Beecher: Is there any evidence in humans that fathers discriminate between their children on the basis of physical resemblance?

Wilson: No. We have hypothesized that physical resemblance to self or family should affect paternal attachment and not maternal attachment, but there isn't any good evidence that supports this.

Buss: Would one expect an evolved mechanism of that sort? It is possible that relatives could judge the similarity of an infant to the father, but if mirrors were not a feature of our ancestral environment then you wouldn't necessarily expect the self to be able to make that similarity judgement.

Wilson: It is possible that they could have had an image of their face from looking in water. However, say for argument's sake that they never had an image of their faces, this is not a problem because there's lots of evidence in the anthropological literature that people use other cues. The question of what is the specificity of our hypothesis is interesting. It's not clear that the baby should look like the father, because from the father's point of view there should be a cue that it's his child, but that's not necessarily the case from the mother's point of view or even the child's point of view. Therefore, the actual prediction is uncertain. A study that was published in *Nature* last year by Christenfeld and Hill at San Diego claimed that when you gave people three pictures of men, one of which was the father, and three pictures of women, one of which was the mother, when the child was one year of age the people were just over 50% accurate in attributing the father to that child, but this wasn't the case for mothers or when the child was a 10-year old or a 20-year old (Christenfeld & Hill 1995).

Scheib: There have been reports on parental resemblance in the donor insemination literature. A few of these have studied what potential parents are concerned about when they're choosing a sperm donor. One of the things that I found interesting was that both the physicians and the recipient couples try to match the donor to the woman's partner, with the articulated belief that parent–offspring resemblance will help the couples keep their use of donor insemination a secret. You might also expect them to behave in the same way, even if secrecy were not a concern. In any case, they have the psychology that values father–offspring resemblance. And there are hints of sex differences too. For example, one study by Klock & Maier (1991) found that women were more concerned about the physical resemblance of the child to the father, but the women's partners were more concerned about the child's personality resemblance to themselves.

Cosmides: David Haig in his plenary address to the Seventh Annual Meeting of the Human Behaviour and Evolution Society (University of California, Santa Barbara 1995) talked of his studies on how women in short-term versus long-term relationships have different risks of various pregnancy-related disorders. It would be interesting to see whether the children of women in long-term relationships bore more resemblance to the father, and whether this was because parts of the paternal genome were being more highly expressed.

Wilson: There was a study published in *Epidemiology* last Spring that is really interesting in the light of David Haig's work on maternal–fetal conflict (Trupin et al 1996). It was an epidemiological study of pre-eclampsia in the US involving 1000s of cases and controlling for age and socioeconomic conditions. The risk of pre-eclampsia normally decreases by half from first birth to second birth or later, but they found that for the second birth or later, if there was a change of paternity pre-eclampsia rates were twice as high. Therefore, there could be some subtle factors that affect the maternal valuation and assessment phase in ways that we could never have imagined before.

Dawkins: What do you mean by change of paternity?

Daly: The alleged father of the second child was different from the alleged father of the first.

Tooby: From looking at the animal literature one might expect there to be precise relationship-specific mechanisms at a level of detail that we're not even beginning to anticipate now. The things that you are talking about now are things that clearly ought to have been studied for the last 25 years. The recruitment of social psychologists into these kind of studies has been remarkably slow, and the number of people trying to come up with empirical results by applying these ideas is remarkably small. What do you think the prospect is for having many more scholars from the relevant fields join this research?

Wilson: The fact that we're all meeting here speaks well for the future, but Dick Nisbett is more in the field of social psychological science so I would like to hear his opinion.

Nisbett: I agree that there has been little penetration of this area into social psychology, yet the two lines of thought have so much to offer one another. Randy Nesse and I just ran an evolution seminar series at the Institute for Social Research, which was the best attended in the history of the Institute, but one of the most poorly attended by social psychologists. Social psychologists mistakenly think that evolutionary theory constitutes a conservative orientation, and the nature of the phenomena they study allows them to remain adherents of the *tabula rasa* position.

Daly: The numbers of evolution-based papers in the *Journal of Personality and Social Psychology* have been increasing over the last three years. Clearly, the 'don't know anything about it and don't want to' antipathy is waning.

Pinker: On the other hand, there are other bodies of scholarship from different disciplines which are obsessed with psychological relationships among family members: the fields of clinical psychology, counselling psychology and psychological social work. The data vary in scientific quality, but there may be relevant information in some of the better-designed studies. These fields have made no contact with scientific social psychology or with evolutionary psychology.

Wilson: We have talked to some of those clinical people. Many of them have good intuitions: they notice things that we only theorize about. However, in general, the family relations literature is dominated by social workers and practitioners, who take a systems approach to flows of information and causal influences rather than focusing on the information processing and decision making of individuals that we as psychologists or evolutionists have. Therefore, they sometimes just seem to miss the boat.

Cronin: What about your own paper 'Whom are newborn babies said to resemble?' (Daly & Wilson 1982), in which you found that the mother's family tries to persuade the father's family that the baby resembles the father?

Wilson: We (Daly & Wilson 1982) and Steve Gaulin (Regalski & Gaulin 1993) have found that mothers and their relatives push much harder in making attributions about the degree of resemblance to the father and father's relatives. It seems as though this is something that people are concerned about, but whether it has any impact is another question. Both studies also found that fathers are more sceptical. They want the truth and are not necessarily persuaded.

Møller: In the data that were presented by Gaulin, there was a decrease in the father's parent's claims with birth order, so that the second child was less likely to be claimed to resemble the father than the first (Gaulin 1997, this volume). This is a rather puzzling result because the paternity data argue exactly the opposite. There is an increased likelihood of having extra-marital second, third or fourth children compared to the first one. Therefore, there's another factor involved in this observed pattern of behaviour. It might be the case that these claims for resemblance to the first child are due to the older generation attempting to mend the bond and once it's established it might not matter that much.

Wilson: We also found that there was more pushing of paternal resemblance for the first child, and we also thought that once the male is committed to the familial unit then it is not necessary to push so hard later. Another thing that happens is that the relatives talk about the subsequent births as looking similar to the prior one rather than to the father.

Pinker: I would like to inject a note of scepticism about the hypothesis that people rely heavily on paternal resemblance to establish paternity. In sexual reproduction there is a mixing of genes, and many children don't look like their fathers. But we do have enormous cognitive machinery for cause-and-effect reasoning, which I suspect plays a much greater role in the psychology of paternity assessment. One way to see how important the cues are is to pit them against each other, as in these thought experiments. First, imagine a man who knew that he hadn't had sex with his wife for nine months, but the offspring resembled him. Now imagine a man who had been close to his wife for the year before, so no one else could have been the father, but the offspring did not particularly resemble him. Which man would be more upset? I suspect that the cognitive calculation of whether the offspring could be his, given the physiological and physical constraints, would trump assessments of resemblance. We tend to think that low level perceptual cues such as smell and physical resemblance would out-weigh pure reasoning, but it might be the other way round—the reason we evolved cause-and-effect reasoning is to apply it to the pressing problems that we face.

Daly: There are hints in the adoption literature, and this should be analysed in more detail, that adoption failures are better predicted by non-resemblance to the father than by non-resemblance to the mother. In this case the adoptive parents have got all their cognitive equipment telling them this is a fictive relationship and still it appears that paternal affection may be facilitated by resemblance in a way that maternal affection isn't.

Sperber: It might be useful, here, to take into account the cultural dimension. There is an important variability across cultures in beliefs about the traits that a child may inherit from the mother and from the father, with extreme cases where the role of one or the other parent in shaping the child is completely denied. Similarly, there is some cross-cultural variability in beliefs regarding the duration of pregnancy. I have done fieldwork in Mauritania, where women seemed to be able to adjust — by several months — claims regarding the duration of a particular pregnancy so as to make it begin at a time when their nomadic husband happened to have been in their company. I have also done fieldwork among the Dorze of Southern Ethiopia, where the duration of pregnancy was claimed to be an even number of months in some clans and odd in others. It could be of course that people's judgements and practices in specific cases are somewhat at odds with their cultural beliefs.

Wilson: My understanding is that the Trobrianders, who articulate that the male plays no role in producing a child, are still concerned about being impolite by implying that a child doesn't resemble its mother's husband. Therefore, there's the whole question of articulation versus the behavioural practice. I agree that it would be interesting to look at cross-cultural variation.

Dawkins: We have talked about paternity certainty and the recognition of resemblances between fathers and offspring, and we've taken it for granted that genetic resemblance can be exploited by an adaptation strategy. However, what if genetic resemblance itself were an adaptation in some sense? One way to look at this would be to consider a strategy of a female deliberately trying to mate with an unusual male, say a male with red hair or a conspicuous character. Then, assuming that she intends to be faithful to him, her offspring are reliably labelled. In a game–theoretical sense it becomes complicated because it depends upon whose benefit we are talking about. This analysis should really be done at the gene level. Another game–theoretical question might be to consider a gene for increasing the facial resemblance between parents and offspring. What will be the fate of that gene and will it flourish or not flourish in the presence of other genes? This in itself could be regarded as a genetically manipulable variable. The result is far from obvious and it is not possible to say intuitively what the result will be because it needs to be set up as a proper model.

Daly: I agree. Some writers have assumed that selection will favour offspring who exhibit paternity cues, but it's not obvious that there wouldn't be selection against such a gene and in favour of a generic child that doesn't reveal paternal phenotypes.

Scheib: There was an article in which Frank Salter (1996) presented a similar argument. The interesting point was that the benefits or the costs of having something like that would depend on whether or not, at least for females, extra-pair copulations were a possibility and whether they would be beneficial. In a context where they were beneficial you certainly wouldn't want to be carrying around a gene that would increase paternal resemblance. However, if extra-pair copulations were not beneficial, and they occurred infrequently, then a gene for increasing paternal resemblance could be useful.

Rogers: The same dynamic would apply for any character that is both visible and heritable. Selection for increasing paternal resemblance amounts to selection for rare types.

Daly: Although males make subtle discriminations in their allocation of paternal effort in response to cues of probability of paternity in various biparental animals, and although there's evidence of various sorts of phenotype matching in other contexts for the allocation of nepotistic investment, there are no non-human examples of males using phenotype matching in the allocation of paternal investment context.

Wilson: In the light of that, in the arms race between cuckoos and their host species with sustained selection you can see evidence of phenotype matching. In contrast, in recently invaded species you don't see it. This suggests that there is far more subtle dynamic conflict between males and females. We don't see any non-human species that are able to use phenotypic cues as paternity cues. Bewick's swans have individual face markings that some people can identify, so the requisite information may be there (Scott 1992). Also, in Mike Beecher's work on colonial-nesting swallows (Beecher 1989, Medvin et al 1993) the information is present in the offspring, i.e. voices and face markings, to allow the parents to identify the offspring once they've been exposed to it. This suggests to me that a sexual conflict is going on.

References

Beecher MD 1989 Evolution of parent–offspring recognition in swallows. In: Dewsbury DA (ed) Contemporary issues in comparative psychology. Sinauer, Sunderland, MA, p 360–380

Christenfeld NJS, Hill EA 1995 Whose baby are you? Nature 378:669

Daly M, Wilson MI 1982 Whom are newborn babies said to resemble? Ethol Sociobiol 3:69–78

Gaulin SJC 1997 Cross-cultural patterns and the search for evolved psychological mechanisms. In: Characterizing human psychological adaptations. Wiley, Chichester (Ciba Found Symp 208) p 195–211

Klock SC, Maier D 1991 Psychological factors related to donor insemination. Fertil Steril 56:489–495

Medvin MB, Stoddard PK, Beecher MD 1993 Signals for parent–offspring recognition: a comparative information analysis of the calls of cliff swallows and barn swallows. Anim Behav 45:841–850

Porter RH 1991 Mutual mother–infant recognition in infants. In: Hepper PG (ed) Kin recognition, Cambridge University Press, Cambridge, p 413–432

Regalski JM, Gaulin SJC 1993 Whom are Mexican infants said to resemble? Monitoring and fostering paternal confidence in the Yucatan. Ethol Sociobiol 14:97–113

Salter F 1996 Carrier females and sender males: an evolutionary hypothesis linking female attractiveness, family resemblance and paternity confidence. Ethol Sociobiol 17:211–220

Scott D 1992 Swans Semper Fidelis. Natural History, July 1992, p 26–33

Trupin LS, Simon LP, Eskenazi E 1996 Change in paternity: a risk factor for preeclampsia in multiparas. Epidemiology 7:240–244

Bird song learning as an adaptive strategy

Michael D. Beecher, S. Elizabeth Campbell and J. Cully Nordby

Departments of Psychology and Zoology, Animal Behavior Program, University of Washington, Seattle, WA 98195, USA

Abstract. Parallels are often drawn between bird song learning and human language learning. The analogies include an early sensitive period for learning, separation of sensory and motor phases of learning, 'innate knowledge' of language or song, and specialized neural systems. Nevertheless, in distinction to human language learning, song learning is usually viewed as a purely auditory process. This view is implied in the typical experimental paradigm for studying song learning, in which the bird is isolated in a sound-proof chamber and exposed only to tape-recorded song. This paradigm remains the major method of studying song learning despite recent demonstrations of the importance of social variables, and it contains the implicit assumption that non-auditory variables, especially social and ecological variables, play only a minor role in this process. I will argue here that understanding the functions of song learning requires that we investigate it in the field, where social and ecological variables have full play. We have done just this for the song sparrow, and when viewed in this perspective song learning in the song sparrow takes on the clear outlines of an adaptive strategy.

1997 Characterizing human psychological adaptations. Wiley, Chichester (Ciba Foundation Symposium 208) p 269–281

Bird song is an attractive system for investigating the evolution of a specialized learning system. Its appeal is enhanced by its many parallels with human language learning, its status as probably the best animal model for studying brain mechanisms of learning, and the vast amount of research on bird song and song learning that we have at our disposal.

Many animals besides the song birds have song, or loud, complex and species-distinctive vocal signals which they broadcast to conspecifics. Although song is widespread, however, the songbirds are unusual in that the form of their song is critically influenced by early learning. A songbird must be exposed to song when young in order to sing normal song as an adult. In most other taxa with song, on the other hand, individuals raised in isolation from conspecifics sing perfectly normal species song as adults. Animals with hard-wired song include many frog species, many insect taxa, some primates, including the lesser apes, and the other major group of passerine birds, the sub-oscines.

Although song in songbirds, or oscine passerines, could be compared with any of these other taxonomic groups, the most interesting comparison is with their closest relatives, the suboscine passerines. These two groups share the same taxonomic order, the Passeriformes or passerine birds. The key point is that song learning has been found to occur in all of the oscine species and none of the suboscine species tested so far, which has led workers in the field to believe that song learning evolved in the original oscine passerine (Kroodsma 1996).

The striking aspect of this contrast is that although song learning is present in the oscines but not the suboscines, song seems to function in essentially the same way in these two groups, to attract mates and/or to repel conspecific rivals, especially in the context of territoriality.

No one has yet offered a plausible explanation of why song learning evolved in the oscines. The diversity of song and song learning patterns in oscines suggests that present song bird species use their song learning abilities to a variety of different ends. Some of the dimensions of this diversity are indicated below.

(1)   When song is learned: from early in life (zebra finch) to throughout the lifetime (canary, thrush nightingale).
(2)   Song repertoire size: from a single species-specific song (white-crowned sparrow) to just a few song types (chaffinch, great tit, <5) to many (marsh wren, nightingale, >100) to very many (brown thrasher, >1000).
(3)   Individual variation (over some defined geographic area): from all birds with the same song type (white-crowned sparrow) to birds with individually distinctive songs (indigo bunting).
(4)   Mimicry: from species that copy only within tight species-specific parameters to mimics that copy virtually anything they hear (mockingbird, thrasher, catbird).
(5)   Copying fidelity: from precise copying (marsh wren) to improvisation (sedge wren).

Identifying the function of song learning, or the original function of song learning, is a daunting task, and we have aimed for a more modest goal, which its to determine its present function in one particular species, *Melospiza melodia*. If we can achieve this goal for a number of well-chosen species, we could perhaps then identify a common denominator, if there is one. In any case, perhaps we could then start to construct a plausible historical scenario of the diversification of song learning strategies in the oscines.

To uncover the present function of song learning in a species we must examine song learning in that species in its natural context, which is to say, we must study it in the field. This is not a popular idea, and there have been only a few such attempts to date. The thought seems to have been that song learning must be studied in the laboratory, where the key variables can be controlled and manipulated. The reigning laboratory paradigm has been the tape–tutor experiment in which we remove from the experiment all aspects of the species- and population-typical song-learning context except for song.

In particular, the young bird is not exposed to other birds, save the songs he hears over the loudspeaker in his isolation chamber. This paradigm has been justified by a particular theoretical view, which is that song learning consists essentially of the interaction of acoustic exposure variables with species-typical (innate) learning programmes and perceptual predispositions.

The problem with this view is that it prevents the investigator from discovering the role of social factors — which have been controlled out of the experiment — or of the many other natural factors that can be observed only in the field. Basically, the approach rarefies song learning by disconnecting it from the behavioural ecology in which it is normally embedded.

The first hint that something was missing in the tape–tutor approach came from species such as zebra finches that simply would not learn from tape: they required a live singing bird. In addition, even those species that will learn from tape–tutors invariably learn better in the laboratory from live tutors (that is, their song repertoires are more like those of wild birds). Realization of the importance of social factors began in earnest with the demonstration by Baptista & Petrinovich (1984, 1986) that different patterns of song learning in white-crowned sparrows are obtained when the standard tape recordings of song are replaced with live, singing birds. Baptista and Petrinovich showed that even for a classic tape–tutor bird such as the white-crowned sparrow, live tutors overwhelm tape–tutors, even if the live tutor is heterospecific and the tape–tutor conspecific, or the live tutor is presented after the putative sensitive period, and the tape–tutor during this sensitive period.

Introducing live tutors in place of tape recordings is, however, only one small step towards a more ecologically valid song-learning context, and it is not automatically even that. For example, in some live tutor experiments the tutor and student remain at unnaturally close quarters over long periods of time. This 'in-your-face' tutor is simply a different kind of unnatural tutor than is a tutor that is heard but never seen. When all is said and done, to understand the functions of song learning we need to study it in the field (Beecher 1996, Kroodsma 1996).

Song learning in a sedentary population of song sparrows

My group has been studying song learning in song sparrows (*Melospiza melodia*). Most of our work has been carried out on a sedentary (non-migratory) population of song sparrows in an undeveloped 200 ha park bordering Puget Sound in Seattle, Washington. We attempt to colour band all (or most) of the males in this population, about 150 birds in an average year.

A male song sparrow typically has six to 10 distinct song types (mode = 8) in his song repertoire. The song types in his repertoire are as distinct from one another as are the song types of different birds. Nevertheless, neighbouring birds will often share some of song types, that is, have similar song types. The song sharing is very local, with birds more than four or five territories away rarely sharing song types. This pattern of neighbour song sharing — which has been observed in many different songbird species—suggests that young birds learn the songs of the neighbourhood to which they immigrate following natal dispersal. Briefly, the key details of song sparrow natural history in our population are as follows (M. D. Beecher, S. E. Campbell & J. C. Nordby, unpublished data, Arcese 1987, 1989). After fledging, young birds leave the natal area at about a month of age and, following dispersal, remain in their new area for the remainder of their lives (Beecher et al 1994, Nordby et al 1997). A young bird usually begins singing subsong and plastic song in the late summer or early fall, but does not sing adult-like song until the following spring. He usually crystallizes his repertoire by early or mid-March, shortly before the breeding season begins in earnest. A song sparrow does not modify his song repertoire after his first breeding season.

I will describe our findings concerning song learning in this population of song sparrows as specific rules the young birds follows in song learning. Collectively, these rules define a 'strategy' of song learning. Because only males sing in song sparrows, and because song occurs most conspicuously in territorial contexts, we have focused on male–male interactions in song learning to date (Beecher et al 1994, J. C. Nordby, S. E. Campbell & M. D. Beecher, unpublished results). We have recently begun to examine possible female influences on song learning (O'Loghlen & Beecher 1997), and will revise our conception of the song-learning strategy of this species as necessary in the future.

The song-learning strategy of the song sparrow

Rule 1: copy complete song types precisely; preserve type and/or tutor

In the field, young song sparrows usually develop near-perfect copies of the songs of their older neighbours (Fig. 1). It is this very fact that made us realize that we could trace song learning in the field. The song similarities are striking, with the differences between tutor and student often being no greater than one normally sees in repetitions of the same song sung by one bird. For the example in Fig. 1, the biggest difference is for the third song of the four shown, which the young bird appears to have simplified the song by dropping the high frequency section near the end.

These field results differ remarkably from the laboratory findings using tape–tutors (Marler & Peters 1987, 1988). In the laboratory setting, song sparrows copy song elements quite precisely, but they frequently combine elements from different songs to form what I will call 'hybrid' songs—songs made up of parts of different song types. That is, songs are often improvised from learned elements.

In our field population, two exceptions to the 'perfect copy' rule actually clarify the rule, and suggest the meaning of the rule. The first exception occurs when the young bird 'blends' two tutors' somewhat different versions of the same song type (vs. copying one or the other). These songs are not true hybrids because although elements are selected from two different tutors, they are selected from the same, or similar, song type. The second exception, which is rare, occurs when the young bird

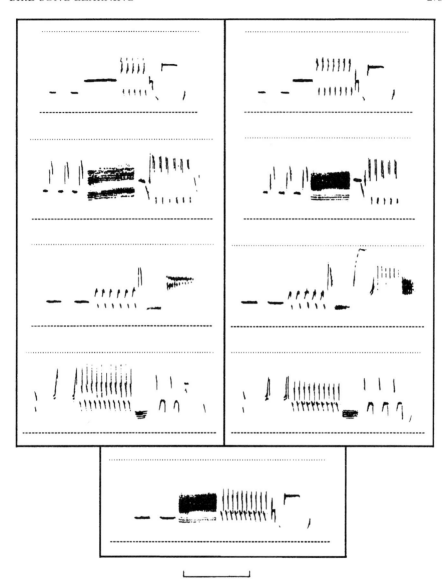

FIG. 1. In the field, young song sparrows learn nearly perfect copies of their tutor songs. Left panel: four of the nine song types of a young bird. Right panel: corresponding songs of one of his tutors. Bottom panel: the bird might develop a song like this if, instead of learning entire song types, he learned elements and re-arranged them into hybrid songs. Note that this hypothetical hybrid song contains one element each from the four tutor songs on the right. Such hybrid songs are common in laboratory experiments, but we do not find them in our field population. Frequency markers at the bottom and top of each sonagram are 0 and 10 kHz, time marker 1 second, bandwidth 117 Hz. Reprinted from Kroodsma (1996) with permission.

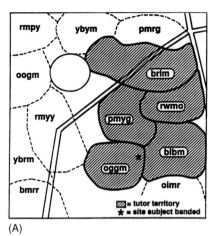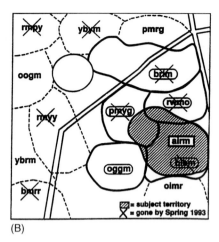

FIG. 2.   Tutors of the young bird are neighbours in the area where the young bird settled post-dispersal in his hatch year, but not all of these tutors survive to the young bird's first breeding season the next spring. (A) Map of territories of 13 adult birds in the area where young male airm settled in his hatch year (1992). The star indicates where airm was banded. The territories of the five birds who were subsequently identified as airm's tutors on the basis of his crystallized repertoire in 1993 are outlined and hatched. (B) The same configuration overlaid by airm's 1993 territory (hatched), and with dead birds crossed out (eight of the 13, including four of the five tutors). Although the actual 1993 territories of birds other than airm are not shown, birds oggm (the sole surviving tutor) and oimr remained in approximately the same place. Also indicated on the map are a pond (circle) and two intersecting paths; open areas were unoccupied (e.g. steep hills, meadows). Reprinted from Kroodsma (1996) with permission.

combines elements from two dissimilar song types of the same tutor. The principle followed by the bird, then, seems to be as follows: combine song elements of different songs only if they are different tutors' versions of the same type, or if they are different song types sung by the same tutor. We summarize this principle by saying the student 'preserves tutor and/or type' in his songs. We have never found a clear example in the field of a bird hybridizing a song type of one singer with a distinctly dissimilar song type of a different singer. My interpretation of this rule is that the goal of the strategy is to give the bird song types he shares with other birds in the neighbourhood. Hybrid songs, being personal inventions, are shared with no one.

Rule 2: learn the songs of several birds who are neighbours

On average, it takes three or four 'tutors' to account for the young bird's entire repertoire of eight or nine song types. Some birds learn from fewer or more tutors. Invariably, these tutors turn out to have been neighbours in the young bird's hatching summer (Fig. 2A). Usually by the following spring (the young bird's first

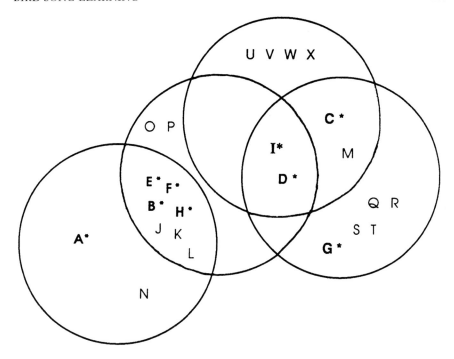

FIG. 3. Young males preferentially learn songs shared by two or more tutors. A Venn diagram of the four tutors (four circles) of subject mxpx. Each song type is indicated by a letter, and shared songs of two or more birds are given the same letter. Shared types are thus in the intersections of the tutor circles (e.g. types I and D were shared by three tutors). The nine song types learned by mxpx are indicated in bold face and starred. He learned seven of the 11 tutor-shared types, but only two of the 13 tutor-unique types. Reprinted from Kroodsma (1996) with permission.

breeding season) some of these tutor neighbours will have died. The young bird appears to commence song learning shortly after he has dispersed from his natal area, and it appears that most or all of song memorization probably occurs in the traditional lab-determined sensitive period, which is roughly the second and third months of life.

Rule 3: attempt to establish a territory near your tutor–neighbours

The young bird usually establishes his territory within the territorial range of his ex-tutors, often replacing a dead tutor (Fig. 2B). There are occasional exceptions to the rule that the young bird sets up his territory amongst his tutors: there are some cases where the young bird sets up on the periphery of the tutor group, perhaps a few territories away, and occasionally a bird goes some distance from his tutors. When a bird does not set up in the midst of his tutors, it invariably turns out that he didn't because he couldn't, i.e. because none of his tutors had died and/or because other young birds moved into this area.

Rule 4: preferentially learn tutor-shared songs

As I noted earlier, in our field population, neighbours typically share a portion of their song repertoires, on average about three or four of their eight to nine song types. Our discovery in the field is that the young bird preferentially learns (or retains) the tutor-shared types. The young bird diagrammed in Fig. 3 (along with his tutors) retained seven types that were shared by two or more of his tutors, and only two that were unique to one of these tutors, despite the fact that in his tutor group there were 11 shared types and 13 unshared types.

My interpretation of this rule is that it represents a 'bet-hedging' strategy designed to guarantee that the young bird has song types he will share with his neighbours in his first breeding season. If instead the bird learned tutor-unique songs, he would have songs 'personalized' for these particular tutor neighbours (i.e. shared with that neighbour only). However, these personalized songs would be good only until the tutor dies or moves, whereas a shared song is good until all the birds having it in the neighbourhood die or move, and even beyond then because other young birds moving into the area also preferentially learn shared types.

Rule 5: preferentially retain song types of tutors surviving to your first breeding season

Birds often have song types that can be traced to tutors that were alive in the young bird's natal summer but died before the next breeding season. Nevertheless, birds generally retain more songs of tutors that survive into the next breeding season than of tutors who do not. I describe this late learning as 'late influence' because it does not appear to be *de novo* learning. We have yet to find one clear example where a young bird learned a new song in the early spring (January–March) that he could not have encountered the previous summer. In those cases where a bird moves away from his natal summer tutors (because none of them die), he does not learn the songs of his new neighbours. He may well retain those song types he memorized earlier that best match those of his new neighbours (we are examining this hypothesis) but his songs still match those of his old tutor neighbours much better than those of his new neighbours.

These results are quite consistent with the two-phase model of song learning of Nelson & Marler (1994), according to which songs are memorized during a sensitive period in the natal summer and songs are kept or dropped on the basis of later social interactions with neighbours of the next breeding season (they describe this later phase as 'selective attrition').

TABLE 1 Correlation of tutor influence with geography

Subject	Number of songs credited to tutor 1	Final territory contiguous to tutor 1?
ARYM	1.41	no
ABGM	2.2	no
GMOY	2.50	no
AIBM	2.50[a]	no[a]
AIRM	2.50[a]	no[a]
WWAM	2.58	no
RMIO	2.83[a]	no[a]
IOOM	3.33	no
ORIM	3.91[a]	no[a]
BARM	4.16	yes
OOIN	4.5	yes
IAYM	4.82	no
YYGM	5.50	yes
OOYM	5.65	yes
GAYM	6.00	yes
IMYY	6.33	yes
YMRR	6.50	yes
GARM	7.00	yes
IIBM	7.00	yes
GAIM	8.00	yes

[a]Tutor did not survive past January 1st.

Rule 6: preferentially retain song types of your primary tutor if you establish a territory next to him

A bird's primary tutor — the bird who had the greatest influence on the young bird's song repertoire — can vary from total influence to only slightly more than the other tutors. We have found a clear correlate for this range of influence: whether or not the young bird and the primary tutor are contiguous neighbours in the bird's first breeding season (Table 1; J. C. Nordby, S. E. Campbell & M. D. Beecher, unpublished data 1996). About half the birds do not wind up next to their primary tutor, and when a bird does not, the influence of his primary tutor is less. That birds do not always wind up strongly influenced by a contiguous neighbour suggests that the commitment to the primary tutor occurs early, probably in the natal summer. If the young bird manages to set up next to that primary tutor, he retains more of that tutor's songs for his final repertoire. If the young bird fails to set up next to the primary tutor, winding up instead next to one of his lesser tutors, or further away yet, he will not show a strong influence from any one tutor.

Summary and conclusions

I have described the song learning programme of the song sparrow in terms of a series of design features. The six rules of song learning, taken together, work to maximize the number of songs the young bird will share with his neighbours, especially his near neighbours, in his first breeding season. I would suggest further, that having songs the young bird shares with his neighbours, tinged with some individuality, may in fact be the 'goal' of this song-learning strategy.

The hypothesis that the song sparrow song-learning strategy is designed to equip the birds with songs they share with their neighbours implies that there is some advantage to such song sharing. An alternative hypothesis is that neighbour song sharing is an incidental effect of birds learning songs in their post-dispersal neighbourhood. Regional dialects are now commonly viewed in this way, as an epiphenomenon. So far, we have two lines of evidence to support the hypothesis that song sharing is advantageous.

The first has to do with how birds use their song types when interacting with their neighbours. We have found that song sparrows preferentially use their shared songs in singing interactions with their neighbours (Beecher et al 1996). That is, if two neighbours share song types A, B and C, but not the rest of the songs in their repertoires, they will use A, B and C in singing interactions. Usually, a bird will reply to his neighbour's song not with the song type, but with another one of the song types he shares with the neighbour (for example if the neighbour sings A, he will reply with B or C). We have called this pattern of song selection 'repertoire matching' because although the bird usually does not match the stimulus song itself, he does match some song in the stimulus bird's repertoire (Beecher et al 1996). Repertoire matching implies knowledge of the singer's repertoire; the bird must know which songs he shares with the singer. We cannot yet say why neighbours counter-sing with shared songs, but this pattern does provide some motivation for a song-learning strategy that equips the bird with songs he shares with his neighbours.

Turning to our second line of evidence, if sharing songs does afford a song sparrow some advantage, this should be reflected in measures of fitness. In a recent interim analysis of a long-term study, we compared the ability of repertoire size and degree of song sharing to predict the territory tenures of a sample of young song sparrows. We found that song sparrows sharing more songs with their neighbours hold their territories longer than birds sharing fewer songs, and that song sharing is a better predictor of years on territory than is repertoire size (M. D. Beecher, S. E. Campbell & J. C. Nordby, unpublished data 1996).

References

Arcese P 1987 Age, intrusion pressure and defence against floater by territorial male song sparrows. Anim Behav 35:773–784

Arcese P 1989 Intrasexual competition, mating system and natal dispersal in song sparrows. Anim Behav 37:45–55

Baptista LF, Petrinovich L 1984 Social interaction, sensitive phases and the song template hypothesis in the white-crowned sparrow. Anim Behav 32:172–181

Baptista LF, Petrinovich L 1986 Song development in the white-crowned sparrow: social factors and sex differences. Anim Behav 34:135–1371

Beecher MD 1996 Bird song learning in the laboratory and the field. In: Kroodsma DE, Miller EL (eds) Ecology and evolution of acoustic communication in birds. Cornell, Ithaca, NY, p 61–78

Beecher MD, Campbell SE, Stoddard PK 1994 Correlation of song learning and territory establishment strategies in the song sparrow. Proc Nat Acad Sci USA 91:1450–1454

Beecher MD, Stoddard PK, Campbell SE, Horning CL 1996 Repertoire matching between neighbouring songbirds. Anim Behav 51:917–923

Kroodsma DE (1996) Ecology of passerine song development. In: Kroodsma DE, Miller EL (eds) Ecology and evolution of acoustic communication in birds. Cornell, Ithaca, NY, p 3–19

Marler P, Peters S 1987 A sensitive period for song acquisition in the song sparrow, *Melospiza melodia*: a case of age-limited learning. Ethology 76:89–100

Marler P, Peters S 1988 The role of song phonology and syntax in vocal learning preferences in the song sparrow, *Melospiza melodia*. Ethology 77: 125–149

Nelson DA, Marler P 1994 Selection-based learning in bird song development. Proc Natl Acad Sci USA 91:10498–10501

Nordby JC, Campbell SE, Beecher MD 1997 Ecological correlates of song learning in song sparrows, submitted

O'Loghlen AL, Beecher MD 1997 Sexual preferences for mate song types in female song sparrows. Anim Behav 53:835–841

DISCUSSION

Shepard: Why is the song useful only whilst the tutor that has the song is still surviving?

Beecher: This only applies to tutor-unique songs. The whole point about learning a song that's shared with several tutors is that it's good as long as any bird that has that song type is in the neighbourhood.

Shepard: But to understand this we have to know something about what the females are doing.

Beecher: I haven't said anything about female preferences. We're working on this at the moment (O'Loghlen & Beecher 1997).

Dawkins: Why it is a good thing to share your song?

Beecher: My point is that song sharing is the most common in birds, so one obvious inference is that it's useful to have shared songs. We wanted to characterize song learning in the field because researchers had already looked at what happens in the laboratory. Firstly, we found that the situation in the field was different from what was observed in the laboratory. Secondly, we observed high levels of song sharing with neighbours in the first breeding season. No one knows the advantages of song sharing, so we've simply tried to obtain some data on what they might be. We also found that neighbours use different songs depending on who they are talking to, using the songs they share with a particular neighbour, but again no one knows the reason for this. If song sharing really is the goal of song learning, one might find the correlation between survival and song sharing that we have in fact have found.

Kacelnik: West & King (1988) looked at song learning in their cow-birds, and they found the response of the females to specific particles in the songs that the juveniles were rehearsing was important for keeping the song in the males' repertoire. We don't know whether in this species the response of females in an area, who are familiar with a certain type of songs, would encourage the males to repeat particular components that lead to the same phenomena you have observed. It's not that your data in any sense are dubious because you haven't looked at the females, it's just that a crucial relationship may exist in this context with the females that is the mechanism producing your results. For example, I worked for about six years on the dawn chorus. I first did a single-sex analysis of why males sing early in the morning, but I did not fully understand the situation until a student of mine (Ruth Msee) decided to look at what the females were doing. It turned out that at least during egg laying the timing of female behaviour was crucial for determining the timing of male behaviour, even though males were singing and responding to each other.

Beecher: I have a couple of comments in response to that. First, you can't do everything at once. If you look at the songbird literature you find that some people are working on male–male competition, some on mate attraction and female choice, and there are a few people working on both. We have only recently begun looking at female responses to different song sparrow songs. We have taken females from the field, brought them into the laboratory and tested their preferences for songs using the copulation solicitation display procedure (O'Loghlen & Beecher 1997). We have found that females prefer mate song to songs of other birds in the area (although this does not necessarily apply to immediate neighbours). Over the last year we've been looking at female responses to shared songs and songs that are not shared, but the data on that are not yet complete. The last point I want to make is that the studies of Meridith West and Drew King are laboratory experiments, not natural observations. Having done studies both in the field and in the laboratory, in my opinion, just bringing birds into the laboratory and trying to re-create the situation that you suppose exists in the field does not guarantee that your situation will be any more natural than a tape–tutor experiment. Female cow-birds do have an influence in the lab, but it remains to be demonstrated that they have an influence in the field. We're presently working on ways of studying female influences in the field.

Hauser: I would like to raise the possibility that the comparison you made between human language acquisition and song learning in birds is a confounded one because we're talking about variation within our species versus variation in lots of different species of birds. A better comparison would be to take species that have a kind of system of learning like our system of learning language and look at the mechanisms that are similar or different within those species.

Beecher: There is currently a lot of interest in within-species comparisons. For example, Nelson et al (1996) have looked at migratory versus non-migratory populations of white-crowned sparrows and we're now following a similar approach with the song sparrows.

Maynard Smith: I have the strong impression that if you were to set up a simulation of a population of males copying song in the way that you describe, and ran it through a 1000 generations or so, you would end up with only one song. You ought to look at this because if this happens then there must be something else that maintains variability in addition to what you have been describing.

Beecher: Yes, and it would be interesting to know how they get from a pattern containing eight or nine songs to a white-crowned pattern which consists of only one song. The simulation would also have to take into account that there are small changes in the songs from generation to generation, i.e. I can't find any of the songs that I was looking at five years ago, although I can find remnants of these songs.

Rogers: I would like to suggest a hypothesis. I'm assuming that birds are using these songs in order to avoid fighting with each other, i.e. while singing the birds are trying to assess each others' fighting ability. If they are singing the same song then perhaps they are better able to assess each others' fighting ability, so they're more likely to avoid combat. In contrast, if they're singing different songs then they may not be able to tell. Therefore, it is possible that birds which sing different songs are less successful at avoiding combat and that's why they disappear from the population. If this hypothesis makes sense then it's possible to understand why birds that sing the same song do better, but it would also be puzzling why there is innovation.

References

Nelson DA, Whaling C, Marler P 1996 The capacity for song memorization varies in populations of the same species. Anim Behav 52:379–387

O'Loghlen AL, Beecher MD 1997 Sexual preferences for mate song types in female song sparrows. Anim Behav 53:835–841

West MJ, King AP 1988 Female visual displays affect the development of male song in the cowbird. Nature 334:244–246

Final general discussion

Hauser: There seem to be at least two ways in which people who work on humans think about how to use the non-human animal data. One of these is to use it as a kind of theoretical inspiration that we can apply to humans, and another is to use it to constrain how we think about the origin of the cognitive process. The latter approach is less common. I'm curious how people here who work on humans use the non-human animal data.

Daly: A third approach may be to illustrate how one might illuminate a relatively proximal function of a capacity or a psychological phenomenon in order to direct the investigation of worthwhile hypotheses about distal functions as well as processes. I'm thinking of Alex Kacelnik's presentation at this symposium (Kacelnik 1997, this volume) in which he provided an explanation for what looked like maladaptively steep discounting in starlings. His explanation may also have relevance for human researchers, who have perhaps been slow to pick up on the possibility of using non-human animal data because people are too easy to talk to. The articulateness of humans is a two-edged sword for studying people: it's so tempting to just ask them questions.

Hauser: There's another approach. If you look at almost all non-human primates, with perhaps the exception of chimpanzees, it is females who engage in community aggression and not the males. Therefore, the selection pressure that caused this important flip needs to be explained. Also, what is it about the ecology of chimpanzees which causes them to be much more like us?

Cosmides: Males who co-operate coalitionally incur risks in doing so; if males who incurred risks can be excluded from the benefits they 'earned' by other, larger males, then it is difficult for coalitional co-operation to evolve. But there are some ecological circumstances that make coalitional aggression more stable. For example, there are fish that fertilize externally which co-operatively drive off other males when a female is releasing eggs. At this point, they dump their sperm on the released eggs. In this situation, it would be difficult (at least compared to species where fertilization is internal) for one individual to exclude from the benefit other males who had helped in driving off the competitors. In contrast, the reason chimpanzees, and not other primates, have evolved human-like coalitional aggression may have more to do with their having certain cognitive prerequisites than with similarities in their ecological situation.

I would also like to ask a general methodological question inspired by the work of Wilson & Daly (1992) on sexual proprietariness. There are mechanisms that make people proprietary about many different things. It is possible that people can be proprietary about almost anything that they value, and so men may simply happen to value fidelity and therefore be proprietary about it. Alternatively, there could be mechanisms that are specialized for sexual proprietariness. These could be different from ordinary proprietariness, so that sexual proprietariness is not merely the confluence of ordinary proprietariness and valuing fidelity. Does anyone have any comments on how we can sort this out?

Dawkins: To try to discover whether there's a separate mechanism for sexual jealousy, for example, as opposed to jealousy about anything else that you might assess, you could look back at some of the older ethology literature. Researchers faced a somewhat similar problem in trying to decide whether, for example, aggression towards conspecifics is the same thing as aggression towards predators. The way that the researchers concluded that they were not the same was by noting that they don't co-vary.

Wilson: Aggression towards conspecifics is often over those resources which are useful in competition for mates. Testosterone, in so far as it's relevant to these aggressive and mate-seeking mechanisms, is going to underlie both of these because the material resources have historically been converted into reproductive opportunities. For example, if a particular species was a seasonal breeder and you're studying the animal in the breeding season then you would see behaviours that reflect both the value of material possessions and the value of females. Therefore, you would need a number of other points of confluence and difference to be able to tease apart the question of specialized mechanisms.

Daly: One of the things that seems most convincing to many people in those cases is to engage in some sort of neurobiological reductionism. How many kinds of aggression are there? Some of the most persuasive arguments have been to isolate brain systems with different neurotransmitters that seem to be involved in different tasks. I introduce this partly because the question of whether psychological science impresses other people as real science in the absence of neurobiological reductionism is a PR problem for psychologists.

Pinker: One has to be careful in looking for or interpreting neurophysiological correlations, because everything is in the brain somewhere. You need other criteria than just finding it in the brain. Neuroscientists themselves have not yet identified the functional units of the brain. Because the brain is an organ of computation, its functional design is not reflected directly in the gross computation, physical size, shape or location. The relevant aspects of the brain are the details of the microcircuitry where the computation takes place. With the tools that we have available it is going to be difficult to identify the psychologically relevant units by looking at the cortex directly, because there are patterns in the neural spaghetti that would require simultaneous recording with a million electrodes to disentangle.

Cosmides: When a phenomenon, such as sexual proprietariness, could be caused by the confluence of two mechanisms, disentangling hypotheses becomes difficult. In experimental manipulations where you are upping the gain on sexual proprietariness, you would have to be careful in identifying what, exactly, you were increasing. For example, in sexual jealousy, increasing the sexual value of the partner to the person won't answer the question because proprietariness should increase whenever you increase the value of a good, whether it is a sexual partner or a diamond. You would want to up the gain on proprietariness *per se* in other domains and see whether that has any effect, while holding the value of the goods in question constant.

Daly: Anders Møller has an operationalization of that in the form of the proximity of and/or attractiveness of your nearest neighbour relative to yourself. You haven't changed the value of the resource, i.e. the mate, to the focal individual, but you've changed the magnitude of the threat and therefore perhaps turned up the evocation of proprietariness towards that resource.

Cosmides: But this could also be the case for food.

Daly: One might find that the guarding of food would also increase when your nearest neighbour is closer, but the fact that he's better looking than you shouldn't have much influence on how much you guard your food.

Dawkins: One possible manipulation could be the consumption of alcohol. Does alcohol raise sexual jealousy but lower other kinds of proprietariness?

Cosmides: I would bet that it does.

Buss: Another approach might be to look at the output of the two types of proprietariness or jealousy. One finding is that when males get jealous in the presence of an intrasexual competitor, then that results in sexual arousal and a keen interest in copulation. In contrast, when someone's property is violated this does not result in increased sexual arousal. Therefore, by looking at the output we might be able to disentangle the two conceptual possibilities.

Thornhill: It seems as though this discussion is about the evidence for special-purpose design, which is the criterion for the demonstration of adaptation. When a priori, i.e. predicted, design features are actually found this provides powerful evidence for adaptation and for the specific effective historical selection responsible for the adaptation.

Cosmides: It's a little tricky because the specialization might be coming from the nature of the thing valued rather than from the proprietariness, e.g. in your work on male sexual coerciveness (Thornhill & Thornhill 1992) it was difficult to disentangle two hypotheses: are there adaptations for rape; or is rape caused by the confluence of adaptations that cause one to use force to get what one wants with adaptations for wanting sex?

Thornhill: If men are sexually aroused by the physical control of the mate in the context of a rape then that would be hard to explain by the by-product hypothesis for rape.

Sherry: As someone who is not an evolutionary psychologist, I would like to go home with a short list of human psychological adaptations that are unequivocally demonstrated by adaptive design features, knowledge of their evolutionary history or perhaps their phylogenetic distribution.

Daly: One could come up with a long list if they were not informed by recent selectionist thinking. Roger Shepard could probably give you a long list of human psychological adaptations — including object recognition, and edge and motion detectors — of which their functionality is well understood. However, those things were discovered without the aid of any explicit selectionism.

Pinker: This is a bit like asking for a list of parts of the body. I would think that the point of evolutionary psychology is not to come up with a list, but to change the way we do psychology in general, namely, to understand all mental phenomena as coming from the operation of psychological adaptations and their by-products. Listing adaptations is equivalent to solving the problems of psychology.

References

Kacelnik A 1997 Normative and descriptive models of decision making: time discounting and risk sensitivity. In: Characterizing human psychological adaptations. Wiley, Chichester (Ciba Found Symp 208) p 51–70

Møller AP 1994 Sexual selection and the barn swallow. Oxford University Press, Oxford

Thornhill R, Thornhill N 1992 The evolutionary psychology of men's coercive sexuality. Behav Brain Sci 15:363–375

Wilson M, Daly M 1992 The man who mistook his wife for a chattel. In: Barkow J, Cosmides L, Tooby J (eds) The adapted mind. Oxford University Press, New York, p 289–322

Index of contributors

Non-participating co-authors are indicated by asterisks. Entries in bold type indicate papers; other entries refer to discussion contributions.

Indexes compiled by Liza Weinkove

Subject index

Other Ciba Foundation Symposia:

No. 194 **Genetics of criminal and anti-social behaviour**
Chairman: Sir Michael Rutter
1996 ISBN 0 471 957219 4

No. 197 **Variation in the human genome**
Chairman: K. M. Weiss
1996 ISBN 0 471 96152 3